T0174787

Ecological Assessment of Selenium in the Aquatic Environment

Other Titles from the Society of Environmental Toxicology and Chemistry (SETAC)

For information about SETAC publications, including SETAC's international journals, Environmental Toxicology and Chemistry and Integrated Environmental Assessment and Management, contact the SETAC office nearest you:

SETAC
1010 North 12th Avenue
Pensacola, FL 32501-3367 USA
T 850 469 1500 F 850 469 9778
E setac@setac.org

SETAC Office
Avenue de la Toison d'Or 67
B-1060 Brussells, Belguim
T 32 2 772 72 81 F 32 2 770 53 86
E setac@setaceu.org

www.setac.org
Environmental Quality Through Science®

Ecological Assessment of Selenium in the Aquatic Environment

Edited by

Peter M. Chapman
William J. Adams
Marjorie L. Brooks
Charles G. Delos
Samuel N. Luoma
William A. Maher
Harry M. Ohlendorf
Theresa S. Presser
D. Patrick Shaw

SETAC Workshop on
Ecological Assessment of Selenium in the Aquatic Environment
Pensacola, Florida, USA

Coordinating Editor of SETAC Books
Joseph W. Gorsuch
Copper Development Association, Inc.
New York, NY, USA

CRC Press is an imprint of the
Taylor & Francis Group, an **informa** business

CRC Press
Taylor & Francis Group
6000 Broken Sound Parkway NW, Suite 300
Boca Raton, FL 33487-2742

First issued in paperback 2019

ISBN-13: 978-1-4398-2677-5 (hbk)
ISBN-13: 978-0-367-38413-5 (pbk)

Visit the Taylor & Francis Web site at
http://www.taylorandfrancis.com

and the CRC Press Web site at
http://www.crcpress.com

SETAC Publications

Books published by the Society of Environmental Toxicology and Chemistry (SETAC) provide in-depth reviews and critical appraisals on scientific subjects relevant to understanding the impacts of chemicals and technology on the environment. The books explore topics reviewed and recommended by the Publications Advisory Council and approved by the SETAC North America, Latin America, or Asia/Pacific Board of Directors; the SETAC Europe Council; or the SETAC World Council for their importance, timeliness, and contribution to multidisciplinary approaches to solving environmental problems. The diversity and breadth of subjects covered in the series reflect the wide range of disciplines encompassed by environmental toxicology, environmental chemistry, hazard and risk assessment, and life-cycle assessment. SETAC books attempt to present the reader with authoritative coverage of the literature, as well as paradigms, methodologies, and controversies; research needs; and new developments specific to the featured topics. The books are generally peer reviewed for SETAC by acknowledged experts.

SETAC publications, which include Technical Issue Papers (TIPs), workshop summaries, newsletter (*SETAC Globe*), and journals (*Environmental Toxicology and Chemistry* and *Integrated Environmental Assessment and Management*), are useful to environmental scientists in research, research management, chemical manufacturing and regulation, risk assessment, and education, as well as to students considering or preparing for careers in these areas. The publications provide information for keeping abreast of recent developments in familiar subject areas and for rapid introduction to principles and approaches in new subject areas.

SETAC recognizes and thanks the past coordinating editors of SETAC books:

A.S. Green, International Zinc Association
Durham, North Carolina, USA

C.G. Ingersoll, Columbia Environmental Research Center
US Geological Survey, Columbia, Missouri, USA

T.W. La Point, Institute of Applied Sciences
University of North Texas, Denton, Texas, USA

B.T. Walton, US Environmental Protection Agency
Research Triangle Park, North Carolina, USA

C.H. Ward, Department of Environmental Sciences and Engineering
Rice University, Houston, Texas, USA

Contents

List of Figures

List of Tables

Acknowledgments

This book presents the proceedings of a Pellston Workshop convened by the Society of Environmental Toxicology and Chemistry (SETAC) in Pensacola, Florida, in February 2009. The 46 scientists, managers, and policymakers involved in this workshop represented 5 countries. We thank all participants for their contributions, both in the workshop and in subsequent discussions resulting in this book.

The workshop and this book were made possible by the generous support of the following organizations (in alphabetical order):

- American Petroleum Institute
- Australian Academy of Technological Sciences and Engineering (partially supported by the International Science Linkages–Science Academies Programme, part of the Australian Government Innovation Statement, Backing Australia's Ability)
- Canadian Industry Selenium Working Group (comprising contributions from members, in alphabetical order: Areva, Cameco, CVRD Inco, Grand Cache Coal, Kemess, Peace River Coal, Prairie Mines, Shell, Sherritt Coal, Teck Coal, Teck Resources, TransAlta, and Western Coal Corporation)
- Canadian Nuclear Safety Commission
- CH2M HILL
- Electric Power Research Institute
- Environment Canada
- Rio Tinto
- U.S. Environmental Protection Agency Office of Water
- U.S. Fish and Wildlife Service
- U.S. Utility Water Act Group

The workshop would also not have been possible without the very capable management and excellent guidance provided by SETAC staff, particularly Greg Schiefer and Nikki Turman. The efforts of Mimi Meredith in the production of this book are gratefully acknowledged.

Peter M. Chapman
William J. Adams
Marjorie L. Brooks
Charles G. Delos
Samuel N. Luoma
William A. Maher
Harry M. Ohlendorf
Theresa S. Presser
D. Patrick Shaw

About the Editors

Peter M. Chapman is a principal and senior environmental scientist with Golder Associates Ltd. in Burnaby, British Columbia, Canada, and has been working on selenium issues since 1995. In 2001 SETAC awarded him the Founders Award, their highest award for lifetime achievement and outstanding contributions to the environmental sciences.

William J. Adams is chief advisor to Rio Tinto on site remediation risk assessment and leads a global team responsible for managing closed mining sites and areas under management. He has conducted research on selenium since 1972 and has published 20 papers on selenium ecotoxicology-related topics.

Marjorie L. Brooks is an assistant professor at Southern Illinois University at Carbondale (Illinois, USA). She is interested in the effects of multiple, abiotic stressors such as UV radiation, metals, and selenium on aquatic biota.

Charles G. Delos is an environmental scientist in the Office of Water of the U.S. Environmental Protection Agency (Washington, DC, USA). He has worked on environmental modeling, risk assessment, and water quality criteria, and is currently responsible for selenium criteria development.

Samuel N. Luoma is with the John Muir Institute of the Environment, University of California (Davis, California, USA) and the Natural History Museum (London), and is an emeritus after 34 years at the U.S. Geological Survey (Menlo Park, California, USA). He has worked on exposure, effects, modeling, and monitoring of selenium since the late 1980s.

William A. Maher is a professor in environmental/analytical chemistry, dean of applied science, and director of the Ecochemistry Laboratory, Institute of Applied Ecology, University of Canberra, Australia. He was awarded the Royal Australian Chemical Institute Analytical Divisions medal in 2002 and their Environmental Chemistry Divisions medal in 2004.

Harry M. Ohlendorf is a technology fellow with CH2M HILL in Sacramento, California, USA. He has conducted evaluations of the ecotoxicology of selenium since 1983, when he began studies of the effects of selenium on birds at Kesterson Reservoir, California.

Theresa S. Presser is a chemist with the National Research Program of the U.S. Geological Survey. She has conducted research on selenium sources and exposure since her initial investigations at Kesterson National Wildlife Refuge and in the Coast Ranges of California in 1983.

D. Patrick Shaw is a senior environmental scientist with Environment Canada in Vancouver, British Columbia, Canada. He has been involved in selenium issues in western Canada since 2003.

Workshop Participants[+]

Workshop chair: Peter M. Chapman, Golder Associates Ltd.
[+]Affiliations were current at the time of the workshop.
*Steering committee member
[†]Provided discussion or commentary presentation in opening plenary

WORKGROUP 1: PROBLEM FORMULATION

William J. Adams*
Rio Tinto
Magna, Utah, USA

John Besser
U.S. Geological Survey
Columbia, Missouri, USA

Keith Finley (Rapporteur)[†]
Duke Energy
Huntersville, North Carolina, USA

William A. Hopkins
Virginia Polytechnic Institute and State
    University
Blacksburg, Virginia, USA

Dianne Jolley[†]
University of Wollongong
Wollongong, New South Wales
Australia

Eugenia McNaughton[†]
U.S. Environmental Protection Agency
San Francisco, California, USA

Theresa S. Presser*
U.S. Geological Survey
Menlo Park, California, USA

D. Patrick Shaw*
Environment Canada
Vancouver, British Columbia
Canada

Jason Unrine
University of Kentucky
Lexington, Kentucky, USA

Terry Young (Chair)
Oakland, California, USA

WORKGROUP 2: ENVIRONMENTAL PARTITIONING

Martina Doblin
University of Technology
Sydney, New South Wales
Australia

Teresa Fan
University of Louisville
Louisville, Kentucky, USA

Simon Foster
University of Canberra
Canberra, ACT
Australia

Reid Garrett
Progress Energy
Raleigh, North Carolina, USA

William Maher*
University of Canberra
Canberra, ACT
Australia

Greg Möller
University of Idaho
Moscow, Idaho, USA

Anthony Roach†
Centre for Ecotoxicology
Lidcombe, New South Wales
Australia

Dirk Wallschläger
Trent University
Peterborough, Ontario
Canada

WORKGROUP 3: TROPHIC TRANSFER

David Buchwalter
North Carolina State University
Raleigh, North Carolina, USA

Nicholas Fisher
Stony Brook University
Stony Brook, New York, USA

Martin Grosell (Rapporteur)
University of Miami
Miami, Florida, USA

Sam Luoma*†
U.S. Geological Survey
Menlo Park, California, USA

Teresa Mathews
Institute for Radiological Protection and
    Nuclear Safety
Cadarache, St. Paul lez Durance Cedex
France

Patricia Orr†
Minnow Environmental
Georgetown, Ontario
Canada

Robin Stewart (Chair)
U.S. Geological Survey
Menlo Park, California, USA

Wen-Xiong Wang
Hong Kong University of Science and
    Technology
Kowloon, People's Republic of China

WORKGROUP 4: TOXIC EFFECTS

Marjorie L. Brooks*
Southern Illinois University
Carbondale, Illinois, USA

Peter M. Chapman*
Golder Associates
Burnaby, British Columbia
Canada

David K. DeForest (Rapporteur)
Windward Environmental
Seattle, Washington, USA

Guy Gilron
Teck Cominco
Vancouver, British Columbia
Canada

Dale Hoff
U.S. Environmental Protection Agency
Duluth, Minnesota, USA

David M. Janz (Chair)†
University of Saskatchewan
Saskatoon, Saskatchewan
Canada

Dennis O. McIntyre†
Great Lakes Environment Center
Columbus, Ohio, USA

Christopher A. Mebane
U.S. Geological Survey
Boise, Idaho, USA

Vincent P. Palace[†]
Department of Fisheries and Oceans
Winnipeg, Manitoba
Canada

Joseph P. Skorupa[†]
US Fish and Wildlife Service
Arlington, Virginia, USA

Mark Wayland
Environment Canada
Saskatoon, Saskatchewan
Canada

WORKGROUP 5: RISK CHARACTERIZATION

Patrick Campbell
West Virginia Department of
   Environmental Protection
Charleston, West Virginia, USA

Steve Canton[†]
GEI Consultants
Littleton, Colorado, USA

Charles G. Delos[*]
U.S. Environmental Protection Agency
Washington, DC, USA

Anne Fairbrother[†]
Exponent
Bellevue, Washington, USA

Nathaniel Hitt
U.S. Geological Survey–Leetown
   Science Center
Kearneysville, West Virginia, USA

Peter Hodson (Chair)
Queens University
Kingston, Onatario
Canada

Lana Miller
University of Lethbridge
Lethbridge, Alberta
Canada

Harry M. Ohlendorf[*]
CH2M HILL
Sacramento, California, USA

Robin Reash (Rapporteur)
American Electric Power
Columbus, Ohio, USA

1 A Pellston Workshop on Selenium in the Aquatic Environment

Peter M. Chapman

CONTENTS

1.1 INTRODUCTION TO THE WORKSHOP

This book is the result of discussions that took place during the Pellston Workshop on Ecological Assessment of Selenium in the Aquatic Environment. The workshop, sponsored by the Society of Environmental Toxicology and Chemistry (SETAC), was held 22–28 February 2009, in Pensacola, Florida, USA.

Selenium (Se) is a naturally occurring metal-like element (a non-metal). It is an essential element required for the health of humans, other animals, and some plants. Specifically, it is necessary for the proper functioning of structural proteins and cellular defenses against oxidative damage. However, Se in excess and in critical chemical species in their diet can cause reproductive failures or abnormalities in egg-laying vertebrates (fish, birds, amphibians, and reptiles).

Selenium has become a contaminant of potential concern (COPC) in North America, Australia, and New Zealand and is likely an unrecognized COPC in other parts of the world. Selenium is a COPC as a result of activities conducted by a wide variety of industrial sectors, including mining (coal, hard rock, uranium, phosphate) and power generation (coal-fired power plants, oil refineries). Selenium is found in organic-rich shales that are source rocks for the above activities. Selenium is also a COPC for agriculture due to discharge of subsurface irrigation drainage waters and to animal husbandry additions of this essential element. Concentrations of Se are increasing in many areas of North America, Australia, New Zealand, and China due to increased mining and power generation activities.

This workshop allowed for a focused discussion regarding the fate and effects of Se in aquatic (freshwater, estuarine, marine) ecosystems. Specifically, this workshop allowed for a forum that determined, as documented in this book, the state-of-the-science and then used this information to provide guidance for assessing and managing the environmental effects of Se. Major areas of uncertainty requiring further

1

research were also documented. The findings of this workshop, as contained in this book, apply equally to areas of the world that currently recognize Se as a COPC and to those that may do so in the future.

1.2   WORKSHOP FORMAT

This workshop brought together a multidisciplinary and international group of 46 scientists, managers, policy makers, and students from Australia, Canada, China, France, and the United States with a common interest in assessment of Se in aquatic environments. During the first day of the workshop, 4 separate series of presentations were given in a plenary session:

- Selenium Past, Present and Future,
- Bioaccumulation and Trophic Transfer,
- Selenium Toxicity Considerations, and
- Risk Characterization.

Each series of presentations included a discussion given by one of the participants, followed by individual commentaries by 2 other participants, and then a plenary discussion. The 3 individuals presenting each series of presentations were encouraged to interact before the workshop such that areas of scientific agreement could be established along with areas of scientific disagreement and reasons for those disagreements. However, it was made clear that the presentations were to be individual, representing each presenter's unique viewpoint. The 4 presentation topics provided the basis for subsequent plenary and working group (WG) discussions and thus, with one exception, do not appear in this book as separate documents. The one exception is a Commentary Paper (Appendix B) on persistence of fish populations in streams with elevated Se concentrations that supplements information provided in the main body of this book.

Daily afternoon plenary meetings during the subsequent 3 days of the workshop provided the opportunity for WG progress review and "cross-fertilization." A final plenary on the last full day of the workshop provided for consensus on areas of agreements and major uncertainties requiring further clarifying research.

Following the initial plenary session, workshop participants moved into 5 separate workgroups, in which each produced a chapter for this book. Following the book's Executive Summary (Chapter 2),

- Chapter 3, Problem Formulation: Context for Selenium Risk Assessment, reviews past and current problems related to Se in aquatic environments, together with lessons learned, and provides a generalized conceptual model.
- Chapter 4, Environmental Partitioning of Selenium, reviews environmental partitioning, in particular Se speciation leading to its entry into the food chain, and provides conceptual models specific to environmental partitioning.
- Chapter 5, Selenium Bioaccumulation and Trophic Transfer, reviews Se bioaccumulation and trophic transfer from the physical environment (i.e.,

water-column particulates), and from primary producers to herbivores to carnivores, including the influence of modifying ecological factors.
- Chapter 6, Selenium Toxicity to Aquatic Organisms, reviews toxic effects from Se, in particular body burdens and their relationship to toxicity.
- Chapter 7, Selenium Risk Characterization, integrates information from Chapters 3 through 6 in a risk assessment format, documenting the state-of-the-science related to risk characterization.

The findings of this SETAC Pellston Workshop, partly outlined above and fully detailed in subsequent chapters of this book, will benefit both scientists and managers. It is hoped that they will assist in preventing future environmental damage due to Se and in providing the basis for focusing management efforts based on good science.

2 Executive Summary

Peter M. Chapman, William J. Adams,
Marjorie L. Brooks, Charles G. Delos,
Samuel N. Luoma, William A. Maher,
Harry M. Ohlendorf, Theresa S. Presser,
and D. Patrick Shaw

This book on Ecological Assessment of Selenium in the Aquatic Environment synthesizes and advances the state-of-the-science regarding this unique metalloid and identifies critical knowledge gaps. Assessment methods appropriate for other metals and metalloids are not always appropriate for selenium (Se). Selenium requires site-specific risk assessments to a much greater extent than do many other contaminants, including adequate quality assurance and quality control of chemical and biological analyses.

Selenium is a growing problem of global concern for which mechanistic, biochemical understanding is required. Aquatic-dependent, egg-laying vertebrates are most at risk; the most sensitive toxicity endpoints are embryo mortality for waterbirds and larval deformities for fish. Aquatic-dependent mammals do not appear to be as sensitive as fish or birds to dietary organic Se exposure. Traditional methods for predicting effects based on direct exposure to dissolved concentrations do not work for Se; site-specific factors are highly important in determining whether Se toxicity will occur.

Selenium uptake is facilitated across most biological membranes (a non-passive, carrier-mediated process), making its partitioning unique among metalloid contaminants. Knowledge of Se speciation is essential to understand its fate and effects. Relating Se sources to risks to sensitive taxa requires measurement of Se accumulation at the base of the food web and in key linkages through the food web. The greatest degree of site-to-site variation in bioaccumulation occurs at the base of the food web but can be predicted by empirically derived site-specific enrichment functions (EFs). Uptake by individual species and in steps of the food web can be described by a trophic transfer function (TTF).

Selenium concentrations in eggs are the best predictors of effects in sensitive egg-laying vertebrates. The vulnerability of a species is the product of its propensity to accumulate Se from its environment as affected by its diet and by site-specific factors controlling the transfer of Se into and within the food web, its propensity to transfer Se from its body into its eggs, and its sensitivity to Se in its eggs.

A notable knowledge gap exists for egg-laying species of amphibians and reptiles, which represent some of the most critically endangered vertebrates. Atmospheric partitioning and dispersion are not well characterized. A better understanding is needed of EFs at the base of food webs and of inter-organ distributions of Se, because both of these are key mediators of toxicity. TTFs need to be documented. Reasons for differential sensitivities among species, even within the same genus, remain to be elucidated, as does the possibility and significance of tolerance (acclimation and/ or adaptation).

3 What You Need to Know about Selenium

Terry F. Young, Keith Finley, William J. Adams,
John Besser, William D. Hopkins, Dianne Jolley,
Eugenia McNaughton, Theresa S. Presser, D.
Patrick Shaw, Jason Unrine

CONTENTS

3.1   INTRODUCTION

In 1976, scientists monitoring North Carolina's Belews Lake were perplexed by the sudden disappearance of the young-of-the-year age class of popular game fish species. This man-made reservoir was fed in part by water from a coal ash settling basin. By 1977, only 3 of the lake's 29 resident species remained. The culprit was determined to be elevated concentrations of selenium (Se) in the food web. Across the country in 1982, federal biologists observed the local extinction of most fish populations in California's Kesterson Reservoir, a wetland area fed by agricultural drainage. They also discovered unnaturally high numbers of dead and deformed bird embryos and chicks. The multiple embryo deformities were sufficiently distinctive to be labeled the "Kesterson syndrome" (Skorupa 1998). Here too, Se was found to be the cause of the devastating impacts to the local ecosystem.

Selenium, however, is not a problem of the past. Se contamination of aquatic ecosystems remains a significant ecological issue of widespread concern, largely because Se is a common by-product of several core economic activities: coal-fired generation of electricity; refining of crude oil; mining of coal, phosphate, copper, and uranium; and irrigated agriculture. Because these industries are likely to continue and grow into the foreseeable future, the potential for large-scale, globally distributed Se contamination of ecological systems is likely to increase.

Since the discovery of the adverse environmental impacts of Se, our ability to identify, quantify, and limit the ecological risk of Se has grown and continues to expand. Starting with the work done at Belews Lake and Kesterson Reservoir, a significant body of research has grown regarding the transport, transformation, and effects of Se in the aquatic environment. We now know that

- Se is distributed globally in organic-rich marine sedimentary rocks,
- most forms of dissolved Se can be transformed and incorporated into food webs,
- organic forms of Se are the most bioavailable,
- the primary route of exposure to Se in consumer animals is via the food web rather than directly from water, and
- maternal transfer of Se to embryos causes reproductive impairment in egg-laying vertebrates.

Although many questions remain, the knowledge we have accumulated during the past 3 decades allows us to assess, predict, and potentially prevent the adverse ecological effects of Se with some confidence.

This chapter 1) provides an overview of the current understanding of Se interactions and impacts, with particular reference to the case studies that are summarized in Appendix A; 2) synthesizes these findings into a conceptual framework that incorporates Se sources, transport and transformation in nature, bioaccumulation and trophic transfer, and effects on ecological systems; 3) uses this conceptual framework to identify strategies for assessing potential Se problems in the field; and 4) recommends key areas for future research. These 4 organizing elements are drawn from the "Problem Formulation" step of the US Environmental Protection Agency (USEPA 1992) Ecological Risk Assessment Guidelines (Text Box 3.1). This chapter provides both an introduction to and a context for the more detailed discussions presented in later chapters.

TEXT BOX 3.1   INITIATING AN ECOLOGICAL RISK ASSESSMENT FOR SELENIUM: PROBLEM FORMULATION

In ecological risk assessments, the Problem Formulation step is designed to help define the nature and extent of the problem, the resources at risk, the ecosystem components to be protected, and the need for additional data to complete the assessment. The Problem Formulation step is often the most important step in the risk assessment process because it identifies the ecosystem attributes to be protected, identifies existing information and data gaps, and provides a means for consensus building between stakeholders for developing an analysis plan. The Problem Formulation step frequently contains 4 main elements, including 1) a synthesis of available information, 2) a conceptual model, 3) assessment endpoints that adequately reflect management goals and the ecosystem they represent, and 4) an analysis plan, which provides the details on data to be collected for risk management decisions (USEPA 1992; Reinert et al. 1998). The conceptual model is intended to identify key features of the ecosystem and resources to be protected and the stressors and the adverse effects that may result. The conceptual model helps identify the hypotheses to be tested during the analysis phase of the assessment.

3.2   WHAT IS SELENIUM (Se)?

The element Se is in the 4th period of group 16 (chalcogen group) of the periodic table. It has an atomic number of 34 and an atomic mass of 78.96 (Lide 1994). Se is chemically related to other members of the chalcogen group, which includes oxygen, sulfur, tellurium, and polonium. Selenium is classified as a non-metal, but elemental Se has several different allotropes that display either non-metal (red Se, black Se) or borderline metalloid or non-metal behavior (grey Se, a semiconductor) (McQuarrie and Rock 1991; Lide 1994). Unlike metals or transition-metals, which typically form cations in aqueous solution, Se is hydrolyzed in aqueous solution to form oxyanions, including selenite (SeO_3^{-2}) and selenate (SeO_4^{-2}). Oxyanions typically have increased solubility and mobility with increasing pH, in contrast to metals, which show the opposite behavior.

Recognizing the non-metallic behavior of Se is one of the keys to a better understanding of its geochemical behavior, but biologically mediated reactions dominate in ecosystems, where Se effects can be beneficial and detrimental (Text Box 3.2). Speciation and biotransformation are widely recognized as playing important roles in determining Se's fate and effects in the environment. Given the richness of biochemical pathways through which Se may be metabolized, it is important to understand the Se biotransformations that may occur in organisms and how they relate to bioavailability, nutrition, and toxicity.

TEXT BOX 3.2   SELENIUM ESSENTIALITY AND TOXICITY

Swedish chemist Jöns Jacob Berzelius is credited with discovering Se in 1818 as a by-product of sulfuric acid production. Berzelius hypothesized that symptoms of toxicity presented by workers in his sulfuric acid factory were due to an impurity present in the pyrite ore used as a production feedstock. Ultimately Berzelius demonstrated that this impurity was an unknown chemical element and named it selenium, from "selene," the ancient Greek word meaning moon (Lide 1994; Wisniak 2000).

In the western United States during the 1930s, Se was identified as the toxic factor of alkali disease in cattle and livestock (Trelease and Beath 1949; Anderson et al. 1961). The US Department of Agriculture conducted both controlled experiments and broad geographic surveys of soil and plant Se to assess the toxic hazards and risks associated with environmental Se. Open-range forage plants included Se accumulator plants of the genus *Astragalus* growing on the Peirre Shale that contained Se concentrations of up to 10,000 mg/kg dw (Trelease and Beath 1949; Anderson et al. 1961). Yang et al. (1983) described an endemic Se intoxication discovered in 1961 in Enshi County, Hubei Province of China. Selenium from a stony coal entered the soil by weathering and was available from alkaline soils for uptake by crops.

In 1957, Se was identified as an essential trace element (or micronutrient) in mammals (Schwarz and Foltz 1957). Proteins containing Se were found to be essential components of certain bacterial and mammalian enzyme systems (e.g., glutathione peroxidase) (Stadtman 1974). Several Se deficiency disorders were identified, including white muscle disease in sheep and mulberry heart disease in pigs (Muth et al. 1958). In the early 1970s, Chinese researchers identified the first major human Se deficiency disease as a childhood cardiomyopathy (Keshan disease; Chinese Medical Association 1979). Thus, Se deficiency as well as toxicity can cause adverse effects in animals.

One of the most important features of Se ecotoxicology is the very narrow margin between nutritionally optimal and potentially toxic dietary exposures for vertebrate animals (Venugopal and Luckey 1978; Wilber 1980; NRC 1989; USDOI 1998). Selenium is less toxic to most plants and invertebrates than to vertebrates. Among vertebrates, reproductive toxicity is one of the most sensitive endpoints, and egg-laying vertebrates have the lowest thresholds of toxicity (USDOI 1998). The most dramatic effects of Se toxicity are extinction of local fish populations and teratogenesis in birds and fish (see Appendix A). Other effects from Se include mortality, mass wasting in adults, reduced juvenile growth, and immune suppression (Skorupa 1998).

Selenium biogeochemistry and the mechanism of entry into living cells are complex (Stadtman 1974, 1996). Se occurs in chemical forms that are analogous to forms of sulfur (S) (Sunde 1997; Fan et al. 1997, 2002; Suzuki and Ogra 2002; Kryukov et al. 2003; Moroder 2005; Unrine et al. 2007). Chief among these are elemental

Se (Se^0), selenide (Se^{-2}), selenite, and selenate, as well as methylated forms Se$_x$(CH$_3$)$_x$. Selenate and selenite can be taken up by plants and converted to organic forms. These organic forms are usually analogues to S-containing biomolecules, especially amino acids. This conversion occurs through either nonspecific isosteric substitution for S in amino acids (selenocysteine or selenomethionine), or through co-translational conjugation of selenophosphate (SePO^{3-}) to serine mediated by selenocysteine tRNA and selenocysteine synthase. In the latter case, selenocysteine is incorporated into genetically encoded selenoproteins (i.e., those proteins whose encoding DNA sequences have a UGA codon and a selenocysteine insertion sequence). In addition, some other metabolites, such as seleno-sugars, are known to occur.

Many enzymes and other proteins have been identified and characterized that require Se for their activity (selenoproteins). In 1973 the first functional selenoproteins were identified: glutathione peroxidase in mammals (Flohé et al. 1973; Rotruck et al. 1973) and formate dehydrogenase and glycine reductase in bacteria (Andreesen and Ljungdahl 1973; Turner and Stadtman 1973). Glutathione peroxidases are part of a large family of proteins that serve a variety of antioxidant and other functions that vary among species and specific tissues (Pappas et al. 2008). These discoveries confirmed Se as an essential nutrient and indicated a role in defense against oxidative injury. It was another decade before a second mammalian selenoprotein was identified as selenoprotein P (Motsenbocker and Tappel 1982). Selenoprotein P (SelP) is now one of the most well-documented selenoproteins. The gene sequence for SelP is highly conserved in bacteria, mammals, and fish (Tujebajeva et al. 2000). The amino acid sequence is rich in selenocysteine, histidine, and cysteine residues, suggesting a function in metal binding or chelation. In fact, SelP has been found to complex with Hg, Ag, Cd, Zn, and Ni (Yoneda and Suzuki 1997a, 1997b; Mostert et al. 1998; Sasaku and Suzuki 1998; Yan and Barrett 1998; Mostert 2000), which supports earlier reports of Se-detoxifying the effects of Hg and Cd in humans and marine mammals (Kosta et al. 1975; Hodson et al. 1984; Pelletier 1985; Osman et al. 1998).

While the glutathione peroxidases and SelP are among the best-known selenoproteins, there are many others. It is now known that the human genome contains 25 genes that encode for selenoproteins (Kryukov et al. 2003). Selenocysteine is genetically encoded by the UGA codon when it occurs with a selenocysteine insertion sequence (SECIS) in the 3′ untranslated region of the DNA sequence (Sunde 1997).

Proteins that contain selenoaminoacids that are nonspecifically incorporated into proteins during translation (i.e., not encoded by a UGA codon and a SECIS) are known as "Se-containing proteins." Selenomethionine, the Se-containing analog of methionine, can be nonspecifically incorporated into peptides because methionyl-tRNA acylase, the enzyme that charges methionyl-tRNA, does not discriminate between methionine and selenomethionine to any great extent (Moroder 2005). A few studies have suggested or demonstrated nonspecific charging of cisteinyl-tRNA with selenocysteine, which could be detrimental for proteins that require cysteine for their structure and function (Wilhelmsen et al. 1985; Müller et al. 1998; Unrine et al. 2007; Garifullina et al. 2008). Analytical identification and quantification of selenocysteine is difficult, which makes it hard to demonstrate nonspecific incorporation into proteins based on analytical data alone (Unrine et al. 2007).

3.3   SOURCES OF Se ENTERING AQUATIC ENVIRONMENTS

Selenium is widely distributed globally and is cycled through environmental compartments via both natural and anthropogenic processes (Nriagu 1989; Haygarth 1994). Ancient organic-rich depositional marine basins are linked to the contemporary global distribution of Se source rocks (Presser et al. 2004a). Figure 3.1 shows a global distribution of phosphate deposits (o) overlain onto that of productive petroleum (a continuum of oil, gas, and coal) basins (+) to generate a global plot of organic-carbon enriched sedimentary basins. The depositional history of these basins and the importance of paleo-latitudinal setting in influencing the composition of the deposits indicate that bioaccumulation may be the primary mechanism of Se enrichment in ancient sediments (Presser 1994; Presser et al. 2004a).

Selenium source rocks in the western United States (Figure 3.2 adapted from Seiler et al. 2003) encompass a wide range of marine sedimentary deposits, from shales mildly enriched in organic carbon to oil shales strongly enriched in organic matter, biogenic silica, phosphate, and trace elements (Presser et al. 2004a). These fine-grained sedimentary rocks provide enriched but disseminated Se sources as 1) bedrock soils for agricultural development or 2) source sediment for alluvial fans (Presser 1994). The areal extent of these rocks in the 17 western states is Upper Cretaceous, approximately 77 million hectares or 17% of the total land area, and Tertiary (mainly Eocene and Miocene), 22 million hectares or 4.6% of the total land area. Depending on their history, Tertiary continental sedimentary deposits may be seleniferous, and these deposits encompass approximately 94.7 million hectares or 20% of the total land area.

Environmental contamination by Se often is associated with particular local Se-enriched geologic formations, as, for example, the Upper Cretaceous–Paleocene Moreno and Eocene–Oligocene Kreyenhagen Formations in the Coast Ranges of California, USA (Presser 1994), the Permian Phosphoria Formation in southeast Idaho

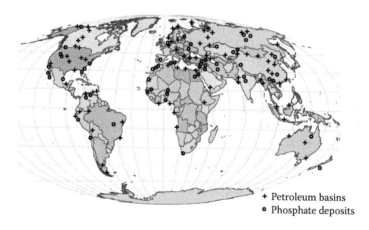

+ Petroleum basins
o Phosphate deposits

FIGURE 3.1   Worldwide distribution of Se-rich geologic formations composed of organic-carbon enriched sedimentary basins. (Adapted from Figure 11-5 in Presser et al. 2004a; http://wwwrcamnl.wr.usgs.gov/Selenium/index.html.)

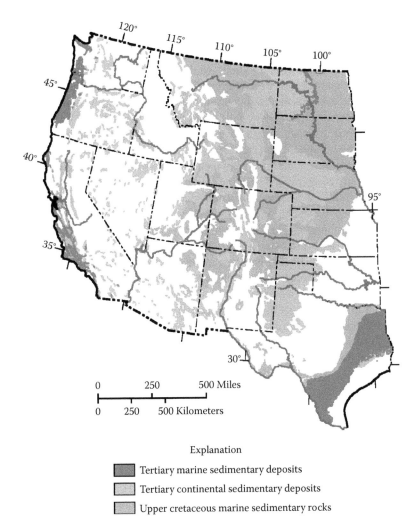

FIGURE 3.2 Selenium source rocks in the western United States. (Adapted from Seiler et al. 2003; http://pubs.usgs.gov/pp/pp1655/.)

(Presser et al. 2004b), the Cretaceous Mist Mountain Formation in Southeastern British Columbia, Canada (Lussier et al. 2003), and the Permian Maokou and Wujiaping shales in south-central China (Zhu et al. 2008). Selenium in these deposits may be present as organic and inorganic forms (Yudovich and Ketris 2006). Selenium is also associated with various sulfide ores of copper, silver, lead and mercury, and uranium (Wang et al. 1993).

Selenium is mobilized through a wide array of anthropogenic activities typically involving contact of a Se-containing matrix with water. In some cases, the contamination will be restricted to local environments, but in other instances Se can be transported a considerable distance from the place of origin.

Selenium in irrigation wastewaters is a significant environmental concern in arid and semi-arid regions (Outridge et al. 1999; Seiler et al. 2003). In areas of seleniferous soils (Figure 3.2), irrigation waters can mobilize dissolved Se predominantly in the form of selenate (Seiler et al. 2003). In these areas, drainage systems often are installed to prevent root zone waterlogging. The resulting oxic drainage water has an alkaline pH and contains elevated concentrations of salts, nitrogenous compounds, and trace elements, including Se (up to 1400 μg Se/L) (Presser and Ohlendorf 1987). Such Se-enriched drainage waters have entered aquatic ecosystems and have been associated with widespread adverse effects (Appendix A).

Although natural weathering slowly mobilizes Se from host rock sequences, this process is greatly accelerated by mining activities that expose the ore and waste rock to oxidation. Oxidized Se and associated metals can infiltrate and leach into the surrounding soils, surface water, and groundwater. Selenium release is of particular concern in coal, phosphate, and uranium mines (Ramirez and Rogers 2002; Presser et al. 2004a, 2004b; Muscatello et al. 2006). Open-pit coal (Dreher and Finkelman 1992; Lussier et al. 2003) and phosphate mines (Hamilton and Buhl 2004) are a significant source of Se because large volumes of rock overlying the target ore seams are left behind in surface waste rock dumps. Selenium is dispersed throughout these deposits but may achieve its highest concentrations in waste-shale zones that occur between the ore zones. In regions where mountaintop mining for coal is practiced, these fresh rock wastes are deposited as "valley fill," providing ideal conditions for both leaching and direct transport of Se-enriched waters into regional ponds, reservoirs, lakes, and rivers (Appendix A).

Selenium release from coal burning for power generation is a major anthropogenic source to the environment, either directly during combustion (Wen and Carignan 2007) or indirectly from disposal of solid combustion waste (coal ash) (Cherry and Guthrie 1977; Johnson 2009). Burning coal oxidizes the organic matter and creates residual wastes, both particulate "fly ash" and larger molten "bottom ash." The fly ash is of particular concern because of its high surface area–to–volume ratio, which facilitates adsorption of mobile trace elements (Jankowski et al. 2006). The resulting Se concentration in waste products may be 4 to 10 times greater than the parent feed coal (Fernández-Turiel et al. 1994). The potential ash waste volumes can be large. More than 400 coal ash disposal sites are designated in the United States. In 2007, about 131 million tons of ash waste was generated, and about 21% of this total was discharged to surface impoundments (Breen 2009). Thermal, pH, and redox conditions during coal combustion help generate predominantly selenite in the ash waste collected on electrostatic precipitators (Yan et al. 2001; Huggins et al. 2007). Selenium is readily solubilized in the alkaline conditions of aquatic fly ash settling basins or fly ash reservoirs (Wang et al. 2007). Clarified ash sluice water or sluice water return flows make their way to local receiving waters as a permitted wastewater discharge or through groundwater seepage. Selenium contamination can occur accidentally due to overfilling events or failures of containment systems. Spectacular events occur as well, such as the catastrophic December 2008 spill of 5.4 million cubic yards of ash from a Tennessee Valley Authority coal-fired power plant (TVA 2009).

The worldwide anthropogenic Se flux to the atmosphere has been estimated at 6.4 M kg/y (Mosher and Duce 1987). Approximately 50% is from coal combustion. Smelting of non-ferrous metal ores involves intense heating to mobilize and isolate the metal of interest; the associated Se and sulfides are volatilized and released in stack gases. Up to 30% of the Se present in feed coal is emitted as a vapor phase, and about 93% of that is returned in the form of elemental Se (Andren and Klein 1975). Roughly 80% of atmospheric Se returns to the ground as wet deposition (Wen and Carignan 2007), mostly near emission sources (Wang et al. 1993). However, depending on atmospheric conditions, stack gases can be carried considerable distances. Seleniferous stack gas from a large copper smelter in Sudbury (ON, Canada) has contaminated lakes up to 30 km away (Schwarcz 1973; Nriagu and Wong 1983).

Crude oil is formed in organic carbon-enriched basins and is a source of Se to the environment. A fraction of Se in crude oil partitions to wastewaters during refining and can be discharged to the environment. Heavy crude oils produced in the San Joaquin Valley and processed at refineries that surround the northern reach of the San Francisco Bay contained 400 to 600 µg Se/L (Cutter and San Diego-McGlone 1990). The northern reach of the bay was listed as impaired by Se discharged from refineries, and control strategies were implemented to reduce Se loads to the bay in 1989 (Presser and Luoma 2006) (Appendix A).

Production and use of Se as a commodity also result in discharge of Se to aquatic systems. More than 80% of the world's production of commercially available Se is derived from anode slimes generated in the electrolytic production of copper (USGS 2000). Processing of the slimes can result in aqueous discharges of Se to surface waters (Naftz et al. 2009). Refined Se is used 1) in electronic components such as rectifiers, capacitors, and photocopy or toner products; 2) in a wide array of industrial applications, such as glass tinting, coloring of plastics, ceramics and glass; 3) as a catalyst in metal plating; and 4) in rubber production (George 2009).

Pharmaceutical applications include dietary Se supplements, anti-fungal treatments, and anti-dandruff shampoos. Each of these uses can result in Se discharges to surface waters and sewage treatment plants. Municipal landfills can generate leachates containing Se that can reach groundwater (Lemly 2004).

In some areas of the world, Se concentrations in soils are below levels adequate to produce feed and forage with sufficient Se to satisfy essential (or optimal) dietary requirements for livestock (Oldfield 1999). Selenium deficiency can be remedied by supplementing Se in feed, some of which may be excreted. Runoff from large feedlot operations where these dietary supplements are used is of particular concern because the Se is in the form of highly bioaccumulative selenomethionine (Lemly 2004). In other cases, fertilizers with nutritional Se amendments (e.g., selenate salts) are applied to lands to rectify this deficiency and enhance production (Watkinson 1983). Under some conditions, application to thin soils having low organic matter has produced short-term elevation of Se concentrations in runoff (Wang et al. 1994), which may be of concern in some receiving environments.

3.3.1 Future Sources of Se

Rapid progress in nanotechnology will likely benefit nearly every sector of science and industry, and consumer products containing nanomaterials are presently entering the market at the rate of 2 to 3 products per week (http://www.nanotechproject.org/). These benefits, however, come with associated risks. Selenium is a key component of nanomaterials such as CdSe or PbSe quantum dots. Quantum dots are nanometer-scale crystallites that function as semiconductors because of quantum confinement effects that occur when the size of the particles approaches the wavelength of their electrons (Reiss et al. 2009). These materials are useful in optoelectronic devices such as light-emitting diodes and photovoltaics. In addition to potential toxicity resulting from degradation of these materials and associated release of Se, emergent properties of the solid-state materials could also elicit toxic responses. For example, active electronic sites that arise from defects in crystal planes and electron hole pairs excited by ultraviolet light could lead to the generation of reactive oxygen species eliciting toxicity (Hardman 2006; Nel et al. 2006). Widespread use of Se-containing nanomaterials could lead to environmental Se contamination, and the environmental consequences may be different from those resulting from current Se sources.

3.3.2 Selected Se Problem Sites

Case studies documented in Appendix A represent a variety of site-specific conditions and include both freshwater and marine sites. Case studies include the following:

- Belews Lake, North Carolina, USA
- Hyco Lake, North Carolina, USA
- Martin Creek Reservoir, Texas, USA
- D-Area Power Plant, Savannah River, South Carolina, USA
- Lake Macquarie, New South Wales, Australia
- Elk River Valley, Southeast British Columbia, Canada
- Areas of the Appalachian mountains affected by mountaintop mining and valley fills
- Kesterson Reservoir, San Joaquin Valley, California, USA
- Terrestrial and aquatic habitats, San Joaquin Valley, California
- Grassland Bypass Project, San Joaquin Valley, California
- San Francisco Bay–Delta Estuary, California
- Phosphate mining in the Upper Blackfoot River watershed, Idaho

Each study compiles information on sources, fate and transformation, effects, and lessons learned. Each case study is distinct with respect to biological receptors; attributes of water, sediment, particulates; food-web pathways; community complexity; and the extent of bioaccumulation and observed effects. A synopsis of 12 case studies, representing diverse Se sources, was previously provided by Skorupa (1998).

A variety of Se contamination events have occurred over the past 40 years in aquatic systems. There have been a number of investigated case studies where elevated environmental Se was attributed to disposal of power plant coal-combustion wastes. These cases (Appendix A) include situations where fly ash was released directly into a nearby water body (e.g., D-Area power plant at Savannah River) or more commonly held in ash settling ponds and the pond effluent released into lakes or reservoirs (e.g., Belews and Hyco Lakes, Martin Reservoir, and Lake Macquarie). In particular, the Belews and Hyco Lakes case studies provided some of the earliest and best-documented evidence of Se effects in the aqueous environment. In some of these cases, confounding factors such as release of other co-occurring contaminants or lack of sufficient information about ecosystem conditions prior to Se addition have made it difficult to ascribe adverse impacts specifically to Se, even though Se toxicosis is well established.

In the now classic study of Belews and Hyco Reservoirs in North Carolina, fly ash pond effluents containing high concentrations of Se were released into the reservoirs for a decade. Both reservoirs experienced reproductive failure of fish populations, transforming formerly diverse fish communities to communities dominated by a few Se-insensitive fish species (Cumbie and Van Horn 1978; Lemly 2002). Fly ash wastewater discharges were later curtailed and a diverse fish community, including Se-sensitive species, was re-established in both waterbodies within several years. However, at each location, more than 20 years later, Se bioaccumulation remains elevated relative to reference sites.

The most well-known case of Se bird poisoning in a field environment is the impoundment of Se-enriched agricultural drainage water in Kesterson Reservoir in the San Joaquin Valley of California. High levels of dissolved Se in drainwater were taken up into the food web, affecting aquatic-dependent wildlife (birds) that showed signs of Se poisoning in adults, as well as reproductive failure due to embryo teratogenesis and failure to hatch (Ohlendorf et al. 1986; Presser 1994). Inputs of irrigation drainwater were halted in the late 1980s, and the reservoir was filled and capped to reduce contact of water with Se-contaminated sediments. Monitoring of ephemeral ponds in the Kesterson area since then shows Se concentrations ranging from 15 to 247 μg Se/L. Aquatic invertebrates collected from these ponds have Se body burdens ranging from 8 to 190 mg/kg dry weight (dw), but Se-induced toxicity has not been observed in aquatic birds (Skorupa 1998). After the capping of Kesterson Reservoir, additional sites receiving agricultural irrigation water were assessed (see case studies in California, Appendix A).

Following the findings at Kesterson Reservoir, the United States Department of the Interior (USDOI) in 1985 initiated the National Irrigation Water Quality Program. Reconnaissance monitoring, or field-level screening, took place at 39 areas in the western United States, where wildlife populations were considered potentially at risk due to agricultural irrigation practices in areas of known seleniferous geological deposits (Presser et al. 1994; Seiler et al. 2003). By 1993, results had confirmed that Se was the contaminant of primary concern at the National Irrigation Water Quality Program study sites, and the receptors generally at greatest risk were water birds (Seiler et al. 2003). Seiler et al. (2003) identified the following sites for further study or remediation planning because these areas were classified as embryotoxic on the basis of Se concentrations in bird eggs:

- Tulare Basin, San Joaquin Valley, California
- Salton Sea, California

- Middle Green River Basin, Utah
- Stillwater Management Area, Nevada
- Kendrick Reclamation Project, Wyoming
- Gunnison-Grand Valley Project, Colorado
- San Juan River area, New Mexico
- Sun River area, Montana
- Riverton Reclamation Project, Wyoming
- Belle Fourche Reclamation, South Dakota
- Dolores-Ute Mountain Area, Colorado
- Lower Colorado River Valley, Texas
- Middle Arkansas Basin, Colorado-Kansas
- Pine River area, Colorado

In some cases, a combination of Se sources has been identified as contributing to elevated levels of Se in ecosystems. For example, the San Francisco Bay–Delta Estuary case study addresses both agricultural drainage-driven inputs plus industrial wastewater contributions. In such instances, an accurate picture of the relative contribution of the multiple sources (e.g., independent characterization of source Se loading and speciation) is useful to conceptually or mechanistically model the ecosystem.

Studies demonstrating the growing potential of Se-related impacts relating to mining activities include coal mining and phosphate mining (Appendix A). Open pit mining practices have in the past produced "pit lakes" with elevated Se concentrations when mining activities were terminated. Mining in areas with productive coal bed or ore deposits results in the weathering of Se from mining overburden material and, in some areas, contamination of groundwater that subsequently seeps into surface water areas. In the Elk River Valley of southeastern British Columbia, open coal pit mining over the past decades has resulted in sharply increasing surface water Se concentrations. Selenium concentrations in discharges (primarily selenate) often exceed 300 µg/L. Downstream of the mines, lotic, lentic, and marsh areas are receiving substantial Se loads, leading to bioaccumulation in macrophytes, benthic macroinvertebrates, and a variety of secondary consumers. Individual-level early life stage effects have been observed in 2 fish species, marsh and water birds, and frogs, but population-level effects linked to Se have been more difficult to establish in field studies (Harding et al. 2005; Orr et al. 2006; Canton et al. 2008).

3.4 SELENIUM CYCLING AND BIOACCUMULATION IN AQUATIC ECOSYSTEMS

Figure 3.3 is a conceptual model of Se dynamics and transfer in aquatic ecosystems. The model illustrates the steps that determine Se effects in ecosystems. Those steps are described in detail below.

3.4.1 SELENIUM SPECIATION IN WATER, PARTICULATES, AND BIOTA

Selenium from natural and anthropogenic sources typically enters aquatic ecosystems as the oxidized inorganic anions, selenate (Se^{+4}) and selenite (Se^{+6}), although

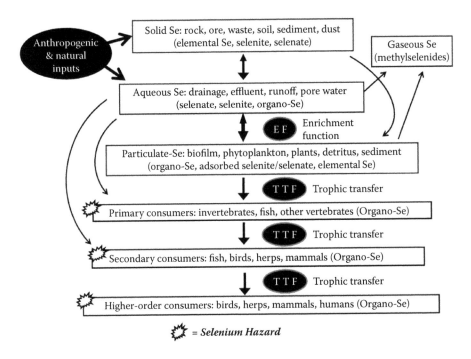

FIGURE 3.3 Conceptual model of Se dynamics and transfer in aquatic ecosystems.

small amounts of dissolved organic Se compounds (Se^{-2}) also can be present in water due to biological activity. Selenate and selenite can be the predominant species present in the water columns of aquatic ecosystems (Figure 3.3). While the aqueous phase is operationally defined as materials passing through a filter with 0.45 μm or smaller pore diameter, colloidal (non-dissolved) Se may be present in this fraction. In terms of mass balance, transport of Se via sediment is usually a lesser route of entry for Se into aquatic ecosystems. However, in terms of biological reactivity, suspended material in an ecosystem plays an important role determining the effects of Se.

The biogeochemical cycling of Se in aquatic systems is characterized by the predominance of biologically mediated reactions over thermodynamically driven reactions (Stadtman 1974, 1996; Oremland et al. 1989, 1990). Both selenate and selenite anions can be actively taken up by microbes, algae, and plants and converted to organic Se compounds, including Se analogues of sulfur-containing biomolecules (Fan et al. 1997, 2002; Stadlober et al. 2001). Selenium is sequentially reduced to Se^{-2} before it is ultimately incorporated into the amino acids selenocysteine and selenomethionine (Sunde 1997). Selenomethionine is the primary organic form of Se at the base of aquatic food webs. Selenocysteine is primarily present in selenoproteins in which the selenocysteine is genetically encoded. Selenocysteine is readily oxidized, indicating that it should not be persistent under ambient conditions outside of organisms. Selenocysteine typically accounts for a relatively small proportion of total Se in most plants with elevated Se concentrations, where excess Se accumulates as selenomethionine (Wu 1998). For these reasons, selenomethionine is thought to be the primary organic form of Se relevant to bioaccumulation and toxicity in food webs (Fan et al. 2002).

For example, Se often enters a stream as selenate. If that stream flows into a wetland and is retained there with sufficient residence time, then recycling of Se may occur. During recycling, particulate Se is generated from dissolved Se species. The transformed reduced species are then returned to the water as these organisms die and decay. The more recycling, the more organo-Se and selenite are produced. Neither of these latter forms can be easily reoxidized to selenate because that reaction takes hundreds of years (Cutter and Bruland 1984). The net outcome of recycling in a watershed is a gradual build-up of selenite and organo-Se in the system. Thus, biologically mediated reactions drive conversions among dissolved species and transformation of dissolved Se to particulate species.

Bacterially mediated reactions can also produce volatile methylated Se species, which are rapidly lost to the atmosphere, or insoluble elemental Se(0), which tends to accumulate in anaerobic sediments (Fan et al. 1998; Turner et al. 1998; Peters et al. 1999; de Souza et al. 2001).

3.4.2 SELENIUM UPTAKE AND TRANSFER IN AQUATIC FOOD WEBS

Fine particulate organic matter, composed of living and dead biotic material and some associated inorganic particles, may contain varying proportions of inorganic and organic Se species. Consumption of these particles by primary consumers, typically invertebrates and small fish, is the primary pathway for Se entry into aquatic food webs (Figure 3.3).

Partitioning between water and particulates is a dynamic biogeochemical process that is difficult to model because equilibrium geochemical modeling fails to describe major biological processes. However, Se partitioning for any location and time can be described by a distribution coefficient or enrichment function (EF), which describes the relationship between Se concentrations in particulate and dissolved phases:

EF = Se concentrations in particulates (micrograms/kg dw)/Se concentrations in water (µg/L)   (1)

The EF usually refers to a simple ratio, as described here, but can be elaborated into a more complex enrichment function that describes variation in Se uptake in response to different environmental factors. Presser and Luoma (2010) compiled data from 52 field studies in which both water-column and particulate Se concentrations were determined. They calculated EFs, which they termed the partitioning coefficient, K_d. The K_ds across the variety of ecosystems (ponds, rivers, estuaries) vary by as much as 2 orders of magnitude (100 to 10,000) and measure up to 40,000. Most rivers and creeks show K_ds of >100 and <300 (e.g., San Joaquin River [CA, USA] at 150). Lakes and reservoirs usually have K_ds > 300, with many in the 500 to 3,000 range (e.g., Belews Lake [NC, USA] at 3,000). Those K_ds >3,000 are usually associated with estuary and ocean conditions (e.g., San Francisco Bay [CA, USA] at 10,000 to 40,000). Exceptions from this categorization can occur as a result of speciation effects and other site-specific conditions.

The EF represents the outcome of Se transformations occurring in a specific eco-system, but it does not differentiate those processes. There have been few attempts to develop biogeochemical models to quantify these processes (Meseck and Cutter 2006). For ecosystem-scale modeling, EF is estimated from field determinations of dissolved Se concentrations and Se concentrations in one or more types of particles. It is recognized that this operational EF will vary widely among environments. An important part of the methodology is to use the characteristics of the environment in question to narrow the potential variability. Hence, it is critical for site-specific Se assessments to quantify Se concentrations in particulates forming the base of the food web.

Bioaccumulation of Se from particulates by primary aquatic consumers is a key determinant of dietary Se exposure and, therefore, of the risk of Se toxicity to higher-order aquatic consumers (e.g., predatory fish and aquatic birds) (Figure 3.3; Wang 2002; Luoma and Rainbow 2005, 2008). Biodynamic models, which characterize the balance between gross Se influx rate and the gross efflux rate, can be the basis for modeling Se bioaccumulation and trophic transfer in aquatic ecosystems (Presser and Luoma 2010). For primary consumers, biodynamic experiments indicate that uptake of dissolved Se is negligible compared with Se uptake from diets of fine particulates (Luoma et al. 1992). With simplifying assumptions (i.e., no uptake of dissolved Se and no growth), the exposure equation for consumers is

$$C_{\text{consumer}} = [(\text{AE})(\text{IR})(C_{\text{diet}})]/[k_e] \qquad (2)$$

The species-specific information in this equation (ingestion rate [IR], assimilation efficiency [AE], and efflux rate constant [k_e]) can be determined from kinetic experiments with invertebrates that serve as the basis of many important food webs (see Chapter 5). These parameters can be combined to calculate a trophic transfer factor (TTF) for Se. The modeled TTF characterizes the potential for a consumer to bioaccumulate Se from its diet, based on the balance of Se influx and efflux. Because TTF is defined as the Se concentration in a consumer (mg/kg dw) divided by Se concentration in diet (mg/kg dw), the preceding equation can be expressed as

$$\text{TTF} = (\text{AE})\,(\text{IR})/k_e \qquad (3)$$

Selenium TTFs determined for marine and freshwater invertebrates vary widely, from 0.6 for amphipods to 23 for barnacles (Presser and Luoma 2010; Chapter 5). This variation in TTF is propagated by trophic transfer, making some food webs and some predatory taxa more vulnerable to Se bioaccumulation and toxic effects.

Biodynamic models have been developed primarily for invertebrates feeding on particulate organic matter, but the same modeling approach can also be applied to higher-order consumers, such as fish feeding on invertebrates or other fish (Baines et al. 2002). Selenium TTFs for predatory fish are less variable (range, 0.6 to 1.7; mean 1.1) than those for invertebrates (Presser and Luoma 2010). The conceptual model (Figure 3.3) summarizes Se transfer from water to organic particulates at the base of the food web to primary consumers and predators. Food web modeling based on EFs and TTFs is illustrated in more detail by Presser and Luoma (2010) and in Chapter 5.

3.4.3 Food-Web Exposure and Toxicity Risks

Biodynamic modeling can provide insight into variability of Se exposures among different ecosystems and different trophic levels. Selenium TTFs are useful metrics for understanding this process because they describe the bioaccumulation in animals across each trophic linkage. Contaminants that biomagnify would be expected to have TTFs >1.0 at each trophic linkage within a food chain. Although Se TTFs are variable among different ecosystems, they tend to be similar within groups of related species or species with similar trophic status. It is clear that the majority of food chain enrichment with Se occurs at the lower trophic levels and that less enrichment occurs at higher trophic levels. A compilation of TTFs for Se indicates that, for freshwater primary consumers, TTFs range from 0.9 for amphipods to 7.4 for zebra mussels; TTFs for fish average 1.1 (Presser and Luoma 2010). These observations have important implications for problem formulation and risk assessment. Unlike contaminants that strongly biomagnify in higher trophic levels (e.g., DDT and Hg), for Se, secondary and tertiary consumers may not experience substantially higher Se exposure than lower trophic levels, because enrichment of Se in aquatic food webs primarily occurs in particulates and primary consumers. For example, a recent study suggests that amphibian larvae that primarily graze periphyton actually bioaccumulate higher Se concentrations than do predatory fish in the same system (Unrine et al. 2007).

However, to establish risk, Se exposure and the magnitude of Se bioaccumulation must be considered along with an animal's sensitivity to Se. Birds and fish (predators) are the 2 taxa of animals most sensitive to aquatic Se contamination (i.e, they are the first to express the effects of Se within ecosystems), with embryonic and larval life-stages being of particular concern. Invertebrates, on the other hand, are relatively insensitive to Se (Lemly 1993a; Presser and Luoma 2006). Thus, the organisms that are most at risk are higher-order predators.

Risks of toxicity to aquatic organisms may be driven by differences in Se exposure mediated by food-web transfer. In a toxicological sense, Se sensitivity is an inherent property of the species. However, differences in Se exposure among ecosystems may be more significant than differences in the toxicological sensitivity among species. Trophic structure (who is eating whom) is as important as trophic position (food chain length) in determining Se bioaccumulation within food webs (Stewart et al. 2004; Presser and Luoma 2010). Combining site-specific estimates of EFs with generic TTFs for different taxonomic groups or species of invertebrates, fish, and birds can help explain how environmental Se concentrations will differ among ecosystems that exhibit differing ecological and biogeochemical characteristics.

3.5 ADVERSE EFFECTS OF Se

Risk assessment protocols for most contaminants consider 2 thresholds: 1) concentrations that cause adverse effects following short-term exposures (acute toxicity) and 2) concentrations that cause adverse effects following long-term exposure (chronic effects). Because adverse effects due to Se exposure are predominantly related to food web exposure, the standard concept of acute Se toxicity based on aqueous exposures has limited applicability in nature.

Chronic dietary toxicity from Se is manifested primarily as reproductive impairment due to the maternal transfer of Se, leading to embryotoxicity and teratogenicity (Gillespie and Bauman 1986; Lemly 1993b, 1998; Skorupa 1998; Ohlendorf 2003). This is particularly true for egg-laying vertebrates because Se is incorporated into egg yolk proteins (Kroll and Doroshov 1991; Davis and Fear 1996; Unrine et al. 2006). In addition to reproductive impairment, Se has a variety of other sublethal effects, including reductions in growth and condition index (Sorenson et al. 1984; Heinz et al. 1987; Ohlendorf 2003), tissue pathology (Sorenson et al. 1982a, 1982b, 1983a, 1983b, 1984; Sorenson 1988), and induction of oxidative stress (Spallholz and Hoffman 2002; Palace et al. 2004). Selenium can be lethal to adult organisms (Ohlendorf 1989, 2003; Heinz 1996) as demonstrated by mass mortalities of adult coots (*Fulica americana*) that occurred in agricultural drainwater habitats in California (USA) (Skorupa 1998). However, most aqueous and dietary concentrations of Se encountered by wildlife are not high enough to be lethal to adults.

Chronic toxicity to birds and fish is strongly associated with concentrations of the Se-substituted amino acid, selenomethionine, in diets and tissues of exposed biota. Studies with mallards (*Anas platyrhynchos*) have demonstrated that diets containing the naturally occurring form of selenomethionine (L-selenomethionine) were more toxic than diets containing either the synthetic enantiomeric mixture, D,L-selenomethionine, or inorganic Se (as selenite) (Heinz et al. 1988; Hoffman et al. 1996). Hamilton et al. (1990) demonstrated that toxic effects of artificial diets spiked with selenomethionine fed to Chinook salmon (*Onchorhynchus tshawytscha*) were similar to effects of diets prepared from wild mosquitofish (*Gambusia affinis*) collected from Se-contaminated habitats.

The sensitivity of aquatic taxa to Se toxicity, expressed in relation to Se concentrations in tissues or diets, varies widely among fish and aquatic-dependent birds (Staub et al. 2004). Concentrations of Se that cause adverse effects may differ substantially even between closely related species, such as rainbow trout (*Onchorhynchus mykiss*) and cutthroat trout (*O. clarki*; see Chapter 6). Similarly, 2 species of wading birds in the family Recurvirostridae showed widely differing effect concentrations for embryo hatchability and teratogenicity, with the black-necked stilt (*Himantopus mexicanus*) being much more sensitive than the American avocet (*Recurvirostra americana*) (Skorupa 1998).

The effects of Se on the survival and reproduction of individuals can lead to adverse changes to populations and community structure (Figure 3.4) (Garrett and Inman 1984; Lemly 1993a). Population and community-level effects have been primarily documented in aquatic systems where movement of organisms (emigration and immigration) is restricted. In the classic example of Belews Lake (NC, USA), 26 of 29 resident fish species experienced local extinction (Appendix A) due to reproductive failure caused by Se (Lemly 1993b, 1998).

Elimination of species from communities, particularly those taxa that exert strong top–down (some predators) or bottom–up (some microbes or benthic invertebrates) effects may have ecosystem-wide repercussions, particularly when sufficient functional redundancy is absent in the system. Se-induced shifts in community composition due to declines of certain invertebrate or forage fish species could result in reduced quality and/or quantity of food resources for higher trophic-level consumers.

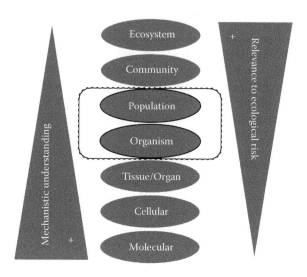

FIGURE 3.4   Hierarchy of effects across levels of biological organization.

Most of what we know about Se bioaccumulation and toxicity comes from studies of birds and fish, but relatively little is known about Se toxicity in other vertebrates. The process of maternal transfer of Se in viviparous vertebrates (i.e., mammals and some herpetofauna) is poorly understood, but it appears that the margin between essentiality and toxicity of Se is much broader for placental mammals than for egg-layers (NRC 1980; see Chapter 6). Thus, among vertebrates, the most notable knowledge gap regarding Se exposure and toxicity is for oviparous species of amphibians and reptiles. This knowledge gap prevents phylogenetic comparisons regarding Se sensitivity.

Amphibians and reptiles are among the most critically endangered vertebrates (Gibbons et al. 2000; Stuart et al. 2004; Wake and Vredenburg 2008). Collectively referred to as "herpetofauna," they are also ecologically important in both aquatic and terrestrial ecosystems. As ectotherms with low energy requirements, herpetofauna can achieve high biomasses compared with mammals and birds occupying similar trophic levels (Hopkins 2006, 2007). In numerous ecosystems, where vertebrate numbers and biomass have been carefully calculated, salamanders, frogs, lizards, and snakes have been shown to be far more abundant than most other vertebrates (Burton and Likens 1975; Roughgarden 1995; Rodda et al. 1999; Petranka and Murray 2001; Gibbons et al. 2006). Thus, herpetofauna greatly influence the cycling of energy and nutrients in many ecological systems (Seale 1980; Wyman 1998; Bouchard and Bjorndal 2000; Beard et al. 2002; Ranvestel et al. 2004; Gibbons et al. 2006; Regester et al. 2006) and may play significant roles in the cycling of contaminants, including Se, in food webs (Hopkins 2006, 2007; Hopkins and Rowe 2010).

In a system contaminated with coal combustion wastes in South Carolina (USA), water snakes (*Nerodia fasciata*) accumulated elevated concentrations of Se from the fish and amphibians they ingested (Hopkins et al. 1999). Based on indirect evidence from long-term controlled feeding studies (Hopkins et al. 2001, 2002a) and additional field studies on amphibians (Roe et al. 2005; Hopkins et al. 2006), it appears that the elevated Se concentrations in snakes were more likely due to ingestion of amphibians than fish (Hopkins 2006). No studies have evaluated the importance of amphibian and reptilian prey as pathways of Se exposure to fish, birds, or mammals that commonly ingest them. Nor have any studies rigorously examined bioaccumulation and effects of Se in top trophic-level reptiles such as snapping turtles and alligators, despite many traits that make these species desirable for ecotoxicological studies (Hopkins 2000, 2006; Roe et al. 2004; Bergeron et al. 2007).

Like birds and fish, reptiles and amphibians partition significant quantities of the Se they accumulate into their ovaries, with subsequent maternal transfer to their eggs. Turtles, alligators, snakes, lizards, and frogs have all been shown to maternally transfer Se (Nagle et al. 2001; Hopkins et al. 2004a, 2005a, 2005b, 2006; Roe et al. 2004). In controlled feeding studies with lizards and field surveys of frogs, 33% to 53% of a female's total body burden of Se prior to oviposition was transferred to her follicles or eggs (Hopkins et al. 2005a, 2005b, 2006). Spinal deformities in Columbia spotted frog embryos with Se concentrations up to 20 mg/kg dw were documented in the Elk River Valley (BC, Canada) watershed (Appendix A).

The reproductive effects and developmental consequences of Se deposition into reptilian eggs remain largely unexplored. A field study with adult amphibians demonstrated that females that transferred excessive concentrations of Se and Sr to their eggs also experienced significant reproductive impairment, including teratogenic effects characteristic of Se toxicity (Hopkins et al. 2006; discussed in more detail in Chapter 6). Additional field studies and controlled dietary exposures linked to adverse reproductive outcomes, much like those conducted on birds and fish, are needed for these diverse and threatened group of vertebrates.

3.6   ECOSYSTEM RECOVERY FOLLOWING Se CONTAMINATION

A limited number of examples are available which document the recovery of impacted aquatic populations in Se-contaminated ecosystems. The recovery of the warm water fish community in Belews Lake represents the most comprehensive example currently available. Prior to being impacted by coal ash effluent, the Belews Lake fish community was diverse, with 29 species. The lake began receiving Se-laden ash pond effluents in 1975. The changes in the warmwater fish community in Belews Lake were documented by sampling lake coves during the period 1977 to 1984, coupled with muscle tissue Se measurements in selected taxa collected from trap nets or by electrofishing (Barwick and Harrell 1997). Monitoring showed significantly reduced fish diversity and biomass during 1977 to 1981, as the lake continued to receive some Se-laden ash pond effluents. In 1978 only 7 taxa were present; in 1979 only 3 were collected. By the mid-1980s, all seleniferous loading to the lake from ash pond effluent was curtailed. Fishery monitoring in successive years indicated a gradual reestablishment of a diverse community, as the range of species successfully expanded downlake from a relatively unimpacted headwater area (Lemly 1997; Barwick and Harrell 1997). By 1985, as median Belews Lake Se water column concentrations decreased to <5 μg/L, 21 fish species had returned to the main body of Belews Lake (1984 and 1985 data; Barwick and Harrell 1997). By 1990, within

5 years of termination of ash pond effluents, 26 fish taxa (combined 1984–1990 data) had been documented (Barwick and Harrell 1997).

Compared with these population-level responses that indicated recovery of the system over a 5-year period, Se residues in monitored taxa, including catfish (*Ameiurus* spp. and *Ictalurus* spp.), green sunfish (*Lepomis cyanellus*), and bluegills (*L. macrochirus*), were slow to decrease. Muscle Se concentrations in these taxa decreased from average concentrations (converted to dry weight from wet weight, using an estimated 75% moisture content) of 42 mg/kg in catfish and 87 mg/kg in green sunfish during 1983–1987, to levels between 4.0 and 15 mg/kg, respectively, by 1992. Those concentrations remained well above reference-site fish residues, however, and low frequencies (up to 6%) of malformed fish larvae continued to be reported as late as 1996 (Lemly 1997). A continuing decline in fish Se concentrations has been closely linked with gradually declining Se concentrations in sediment and benthic food webs in Belews Lake.

Following the termination of drainwater inputs and the filling of the ponds at Kesterson National Wildlife Refuge, monitoring and modeling indicated that reduced, but persistent, Se exposures from the terrestrial habitat and ephemeral pools would continue to present a low level of risk to wildlife (Ohlendorf 2002). Although Se concentrations in specific food webs remained above toxicity levels of concern and slightly elevated with respect to reference sites, Ohlendorf (2002) concluded that Se concentrations in terrestrial and aquatic wildlife did not pose substantial risk of adverse effects on reproductive or other responses.

Under some conditions, recovery of populations of a specific receptor species may not take place. For example, Se amendments made to a series of Swedish lakes with elevated levels of mercury in the 1980s is thought to have resulted in the local extirpation of perch (*Perca fluviatilis*) from several lakes isolated from source populations (Paulsson and Lundbergh 1989; Skorupa 1998).

In summary, these cases indicate that some aquatic populations may recover in the several years following the cessation of aqueous Se inputs. However, aquatic communities commonly include important benthic food webs. Selenium concentrations in sediment typically decline more slowly than water column concentrations. Therefore, natural attenuation of Se in food webs may require several years or even decades.

3.7   STRATEGIES FOR ASSESSING THE RESOURCE TO BE PROTECTED

3.7.1   System Characteristics

Source, habitat, and food web characteristics, along with other stressors, influence Se's overall effect on an ecosystem (Figure 3.5). These characteristics are important in developing a strategy to assess an ecosystem that may be at risk from Se contamination.

Both the amount and the chemical form of Se discharged into an ecosystem help to determine its fate and effects. Most often, Se enters aquatic systems as a highly water-soluble oxyanion (i.e., selenate or selenite). In typical coal combustion wastewaters, for example, most of the Se enters the ecosystem as selenite. The efficiency of uptake

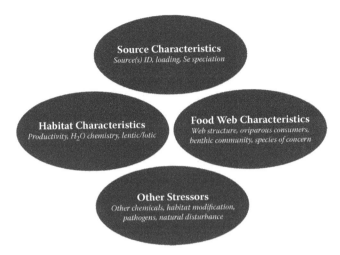

FIGURE 3.5 Ecosystem characteristics that influence Se cycling, bioavailability, and effects.

by plankton from the water column is greater for selenite than for selenate, resulting in a rapid flux of Se into the aquatic food web (Besser et al. 1989, 1993; Riedel et al. 1996). Relative to selenate, selenite is also more readily complexed and precipitated from the water column via non-biological pathways (e.g., by co-precipitation with metal hydroxides: NAS 1976; Simmons and Wallschläger 2005). These properties tend to favor incorporation of selenite-Se into particulates, which facilitates a benthic exposure pathway for consumers. Increased severity and rate of manifestations of selenite-Se-induced toxicity observed in biota (e.g., at Belews Lake), relative to eco-systems receiving a similar or greater concentrations of selenate-Se (e.g., Kesterson Reservoir), have been attributed to these differences in source speciation (Skorupa 1998; Appendix A). Source characterization should include temporal analyses both as a means to accurately assess loading rates and to confirm Se speciation over time.

The conditions within a receiving water body are important factors contributing to Se accumulation within components of food webs. The most severe Se toxicity problems documented to date have occurred in lentic systems with elevated Se inputs and comparatively long residence time. High biological productivity tends to increase the rate of incorporation of dissolved inorganic Se into biota, resulting in high concentrations of bioavailable Se in biota and organic detritus (Orr et al. 2006). Systems with lower productivity and shorter residence times result in less accumulation of Se. High levels of microbial activity are typical of high-productivity lentic and wetland habitats that are most often also associated with high levels of Se bioavailability. This is not surprising, because microbially mediated reactions are involved in many of the transformations that affect Se fate and bioavailability, including reduction of selenate (least bioavailable) to selenite (more bioavailable) and reduction of these inorganic species to organic selenides (most bioavailable) (Riedel et al. 1996). Microbial activity can also lead to reduced Se bioavailability, for example, by formation of elemental Se, an insoluble form that tends to accumulate in sediments, or loss from the aquatic system by formation of volatile methylselenide species (Fan et al. 1998).

In aquatic ecosystems with low productivity and short residence times the accumulation of Se in the food web is expected to be reduced. For example, Adams et al. (2000) demonstrated significant differences in the accumulation of Se in fish from lotic versus lentic systems. However, the fate of Se in some localized habitats may vary widely. In sites such as marginal wetlands, side channels, and seasonally flooded areas, local hydraulic residence time is longer and productivity is higher than in main-channel habitats. This leads to greater Se accumulation in organic detritus and organic-rich sediments, greater biotransformation of inorganic Se, and greater Se bioaccumulation.

Hydrology, productivity, and microbial activity of aquatic habitats influence the quantity and type of fine particulate organic matter available at the base of aquatic food webs. These differences are reflected in the speciation and bioavailability of particulate Se (Presser and Luoma 2010). Operationally defined EFs characterize Se partitioning between water and particulate matter for aquatic systems. Systems with relatively low EFs (<500) are streams, whereas systems with the highest EFs (>2000) tend to be dominated by highly productive wetlands and estuaries.

The magnitude of EFs for primary producers is an important determinant of the potential for Se bioaccumulation in food webs. The fate of Se entering aquatic food webs, however, is further modified by differences in food web structure among aquatic ecosystems. The Se exposure of higher-order predators is predominantly determined by the specific taxa that comprise these links rather than the number of trophic links in their food webs. Predators that consume aquatic taxa such as marine bivalves, which have exceptionally high TTFs (range: 1.4 to 23), may experience greater Se exposure than other predators in the same ecosystems (Presser and Luoma 2010).

Food web linkages to the top oviparous consumers should be included in Se site assessments. Reproductive impairment and early life stage malformations in high trophic-level egg-laying (oviparous) vertebrate species, including fish and aquatic-dependent birds, are the most frequently documented manifestations of Se toxicity. Understudied oviparous species, such as reptiles and amphibians, can make up a substantial fraction of biomass and are critical components in system energy transfer and ecology.

Food web structures and hence the potential for dietary Se exposure of top predator species are commonly highly complex. Consumers utilize a wide variety of food sources that are influenced by season, migratory patterns, or life stage–dependent factors. Temperate lentic habitats, when provided with sufficient soluble nutrients, support a robust but seasonally variable food web. Partitioning of water-column Se in particulates is efficient, as reflected in higher EFs in lentic versus lotic systems. Benthic organisms comprise an important component of both lentic and lotic food webs, but lentic sediment is typically comprises fine particulates, including biogenic particulate organic material. The organic component (total organic carbon) of sediments has been associated with higher Se concentrations and, further, strongly appears to influence the magnitude of Se bioaccumulation in benthic invertebrates. In lotic systems, substrates and stream velocities are less amenable to accumulation in fine particulate material and detritus, except in backwater areas, which are essentially lentic habitats.

The length of the food web and number of trophic levels represented may not reflect the magnitude of the risks posed by environmental Se contamination to species of concern. In San Francisco Bay (Appendix A), white sturgeon (*Acipenser transmontanus*), an exceptionally long-lived top predator, consume great quantities of an invasive clam species (*Potamocorbula amurensis*). While Se concentrations remain relatively low in both the water column (<1 μg/L) and suspended particulates (0.5 to 1.5 mg/kg dw), Se is bioaccumulated efficiently to potentially problematic Se concentrations by sturgeon because of the approximate 6-fold trophic transfer from particulate to clam. In the same ecosystem, juvenile striped bass (*Morone saxatilis*) utilize a slightly longer food web comprising first- and second-order crustacean consumers (zooplankton and mysid shrimp). The bass do not accumulate Se to problematic concentrations because trophic transfer is less than 2-fold. In mechanistic terms, the key difference between the 2 food webs, and therefore the exposures of predatory fish, is the very low efflux rate of Se from clam tissue relative to the crustacean food items (Stewart et al. 2004).

Se toxicity may be enhanced by other ecological variables normally encountered by animals in nature. For example, Chapter 6 discusses potential temperature effects on Se toxicity (e.g., the Lemly [1993c] "Winter Stress Syndrome"). Selenium-induced shifts in community composition due to declines of certain invertebrate or forage fish species could result in reduced quality and/or quantity of food resources for higher trophic-level consumers. Such indirect effects mediated through nutritional deficits are widespread in systems contaminated by other contaminants (Fleeger et al. 2003), including complex waste mixtures containing Se (Hopkins et al. 2002b, 2004b; Roe et al. 2006). Possible interactions between Se and abiotic variables (e.g., temperature, salinity, climate), life history events (e.g., migration, metamorphosis), and other anthropogenic factors (e.g., eutrophication, habitat modification, interactions with other contaminants) are also knowledge gaps that need to be addressed to better inform future risk of Se to aquatic biota.

Another major challenge to evaluating Se toxicity is its well-documented interaction with other constituents of aquatic environments. For example, sulfate inhibits uptake of selenate by plants and has an antagonistic effect on the acute toxicity of selenate (dissolved route of exposure only) to invertebrates and fish (Brix et al. 2001). However, sulfate–selenate interactions have not been shown to influence Se transfer via trophic transfer, which is the primary exposure mechanism for chronic toxicity (Besser et al. 1989; Skorupa 1998; Presser and Luoma 2010). A more significant challenge to evaluating Se toxicity in the field is its common co-occurrence with other contaminants. Many of the industrial sources of Se also emit additional trace elements and in some cases organic contaminants. For example, coal combustion produces solid waste containing elevated concentrations of more than a dozen potentially toxic trace elements (Rowe et al. 2002). This complication is not unique to Se, because all habitats on the planet contain measurable concentrations of other contaminants. However, for Se this may become a major source of uncertainty because it is well known that Se interacts with other contaminants such as Hg and As (Cuvin-Aralar and Furness 1991; Yoneda and Suzuki 1997a, 1997b; Heinz and Hoffman 1998; Hopkins et al. 2006, 2007). Synergistic, additive, and/or antagonistic interactions are likely in some Se-contaminated systems. These interactions are complex and are likely to be site specific. Revealing the molecular mechanisms behind these interactions with Se also may allow better predictive power in these situations.

Investigation of population-, community-, and ecosystem-level responses to Se contamination also may be complicated by the presence of other stressors such as habitat modification, altered hydrology, species introductions, diseases, and the like. Each of these factors would be relevant for establishing hypothetical or actual reference site conditions, as would consideration of natural successional stages.

3.7.2 ASSESSMENT ENDPOINTS AND MEASURES OF EXPOSURE AND EFFECT

When episodes of Se contamination occur or are suspected, it is useful to have a method to assess the possible adverse effects on the ecological systems in the field. The Ecological Risk Assessment framework developed by the US Environmental Protection Agency (USEPA 1992) recommends that assessment endpoints and associated measures be used for this purpose. In this context, assessment endpoints represent components that sustain the structure, function, and diversity of an ecological system, or components that may be valued for other reasons (such as a rare species). Assessment endpoints may be identified at any level of biological organization: molecular, cellular, organism, population, community, and ecosystem (Figure 3.4). Once the assessment endpoints are selected, measures of exposure and effects can be identified. These measures reflect the actual types of data that will need to be collected in order to complete the risk assessment. Ideally, they should be able to be measured relatively easily, either indirectly or directly.

Generic assessment endpoints and measures that can be used to determine the effects of Se contamination on an ecological system were derived from the synthesis of Se research presented previously, as well as the conceptual models proposed for exposure pathways and ecological effects. Measures of exposure and effects are categorized in Table 3.1, and the measures of system characteristics are subsumed within the community- and ecosystem-level exposure and effects measures. Data collection on the key measures (in bold text) is recommended for systems where a Se problem is strongly suspected or has been identified. For systems where studies are just beginning and less information exists on whether Se is an influence, the first steps might be to measure Se concentrations in water, particulate phases (including organic carbon content of the sediment), and tissues of primary consumers.

For the purpose of characterizing Se exposure in a particular aquatic ecosystem, the recommended measures are Se concentrations in water and in biogenic particulates (used to calculate the EF) and measurement of Se concentrations in dominant primary consumers. These measures capture much of the site-specific variation in Se enrichment at the base of aquatic food webs. Temporally and spatially matched samples related to specific food webs are valuable given the site-specific nature of Se effects. The most appropriate measure of Se exposures for the purpose of estimating Se hazards to higher-order consumers is Se concentrations in eggs or mature ovaries of vertebrates (fish and/or birds), which are the best predictors of the toxic effects of Se on embryo and larval stages; measurement of Se concentrations in diets, muscle, and whole organisms are less predictive of toxic effects of Se (Chapter 7). Measurement of the biologically active species, Se-methionine, at various levels of

TABLE 3.1

Key Assessment Endpoints and Corresponding Exposure and Effects Measures for Se Risk Assessments in Aquatic Systems. Data Collection for the Key Measures (in Boldface Type) Is Recommended for Systems Where a Se Problem Is Strongly Suspected or Has Been Identified

Level of Organization	Assessment Endpoint	Measures of Exposure	Measures of Effect
Molecular or cellular	Oxidative stress protection	Se in subcellular compartments	Enzyme assays and gene expression
	Normal biomolecule structure and function	Se substitution in biomolecules	
Tissue	Normal tissue structure and function	Total Se and /or selenomethionine in tissue	Pathology of liver, kidney, eyes, gills, blood, gonad
			Relative organ weight
Organism	Survival, growth, and reproduction of egg-laying vertebrates	**Se in female reproductive tissue of oviparous vertebrates**	Survival
			Growth
		Se in whole-body or surrogate tissue	Body condition
			Edema
			Embryo malformation
			Egg hatchability
			Immuno-competence
			Incidence of parasites or disease
Population	Population sustainability	Se in diet	**Reduced abundance**
			Population structure
			Change in genetic diversity
Community	Community structure and function	**Se in water and particulates (enrichment function)**	**Presence or absence of sensitive species** and functional groups
		Se speciation in particulates	Taxa richness and diversity
		Se in primary consumers	
		Trophic transfer factor	
		Food web structure	
Ecosystem	Ecosystem structure and function	Se loading and speciation in ecosystem	Productivity
		Residence time of Se in ecosystem	Nutrient cycling
		Organic carbon in sediment	

organization (particulates, whole-body, tissues, and subcellular components) may also provide insight into differences in Se bioavailability and toxicity among ecosystems and taxa.

The measures of effect that are most reliably diagnostic of Se toxicity in aquatic and associated terrestrial ecosystems are those most directly related to Se reproductive toxicity at the organism level: embryo malformations (terata), embryo-larval edema, and egg hatchability. Reproductive failure can lead to effects at both the population level (reduced abundance, loss of year classes) and the community level (loss of Se-sensitive species); these changes are often the most visible evidence of Se toxicity in aquatic ecosystems. However, these measurements can be difficult to implement because of the need for a large number of samples, specialized equipment, or extensive time and resources; they also may be less diagnostic of Se toxicity because they may reflect effects of other stressors. Measures of effects at tissue and subcellular levels may be diagnostic of Se toxicity (e.g., measures of oxidative stress), but these measures are generally less predictive of effects at higher levels of organization.

3.8 SUMMARY

The ecological effects of Se are mediated by site-specific factors, but certain general patterns emerge from a synthesis of current research. These generalizations address the geochemistry and anthropogenic activities likely to cause risk, Se biochemistry, the cycling of Se in aquatic environments, the uptake and transfer of Se through food webs, and the mechanisms of action for Se toxicity. While recognizing that each site is different, these general patterns not only can be used to assess contaminated sites, but also to predict situations in which potential Se mobilization may cause great risk.

3.8.1 SELENIUM'S BIOCHEMICAL ROLE

Selenium is both an essential element for animal nutrition and a toxicant. In fish and birds there is a narrow margin between essentiality and toxicity. Selenium occurs in a variety of organic and inorganic forms, but selenomethionine has been associated most closely with trophic transfer and toxicity in the environment. In aquatic systems, bacteria, algae, and plants convert inorganic forms of Se into organic forms, including selenomethionine, which is then transferred through food webs and, for egg-laying species, from mother to egg. The confirmed effects of Se on reproductive success in egg-laying vertebrates, including developmental abnormalities, have been linked to vertebrate population extirpations.

3.8.2 SELENIUM AS A GLOBAL PROBLEM

Selenium is distributed globally but not uniformly in organic-rich marine sedimentary rocks. Anthropogenic activities such as coal, phosphate, and metals mining can expose Se-rich strata to greatly enhanced leaching and subsequent transport. Soils derived from weathering and erosion of Se-rich sedimentary rocks can contribute Se through agricultural irrigation runoff and drainage. Selenium also is associated with processing and combustion of fossil fuels such as coal and oil. Coal combustion and oil refinery wastes may contain greatly concentrated Se relative to the raw material, and wastes from these processes can elevate Se concentrations in aquatic environments. These and other human uses of Se-associated products can transport contamination far from sources, potentially generating problems in areas distant from source rocks. Selenium discharges and Se contamination of aquatic ecosystems can be expected when known geologic sources of Se are combined with anthropogenic activities such as mining, irrigation, and coal-fired power plant operation unless appropriate management measures are instituted.

Specific examples of Se contamination from anthropogenic activities are well documented in the literature (Appendix A). In many of these cases, significant adverse effects on biota that are typical of Se toxicity have been documented; in some cases, population- and/or community-level effects also occurred. These case studies also demonstrate that the ecological outcome of Se contamination depends in part on measures of system characteristics such as Se loading, dissolved Se speciation, residence time or flow conditions, productivity, general food web characteristics, including diet and predator linkages, and the presence of other stressors.

Demand for coal, oil, and phosphate ore are expected to continue to increase in the foreseeable future. In addition, certain new technologies that use Se, such as nanotechnology, may have unpredicted impacts. As a result, both localized and landscape-scale Se contamination are global issues that are expected to increase in prominence in the future.

3.8.3 Movement and Transformation of Se

Much has been learned in recent years regarding the transport and transformation of Se in aquatic systems (Figure 3.3). Most important, research has shown that diet is the dominant pathway of Se exposure for both invertebrates and vertebrates. For this reason, traditional methods for predicting toxicity on the basis of exposure to dissolved concentrations do not work for Se. Selenium moves readily from water to primary producers and the other organic particulates that form the base of aquatic food webs. The EF, the ratio of the Se concentration in particulates to the Se concentration in water, describes the initial enrichment step for Se at the base of the food web. The EF measure in natural systems can vary by up to 2 orders of magnitude at different locations, although there is some evidence that EF values cluster more closely among sites with similar characteristics (e.g., lake systems versus river systems). This variability in EF makes it difficult to predict Se exposure and effects from water concentrations alone.

Transfer from particulates to primary consumers is less variable. TTFs (ratio of Se concentration in consumers to Se concentration in diet) for invertebrates are site and species-specific, but generally vary within 0.6 to 23. This dietary pathway is dominant; uptake of Se directly from water by consumers is negligible. Similarly, transfer from invertebrates to fish is from 0.6 to 1.7. For these reasons, the composition of the food web is important in determining bioaccumulation; the length of the food chain does not necessarily predict the level of Se exposure.

3.8.4 Effects of Se on Ecosystems

Acute toxicity from exposure to elevated dissolved Se concentrations has rarely, if ever, been reported in the aquatic environment. Significant chronic effects would be expected at far lower dissolved Se concentrations due to the incorporation of Se into the food web and resulting exposure and toxicity to fish and birds.

Chronic Se toxicity is manifested through reproductive impairment via maternal transfer in egg-laying vertebrates, resulting in embryotoxicity and teratogenicity. Other chronic effects include reductions in growth, tissue pathologies, induction of oxidative stress, and mortality. Sensitivity to chronic Se toxicity may vary widely, even among closely related species. Because estimates of risk are developed from knowledge of exposure and effects, the species that are most sensitive to Se are not always the most exposed to Se in nature. Species-specific feeding habits that result in high exposure levels may also drive toxicity risks. While much has been learned about bird and fish species, far less is known about toxicity in other oviparous vertebrates. A notable knowledge gap exists for egg-laying species of amphibians and reptiles, which include some of the most critically endangered vertebrate species.

Direct effects of Se on the population and community levels of biological organization have been documented at some sites (Appendix A). There is much less information about ecologically relevant indirect effects at the community or the ecosystem levels. Changes in invertebrate community structure caused by Se-induced loss of fish predators could be one example. Interactions between Se and temperature or other stressors also may occur but require further study.

These observations help explain why the behavior and toxicity of Se in ecological systems are highly dependent on site-specific factors. Knowledge of the food web is one of the keys to determining which biological species or other ecological characteristics will be affected. Other important parameters include rates of input of Se into the system, hydraulic residence time, and Se speciation in water and particulates.

It is difficult to generalize about system recovery when Se contamination is reduced or removed. Recovery is a function of the characteristics of the particular ecosystem and the decreases in mass loading of Se. Experience at Belews and Hyco Lakes shows that, once the source is removed, aquatic communities can substantially recover within a few years, although the community composition may be altered. Selenium in sediment may contribute to long (decadal) recovery times of tissue residues and possible long-term persistence of adverse effects in aquatic consumers.

3.8.5 How to Investigate a Potential Se Problem

Key assessment endpoints and corresponding exposure and effects measures at multiple levels of biological organization can be used to diagnose a suspected Se problem (Table 3.1). Similar assessment endpoints and measures also can be used to help predict potential impacts of a future anthropogenic activity.

Based on current knowledge, the endpoints most diagnostic of Se exposure occur at the tissue and organism levels. Table 3.1 presents the key measures recommended for assessing an ecosystem where significant Se contamination is strongly suspected or known. In systems where Se contamination is less certain, a shorter list of initial endpoints is proposed that includes Se concentrations in water, particulates, reproductive tissues from oviparous fish and wildlife, and tissues from primary consumers. In either situation, significant insight into the fate and effects of Se also may be gained by evaluating system characteristics such as Se loading and speciation, hydraulic residence time, ecological productivity, general food web characteristics, and the presence of other anthropogenic or natural stressors.

3.9 PRIORITIES FOR FUTURE RESEARCH

Selenium research has progressed in recent decades and has resulted in significant advances in our knowledge of Se dynamics and effects in aquatic systems. There are still important unknowns, however, and we suggest the following priorities for continued research:

1. Determine the species sensitivity of other egg-laying vertebrates, including reptiles and amphibians.

   Research has confirmed the susceptibility of oviparous fish and birds due to the maternal transfer of Se, and subsequent embryonic effects. There is insufficient toxicity information (in some cases, no toxicity information) on other oviparous species, including reptiles and amphibians.

2. Synthesize information regarding methods for collection of particulate components and develop a database of EF values.

   Particulate Se determines the uptake of Se into the base of the food web and serves as the Se source for primary consumers. There is substantial variability in approaches to particulate matter definition, collection, and analysis.

3. Obtain more information on Se sensitivity of marine species.

   There is insufficient information on Se effects in marine organisms.

4. Expand biodynamic modeling in freshwater systems.

   Collection of additional data regarding relationships among environmental compartments should lead to more reliable predictions of exposure and effects in freshwater systems. This would include more generalizable relationships across systems.

5. Develop additional quantitative surrogates for reproductive endpoints.

   Because it may be difficult or impractical to measure reproductive endpoints directly, alternative approaches would be valuable. For example, if a confirmed, quantitative relationship between diet and a reproductive endpoint is established, data on diet can then be used to predict reproductive toxicity risk.

6. Elucidate the mechanisms of Se toxicity.

   Although selenomethionine appears to be the form of Se that is most closely associated with adverse reproductive outcomes in wildlife, the precise mode of action for these toxic effects is poorly understood.

7. Explore indirect effects of selenium exposure within ecological systems.

   An understanding of changes in ecosystem ecological structure due to Se exposure is needed, including system-wide effects mediated via loss of food resources, disruption of predator–prey relationships, and loss of predators.

8. Identify interactive effects of selenium with other contaminants and stressors. Future studies on Se toxicity should consider the possible interactions between Se and common ecological variables (e.g., temperature, salinity, climate), important events in an animal's life history (e.g., migration, metamorphosis), and other anthropogenic factors (e.g., eutrophication, habitat modification, interactions with other contaminants). Although it is well known that Se interacts with other elements such as Hg, much remains to be known about the molecular mechanisms driving these interactions and their implications for toxicity.

REFERENCES

Adams WJ, Toll JE, Brix KV, Tear LM, DeForest DK. 2000. Site-specific approach for setting water quality criteria for selenium: differences between lotic and lentic systems. Proceedings of the 24th Annual British Columbia Mine Reclamation Symposium, June 21–22, 2000. Williams Lake (BC, Canada): The British Columbia Technical and Research Committee on Reclamation. p 231–240.

Anderson MS, Lakin HW, Beeson KC, Smith FF, Thancker E. 1961. Selenium in agriculture. Handbook No. 200. Washington (DC, USA): US Department of Agriculture.

Andreesen JR, Ljungdahl L. 1973. Formate dehydrogenase of *Clostridium thermoaceticum*: incorporation of selenium-75, and the effect of selenite, molybdate and tungstate on the enzyme. *J Bacteriol* 116:867–873.

Andren A, Klein D. 1975. Selenium in coal-fired steam plant emissions. *Environ Sci Technol* 9:856–858.

Baines SB, Fisher NS, Stewart R. 2002. Assimilation and retention of selenium and other trace elements from crustacean food by juvenile striped bass (*Morone saxatilis*). *Limnol Oceanogr* 43:646–655.

Barwick DH, Harrell RD. 1997. Recovery of fish populations in Belews Lake following selenium contamination. *Proc Ann Conf SE Assoc Fish Wildl Agencies* 51:209–216.

Beard KH, Vogt KA, Kulmatiski A. 2002. Top-down effects of a terrestrial frog on forest nutrient dynamics. *Oecologia* 133:583–593.

Bergeron CM, Husak JF, Unrine JM, Romanek CS, Hopkins WA. 2007. Influence of feeding ecology on blood mercury concentrations in four species of turtles. *Environ Toxicol Chem* 26:1733–1741.

Besser JM, Huckins JN, Little EE, LaPoint TW. 1989. Distribution and bioaccumulation of selenium in aquatic microcosms. *Environ Pollut* 62:1–12.

Besser JM, Canfield TJ, La Point TW. 1993. Bioaccumulation of organic and inorganic selenium in a laboratory food chain. *Environ Toxicol Chem* 12:57–72.

Bouchard SS, Bjorndal KA. 2000. Sea turtles as biological transporters of nutrients and energy from marine to terrestrial ecosystems. *Ecology* 81:2305–2313.

Breen B. 2009. Testimony before the U.S. House of Representatives Subcommittee on Water Resources and the Environment, April 30, 2009. Available from: http://www.epa.gov/ epawaste/nonhaz/industrial/special/fossil/coalashtest409.pdf Accessed 27 Nov 2009.

Brix KV, Volosin JS, Adams WJ, Reash RJ, Carlton RG, McIntyre DO. 2001. Effects of sulfate on the acute toxicity of selenate to freshwater organisms. *Environ Toxicol Chem* 20:1037–1045.

Burton TM, Likens GE. 1975. Energy flow and nutrient cycling in salamander populations in the Hubbard Brook experimental forest, New Hampshire. *Ecology* 56:1068–1080.

Canton SP, Fairbrother A, Lemly AD, Ohlendorf H, McDonald LE, MacDonald DD. 2008. Experts workshop on the evaluation and management of selenium in the Elk Valley, British Columbia, Workshop Summary Report. British Columbia Ministry of Environment, Kootenay Region, Nelson, BC. http://www.env.gov.bc.ca/eirs/epd/.

Cherry DS, Guthrie RK. 1977. Toxic metals in surface waters from coal ash. *Water Resour Bull* 13:1227–1236.

Chinese Medical Association. 1979. Observations on effect of sodium selenite in prevention of Keshan disease. *Chinese Med J* 92:471–476 (Reprinted in 2001in *J Trace Elements Exper Med* 14:221–226).

Cumbie PM, Van Horn SL. 1978. Selenium accumulation associated with fish mortality and reproductive failure. *Proc Ann Conf SE Assoc Fish Wildl Agencies* 32:612–624.

Cutter GA, Bruland KW. 1984. The marine biogeochemistry of selenium: a re-evaluation. *Limnol Oceanogr* 29:1179–1192.

Cutter GA, San Diego-McGlone MLC. 1990. Temporal variability of selenium fluxes in San Francisco Bay. *Sci Tot Environ* 97/98:235–250.

Cuvin-Aralar MLA, Furness RW. 1991. Mercury and selenium interaction — a review. *Ecotoxicol Environ Saf* 21:348–364.

Davis RH, Fear J. 1996. Incorporation of selenium into egg proteins from dietary selenite. *British Poultry Sci* 37:197–211.

de Souza MP, Amini A, Dojka MA, Pickering IJ, Dawson SC, Pace NR, Terry N. 2001. Identification and characterization of bacteria in a selenium-contaminated hypersaline evaporation pond. *Appl Environ Microbiol* 67:3785–3794.

Dreher GB, Finkelman RB. 1992. Selenium mobilization in a surface coal mine, Powder River Basin, Wyoming, U.S.A. *Environ Geol* 19:155–167.

Fan TW-M, Lane AN, Higashi RM. 1997. Selenium biotransformations by a euryhaline microalga isolated from a saline evaporation pond. *Environ Sci Technol* 31:569–576.

Fan TW-M, Higashi RM, Lane AN. 1998. Biotransformations of selenium oxyanion by filamentous cyanophyte-dominated mat cultured from agricultural drainage waters. *Environ Sci Technol* 32:3185–3193.

Fan TW-M, Teh SJ, Hinton DE, Higashi RM. 2002. Selenium biotransformations into proteinaceous forms by foodweb organisms of selenium-laden drainage waters in California. *Aquat Toxicol* 57:65–84.

Fernández-Turiel JL, Carvalho W, Cabañas M, Querol X, López-Soler A. 1994. Mobility of heavy metals from coal fly ash. *Environ Geol* 23:264–270.

Fleeger JW, Carman KR, Nisbet RM. 2003. Indirect effects of contaminants in aquatic ecosystems. *Sci Tot Environ* 317:207–233.

Flohé L, Günzler EA, Schock HH. 1973. Glutathione peroxidase: a selenoenzyme. *FEBS Lett* 32:132–134.

Garifullina G, Owen J, Lindbolm S-D, Tufan H, Pilon M, Pilon-Smits E. 2008. Expression of a mouse selenocysteine lyase in *Brassica juncea* chloroplasts affects selenium tolerance and accumulation. *Physiolog Plantarum* 118:538–544.

Garrett GP, Inman CR. 1984. Selenium-induced changes in fish populations in a heated reservoir. *Proc Ann Conf SE Assoc Fish and Wildl Agencies* 38:291–301.

George MW. 2009. Mineral commodity summaries. Selenium. Washington (DC, USA): US Geological Survey. p 144–145. Available from http://minerals.usgs.gov/minerals/pubs/mcs/2009/mcs2009.pdf. Accessed Jan 2010.

Gibbons JW, Scott DE, Ryan TJ, Buhlmann KA, Tuberville TD, Metts BS, Greene JL, Mills T, Leiden Y, Poppy S, Winne CT. 2000. The global decline of reptiles, déjà vu amphibians. *BioScience* 50:653–666.

Gibbons JW, Winne CT, Scott DE, Willson JD, Glaudas X, Andrews KM, Todd BD, Fedewa LA, Wilkinson L, Tsaliagos RN, Harper SJ, Greene JL, Tuberville TD, Metts BS, Dorcas ME, Nestor JP, Young CA, Akre T, Reed RN, Buhlmann KA, Norman J, Crosawh DA, Hagen C, Rothermel BB. 2006. Remarkable amphibian biomass and abundance in an isolated wetland: implications for wetland conservation. *Conserv Biol* 20:1457–1465.

Gillespie RB, Baumann PC. 1986. Effects of high tissue concentrations of selenium on reproduction in bluegills. *Trans Am Fish Soc* 115:208–213.

Hamilton SJ, Buhl, KJ. 2004. Selenium in water, sediment, plants, invertebrates, and fish in the Blackfoot River Drainage. *Wat Air Soil Pollut* 159:3–34.

Hamilton SJ, Buhl KJ, Faerber NL, Weidmeyer RH, Bullard FA. 1990. Toxicity of organic selenium in the diet to chinook salmon. *Environ Toxicol Chem* 9:347–358.

Harding LE, Graham M, Paton D. 2005. Accumulation of selenium and lack of severe effects on productivity of American dippers (*Cinclus mexicanus*) and spotted sandpipers (*Actitis macularia*). *Arch Environ Contam Toxicol* 48:414–423.

Hardman R. 2006. A toxicologic review of quantum dots: toxicity depends on physicochemical and environmental factors. *Environ Health Persp* 114:165–172.

Haygarth PM. 1994. Global importance and global cycling of selenium. In: Frankenberger WT Jr, Benson S, editors. Selenium in the environment. New York (NY, USA): Marcel Dekker. p 1–27.

Heinz GH. 1996. Selenium in birds. In: Beyer WN, Heinz GH, Redmon-Norwood AW, editors. Environmental contaminants in wildlife: interpreting tissue concentrations. Boca Raton (FL, USA): CRC Pr. p 447-458.

Heinz GH, Hoffman DJ. 1998. Methylmercury chloride and selenomethionine interactions on health and reproduction in mallards. *Environ Toxicol Chem* 17:139–145.

Heinz GH, Hoffman DJ, Krynitsky AJ, Weller DMG. 1987. Reproduction in mallards fed selenium. *Environ Toxicol Chem* 6:423–433.

Heinz GH, Hoffman DJ, Gold LG. 1988. Toxicity of organic and inorganic selenium to mallard ducklings. *Arch Environ Contam Toxicol* 17:561–568.

Hodson PV, Whittle DM, Hallett DJ. 1984. Selenium contamination of the Great Lakes and its potential effects on aquatic biota. In: Nriagu JO, Simmons MS, editors. Toxic contaminants in the Great Lakes. New York (NY, USA): J Wiley. p 371–391.

Hoffman DJ, Heinz GH, LeCaptain LJ, Eisemann JD, Pendleton GW. 1996. Toxicity and oxidative stress of different forms of organic selenium and dietary protein in mallard ducklings. *Arch Environ Contam Toxicol* 31:20–27.

Hopkins WA. 2000. Reptile toxicology: challenges and opportunities on the last frontier of vertebrate ecotoxicology. *Environ Toxicol Chem* 19:2391–2393.

Hopkins WA. 2006. Use of tissue residues in reptile ecotoxicology: a call for integration and experimentalism. In: Gardner S, Oberdorster E, editors. New perspectives: toxicology and the environment. Volume 3, Reptile toxicology. London (UK): Taylor & Francis. p 35–62.

Hopkins WA. 2007. Amphibians as models for studying environmental change. *ILAR J* 48:270–277.

Hopkins WA, Rowe CL. 2010. Interdisciplinary and hierarchical approaches for studying the effects of metals and metalloids on amphibians. In: Sparling DW, Linder G, Bishop CA, Crest S, editors. Ecotoxicology of amphibians and reptiles. 2nd ed. Pensacola (FL, USA): in press, SETAC Pr.

Hopkins WA, Rowe CL, Congdon JD. 1999. Elevated trace element concentrations and standard metabolic rate in banded water snakes (*Nerodia fasciata*) exposed to coal combustion wastes. *Environ Toxicol Chem* 18:1258–1263.

Hopkins WA, Roe JH, Snodgrass JW, Jackson BP, Kling DE, Rowe CL, Congdon JD. 2001. Nondestructive indices of trace element exposure in squamate reptiles. *Environ Pollut* 115:1–7.

Hopkins WA, Roe JH, Snodgrass JW, Staub BP, Jackson BP, Congdon JD. 2002a. Trace element accumulation and effects of chronic dietary exposure on banded water snakes (*Nerodia fasciata*). *Environ Toxicol Chem* 21:906–913.

Hopkins WA, Snodgrass JW, Roe JH, Staub BP, Jackson BP, Congdon JD. 2002b. Effects of food ration on survival and sublethal responses of lake chubsuckers (*Erimyzon sucetta*) exposed to coal combustion wastes. *Aquat Toxicol* 57:191–202.

Hopkins WA, Staub BP, Baionno JA, Jackson BP, Roe JH, Ford NB. 2004a. Trophic and maternal transfer of selenium in brown house snakes (*Lamprophis fuliginosus*). *Ecotox Environ Saf* 58:285–293.

Hopkins WA, Staub BP, Snodgrass JW, Taylor BE, DeBiase AE, Roe JH, Jackson BP, Congdon JD. 2004b. Responses of benthic fish exposed to contaminants in outdoor microcosm—examining the ecological relevance of previous laboratory toxicity test. *Aquat Toxicol* 68:1–12.

Hopkins WA, Staub BP, Baionno JA, Jackson BP, Talent LG. 2005a. Transfer of selenium from prey to predators in a simulated terrestrial food chain. *Environ Pollut* 134:447–456.

Hopkins WA, Snodgrass JW, Baionno JA, Roe JH, Staub BP, Jackson BP. 2005b. Functional relationships among selenium concentrations in the diet, target tissues, and nondestructive tissue samples of two species of snakes. *Environ Toxicol Chem* 24:344–351.

Hopkins WA, DuRant SE, Staub BP, Rowe CL, Jackson BP. 2006. Reproduction, embryonic development, and maternal transfer of contaminants in an amphibian *Gastrophryne carolinensis*. *Environ Health Persp* 114:661–666.

Hopkins WA, Hopkins LB, Unrine JM, Snodgrass J, Elliot J. 2007. Mercury concentrations in tissues of osprey from the Carolinas, USA. *J Wildl Manage* 71:1819–1829.

Huggins FE, Senior CL, Chu P, Ladwig K, Huffman GP. 2007. Selenium and arsenic speciation in fly ash from full-scale coal-burning utility plants. *Environ Sci Technol* 41:3284–3289.

Jankowski J, Ward CR, French D, Groves S. 2006. Mobility of trace elements from selected Australian fly ashes and its potential impact on aquatic ecosystems. *Fuel* 85:243–256.

Johnson J. 2009. The foul side of 'clean coal'. *ChemEngineer News* 87:44–47.

Kosta L, Byrne AR, Zelenko V. 1975. Correlation between selenium and mercury in man following exposure to inorganic mercury. *Nature* 254:238–239.

Kroll KJ, Doroshov SI. 1991. Vitellogenin: potential vehicle for selenium bioaccumulation in oocytes of the white sturgeon (*Acipenser transmontanus*). In: Williot P, editor. Acipenser. Antony (FR): Cemagref. p 99–106.

Kryukov G, Castellano S, Novoselov S, Lobanov A, Zehtab O, Guigo R, Gladyshev V. 2003. Characterization of mammalian selenoproteomes. *Science* 300:1439–1443.

Lemly AD. 1993a. Guidelines for evaluating selenium data from aquatic monitoring and assessment studies. *Environ Monitor Assess* 28:83–100.

Lemly AD. 1993b. Teratogenic effects of selenium in natural populations of freshwater fish. *Ecotoxicol Environ Saf* 26:181–204.

Lemly AD. 1993c. Metabolic stress during winter increases the toxicity of selenium to fish. *Aquat Toxicol* 27:133–158.

Lemly AD. 1997. Ecosystem recovery following selenium contamination in a freshwater reservoir. *Ecotoxicol Environ Saf* 36:275–281.

Lemly AD. 1998. Pathology of selenium poisoning in fish. In: Frankenberger WT Jr, Engberg RA, editors. Environmental chemistry of selenium. New York (NY, USA): Marcel Dekker. p 281–296.

Lemly AD. 2002. Symptoms and implications of selenium toxicity in fish: the Belews Lake example. *Aquat Toxicol* 57:29–49.

Lemly AD. 2004. Aquatic selenium pollution is a global environmental safety issue. *Ecotoxicol Environ Saf* 59:44–56.

Lide D. 1994. CRC handbook of chemistry and physics: a ready-reference book of chemical and physical data. Boca Raton (FL, USA): CRC Pr.

Luoma SN, Rainbow PS. 2005. Why is metal bioaccumulation so variable? Biodynamics as a unifying concept. *Environ Sci Technol* 39:1921–1931.

Luoma SN, Rainbow PS. 2008. Metal contamination in aquatic environments. Cambridge (UK): Cambridge Univ Pr.

Luoma SN, Johns C, Fisher NS, Steinberg NA, Oremland RG, Reinfelder JR. 1992. Determination of selenium bioavailability to a benthic bivalve from particulate and solute pathways. *Environ Sci Technol* 26:484–491.

Lussier C, Veiga V, Baldwin S. 2003. The geochemistry of selenium associated with coal waste in the Elk River Valley, Canada. *Environ Geol* 44:905–913.

McQuarrie D, Rock P. 1991. Chemistry of the main-group elements II. In: McQuarrie D, Rock P, editors. General chemistry. New York (NY, USA): Freeman. p 1083–1097.

Meseck SL, Cutter GA. 2006. Evaluating the biogeochemical cycle of selenium in San Francisco Bay through modeling. *Limnol Oceanogr* 51:2018–2032.

Moroder L. 2005. Isosteric replacement of sulfur with other chalcogens in peptides and proteins. *J Peptide Sci* 11:187–214.

Mosher B, Duce R. 1987. A global atmospheric selenium budget. *J Geophys Res* 92:13289–13298.

Mostert V. 2000. Selenoprotein P: properties, functions, and regulation. *Arch Biochem Biophys* 376:433–438.

Mostert V, Lombeck I, Abel J. 1998. A novel method for the purification of selenoprotein P from human plasma. *Arch Biochem Biophys* 357:326–330.

Motsenbocker MA, Tappel AL. 1982. Selenocysteine-containing proteins from rat and monkey plasma. *Biochim Biophys Acta* 704:253–260.

Müller S, Senn H, Gsell B, Vetter W, Baron C, Bock A. 1998. The formation of diselenide bridges in proteins by incorporation of selenocysteine residues: biosynthesis and characterization of (Se)2-thioredoxin. *Biochemistry* 33:3404–3412.

Muscatello JR, Bennett PM, Himbeault KT, Belknap AM, Janz DM. 2006. Larval deformities associated with selenium accumulation in northern pike (*Esox lucius*) exposed to metal mining effluent. *Environ Sci Technol* 40:6506–6512.

Muth OH, Oldfield JE, Remmert LF, Schubert JR. 1958. Effects of selenium and vitamin E on white muscle disease. *Science* 128:1090–1091.

Naftz DL, Johnson WP, Freeman ML, Beisner K, Diaz X. 2009. Estimation of selenium loads entering the south arm of Great Salt Lake, Utah, from May 2006 through March 2008. U.S. Geological Survey Scientific Investigations Report 2008–5069. Reston (VA, USA): US Department of Interior.

Nagle RD, Rowe CL, Congdon JD. 2001. Accumulation and selective maternal transfer of contaminants in the turtle *Trachemys scripta* associated with coal ash deposition. *Arch Environ Contam Toxicol* 40:531–536.

[NAS] National Academy of Sciences. 1976. Selenium: medical and biological effects of environmental pollutants. Washington (DC, USA): NAS.

Nel A, Xia T, Madler L, Li N. 2006. Toxic potential of materials at the nanolevel. *Science* 311:622–627.

[NRC] National Research Council. 1980. Mineral tolerance of domestic animals. Washington (DC, USA): National Academy Pr.

[NRC] National Research Council. 1989. Irrigation-induced water quality problems: what can be learned from the San Joaquin Valley experience. Washington (DC, USA): National Academy Pr.

Nriagu JO. 1989. Global cycling of selenium. In: Inhat M, editor. Occurrence and distribution of selenium. Boca Raton (FL, USA): CRC Pr. p 327–339.

Nriagu JO, Wong HK. 1983. Selenium pollution of lakes near the smelters at Sudbury, Ontario. *Nature* 301:55–57.

Ohlendorf HM. 1989. Bioaccumulation and effects of selenium in wildlife. In: Jacobs LW, editor. Selenium in agriculture and the environment. Special Publication 23. Madison (WI, USA): American Society of Agronomy and Soil Science Society of America. p 133–177.

Ohlendorf HM. 2002. The birds of Kesterson Reservoir: a historical perspective. *Aquat Toxicol* 57:1–10.

Ohlendorf HM. 2003. Ecotoxicology of selenium. In Hoffman DJ, Rattner BA, Burton GA Jr, Cairns J, editors. Handbook of ecotoxicology, 2nd edition. Boca Raton (FL, USA): Lewis. p 465–501.

Ohlendorf HM, Hoffman DJ, Saiki MK, Aldrich TW. 1986. Embryonic mortality and abnormalities of aquatic birds: apparent impacts of selenium from irrigation drainwater. *Sci Tot Environ* 52:49–63.

Oldfield JE. 1999. Selenium world atlas. Grimbergen (BE): Selenium Tellurim Development Association.

Oremland RS, Hollibaugh JT, Maest AS, Presser TS, Miller LG, Culbertson CW. 1989. Selenate reduction to elemental selenium by anaerobic bacteria in sediments and culture: biogeochemical significance of a novel, sulfate-independent respiration. *Appl Environ Microbiol* 55:2333–2343.

Oremland RS, Steinberg NA, Maest AS, Miller LG, Hollibaugh JT. 1990. Measurement of in situ rates of selenate removal by dissimilatory bacterial reduction in sediments. *Environ Sci Technol* 24:1157–1164.

Orr PL, Guiguer KR, Russel CK. 2006. Food chain transfer of selenium in lentic and lotic habitats of a western Canadian watershed. *Ecotoxicol Environ Saf* 63:175–188.

Osman K, Schutz A, Akesson B, Maciag A, Vahter M. 1998. Interactions between essential and toxic elements in lead exposed children in Katowice, Poland. *Clin Biochem* 3:657–665.

Outridge PM, Scheuhammer AM, Fox GA, Braune BM, White LM, Gregorich LJ, Keddy C. 1999. An assessment of the potential hazards of environmental selenium for Canadian water birds. *Environ Rev* 7:81–96.

Palace VP, Spallholz JE, Holm J, Wautier K, Evans RE, Baron CL. 2004. Metabolism of selenomethionine by rainbow trout (*Oncorhynchus mykiss*) embryos can generate oxidative stress. *Ecotoxicol Environ Saf* 58:17–21.

Pappas AC, Zoidis E, Surai PF, Zervas G. 2008. Selenoproteins and maternal nutrition. *Comp Biochem Physiol Part B: Biochem Molec Biol* 151:361–372.

Paulsson K, Lundbergh K. 1989. Selenium method for treatment of lakes for elevated levels of mercury in fish. *Sci Tot Environ* 87/88:495–507.

Pelletier E. 1985. Mercury-selenium interactions in aquatic organisms: a review. *Mar Environ Res* 18:111–132.

Peters GM, Maher WA, Jolley D, Carroll BI, Gomes VG, Jenkinson AV, McOrist GD. 1999. Selenium contamination, redistribution and remobilisation in sediments of Lake Macquarie, NSW. *Organic Geochem* 30:1287–1300.

Petranka JW, Murray SM. 2001. Effectiveness of removal sampling for determining salamander density and biomass: a case study in an Appalachian streamside community. *J Herpetol* 35:36–44.

Presser TS. 1994. The Kesterson effect. *Environ Manage* 18:437–454.

Presser TS, Ohlendorf HM. 1987. Biogeochemical cycling of selenium in the San Joaquin Valley, California, USA. *Environ Manage* 11:805–821.

Presser TS, Luoma SN. 2006. Forecasting selenium discharges to the San Francisco Bay-Delta Estuary: ecological effects of a proposed San Luis Drain extension. Menlo Park (CA, USA): US Geological Survey Professional Paper 1646. Available at http://pubs.usgs.gov/pp/p1646.

Presser TS, Luoma SN. 2010. A methodology for ecosystem-scale modeling of selenium. *Integr Environ Assess Manage* (in press).

Presser TS, Sylvester MA, Low WH. 1994. Bioaccumulation of selenium from natural geologic sources in western states and its potential consequences. *Environ Manage* 18:423–436.

Presser TS, Piper DZ, Bird KJ, Skorupa JP, Hamilton SJ, Detwiler SJ, Huebner MA. 2004a. The Phosphoria Formation: a model for forecasting global selenium sources to the environment. In: Hein JR, editor. Life cycle of the Phosphoria Formation: from deposition to post-mining environment. New York (NY, USA): Elsevier. p 299–319.

Presser TS, Hardy M, Huebner MA, Lamothe PJ. 2004b. Selenium loading through the Blackfoot River watershed: linking sources to ecosystems. In: Hein JR, editor. Life cycle of the Phosphoria Formation: from deposition to post-mining environment. New York (NY, USA): Elsevier. p 437–466.

Ramirez P, Rogers B. 2002. Selenium in a Wyoming grassland community receiving wastewater from an *in situ* uranium mine. *Arch Environ Contam Toxicol* 42:431–436.

Ranvestel AW, Lips KR, Pringle CM, Whiles MR, Bixby RJ. 2004. Neotropical tadpoles influence stream benthos: evidence for the ecological consequences of decline in amphibian populations. *Freshw Biol* 49:274–285.

Regester KJ, Lips KR, Whiles MR. 2006. Energy flow and subsidies associated with the complex life cycle of ambystomatid salamanders in ponds and adjacent forest in southern Illinois. *Oecologia* 147:303–314.

Reinert KH, Bartell SM, Biddinger GR, editors. 1998. Ecological risk assessment decision-support system: a conceptual design. Pensacola (FL, USA): SETAC Pr.

Reiss P, Protiere M, Li L. 2009. Core/shell semiconductor nanocrystats. *Small* 5:154–168.

Riedel GF, Sanders JG, Gilmour CC. 1996. Uptake, transformation, and impact of selenium in freshwater phytoplankton and bacterioplankton communities. *Aquat Microb Ecol* 11:43–51.

Rodda GH, Perry G, Rondeau RJ. 1999. The densest terrestrial vertebrate. In Abstracts of the Society for the Study of Amphibians and Reptiles. State College (PA, USA): Pennsylvania State Univ. p 195.

Roe JH, Hopkins WA, Baionno JA, Staub BP, Rowe CL, Jackson BP. 2004. Maternal transfer of selenium in *Alligator mississippiensis* nesting downstream of a coal-burning power plant. *Environ Toxicol Chem* 23:1969–1972.

Roe JH, Hopkins WA, Jackson BP. 2005. Species- and stage-specific differences in trace element tissue concentrations in amphibians: implications for the disposal of coal-combustion wastes. *Environ Pollut* 136:353–363.

Roe JH, Hopkins WA, DuRant SE, Unrine JM. 2006. Effects of competition and coal combustion wastes on recruitment and life history characteristics of salamanders in temporary wetlands. *Aquat Toxicol* 79:176–184.

Rotruck JT, Pope AL, Ganther H, Swanson A, Hafeman DG, Hoekstra WG. 1973. Selenium: biochemical role as a component of glutathione peroxidase. *Science* 179:588-590.

Roughgarden J. 1995. Anolis lizards of the Caribbean: ecology, evolution, and plate tectonics. Oxford Series in Ecology and Evolution. New York (NY, USA): Oxford Univ Pr.

Rowe CL, Hopkins WA, Congdon JD. 2002. Ecotoxicological implications of aquatic disposal of coal combustion residues in the United States: a review. *Environ Monit Assess* 80:207–276.

Sasaku C, Suzuki KT. 1998. Biological interaction between transition metals (Ag, Cd & Hg) selenide/sulfide and selenoprotein P. *J Inorg Biochem* 71:159–162.

Schwarcz H. 1973. Sulfur isotope analyses of some Sudbury, Ontario ores. *Can J Earth Sci* 10:1444–1459.

Schwarz K, Foltz CM. 1957. Selenium as an integral part of factor-3 against dietary necrotic liver degeneration. *J Am Chem Soc* 79:3292–3293.

Seale DB. 1980. Influence of amphibian larvae on primary production, nutrient flux, and competition in a pond ecosystem. *Ecology* 61:1531–1550.

Seiler R, Skorupa J, Naftz D, Nolan B. 2003. Irrigation-induced contamination of water, sediment, and biota in the Western United States: synthesis of data from the National Irrigation Water Quality Program. Denver (CO, USA): US Geological Survey Professional Paper.

Simmons DBD, Wallschläger D. 2005. A critical review of the biogeochemistry and eco-toxicology of selenium in lotic and lentic environments. *Environ Toxicol Chem* 24:1331–1343.

Skorupa JP. 1998. Selenium poisoning of fish and wildlife in nature: lessons from twelve real-world examples. In: Frankenberger WT Jr, Engberg RA, editors. Environmental chemistry of selenium. New York (NY, USA): Marcel Dekker. p 315–354.

Sorensen EMB. 1988. Selenium accumulation, reproductive status, and histopathological changes in environmentally exposed redear sunfish. *Arch Toxicol* 61:324–329.

Sorensen EMB, Harlan CW, Bell JS. 1982a. Renal changes in selenium-exposed fish. *Am J Forensic Med Pathol* 3:123–129.

Sorensen EMB, Bauer TL, Bell JS, Harlan, CW. 1982b. Selenium accumulation and cytotoxicity in teleosts folowing chronic, environmental exposure. *Bull Environ Contam Toxicol* 29:688–696.

Sorensen EMB, Bauer TL, Harlan CW, Pradzynski AH, Bell JS. 1983a. Hepatocyte changes following selenium accumulation in a freshwater teleost. *Am J Forensic Med Pathol* 4:25–32.

Sorensen EMB, Bell JS, Harlan CW. 1983b. Histopathological changes in selenium exposed fish. *Am J Forensic Med Pathol* 4:111–123.

Sorensen EMB, Cumbie PM, Bauer TL, Bell JS, Harlan CW. 1984. Histopathological, hematological, condition-factor, and organ weight changes associated with selenium accumulation in fish from Belews Lake, NC. *Arch Environ Contam Toxicol* 13:152–162.

Spallholz JE, Hoffman DJ. 2002. Selenium toxicity: cause and effects in aquatic birds. *Aquat Toxicol* 57:27–37.

Stadlober M, Sager M, Irgolic KJ. 2001. Effects of selenate supplemented fertilisation on the selenium level of cereals: identification and quantification of selenium compounds by HPLC-ICP-MS. *Food Chem* 73:357–366.

Stadtman TC. 1974. Selenium biochemistry. *Science* 183:915–922.

Stadtman TC. 1996. Selenocysteine. *Ann Rev Biochem* 65:83–100.

Staub BP, Hopkins WA, Novak J, Congdon JD. 2004. Respiratory and reproductive characteristics of eastern mosquitofish (*Gambusia holbrooki*) inhabiting a coal ash settling basin. *Arch Environ Contam Toxicol* 46:96–101.

Stewart AR, Luoma SN, Schlekat CE, Doblin MA, Heib KA. 2004. Food web pathway determines how selenium affects aquatic ecosystems: a San Francisco Bay case study. *Environ Sci Technol* 38:4519–4526.

Stuart SN, Chanson JS, Cox NA, Young BE, Rodrigues ASL, Fischman DL, Waller RW. 2004. Status and trends of amphibian declines and extinctions worldwide. *Science* 306:1783–1786.

Sunde RA. 1997. Selenium. In: O'Dell BL, Sunde RA, editors. Handbook of nutritionally essential mineral elements. New York (NY, USA): Marcel Dekker. p 493–556.

Suzuki KT, Ogra Y. 2002. Metabolic pathway for selenium in the body: speciation by HPLC-ICP MS with use of enriched selenium. *Food Addit Contam* 19:974–983.

Trelease SF, Beath OA. 1949. Selenium: its geological occurrence and its biological effects in relation to botany, chemistry, agriculture, nutrition, and medicine. New York (NY, USA): Trelease and Beath.

Tujebajeva R, Ransom DG, Harney JW, Berry MJ. 2000. Expression and characterisation of nonmammalian selenoprotein P in the zebrafish, *Danio rerio*. *Genes Cells* 5:897–903.

Turner DC, Stadtman TC. 1973. Purification of protein components of the clostridal glycine reductase system and characterization of protein A as a selenoprotein. *Arch Biochem Biophys* 154:366–381.

Turner RJ, Weiner JH, Taylor DE. 1998. Selenium metabolism in *Escherichia coli*. *Biometals* 11:223–227.

[TVA] Tennessee Valley Authority. 2009. Environmental assessment: initial emergency response actions for the Kingston fossil plant ash dike failure, Roane County, Tennessee. February, 2009. Available from: http://www.tva.gov/environment/reports/Kingston/pdf/2009-13_KIF_EmergencyResponse_EA.pdf. Accessed 27 Nov 2009.

Unrine J, Jackson B, Hopkins W, Romanek C. 2006. Isolation and partial characterization of proteins involved in maternal transfer of selenium in the western fence lizard (*Sceloporus occidentalis*). *Environ Toxicol Chem* 25:1864–1867.

Unrine JM, Jackson BP, Hopkins WA. 2007. Selenomethionine biotransformation and incorporation into proteins along a simulated terrestrial food chain. *Environ Sci Technol* 41:3601–3606.

[USDOI] United States Department of the Interior. 1998. Guidelines for interpretation of the biological effects of selected constituents in biota, water, and sediment. Denver (CO, USA): National Irrigation Water Quality Program, USDOI, Bureau of Reclamation. p 139–184.

[USEPA] United States Environmental Protection Agency. 1992. Framework for ecological risk assessment. EPA/630/R-92/001. Washington (DC, USA): Office of Research and Development.

[USGS] United States Geological Survey. 2000. Minerals yearbook: metals and minerals. Volume 1. Washington (DC, USA): USGS.

Venugopal B, Luckey TD. 1978. Metal toxicity in mammals. Volume 2. New York (NY, USA): Plenum Pr.

Wake DB, Vredenburg VT. 2008. Are we in the midst of the sixth mass extinction? A view from the world of amphibians. *Proc Nat Acad Sci* 105:11466–11473.

Wang D, Alfthan G, Aro A. 1993. Anthropogenic emissions of Se in Finland. *Appl Geochem* Suppl Issue 2:87–93.

Wang D, Alfthan G, Aro A, Lahermo P, Väänänen P. 1994. The impact of selenium fertilisation on the distribution of selenium in rivers in Finland. *Agricult Ecosyst Environ* 50:133–149.

Wang T, Wang J, Burken JG, Ban H, Ladwig K. 2007. The leaching characteristics of selenium from coal fly ashes. *J Environ Qual* 36:1784–1792.

Wang W-X. 2002. Interactions of trace metals and different marine food chains. *Mar Ecol Progr Ser* 243:295–309.

Watkinson J. 1983. Prevention of selenium deficiency in grazing animals by annual top-dressing of pasture with sodium selenate. *New Zealand Vet J* 31:78–85.

Wen H, Carignan J. 2007. Reviews on atmospheric selenium: emissions, speciation and fate *Atmospher Environ* 41:7151–7165.

Wilber CG. 1980. Toxicology of selenium: a review. *Clin Toxicol* 17:171–230.

Wilhelmsen E, Hawkes W, Tappel A. 1985. Substitution of selenocysteine for cysteine in a reticulocyte lysate protein synthesis system. *Biol Trace Elem Res* 7:141–151.

Wisniak J. 2000. Jons Jacob Berzelius: a guide to the perplexed chemist. *Chem Educator* 5:343–350.

Wu L. 1998. Selenium accumulation and uptake by crop and grassland plant species. In: Frankenberger WT Jr, Engberg RA, editors. Environmental chemistry of selenium. New York (NY, USA): Marcel Dekker. p 657–686.

Wyman RL. 1998. Experimental assessment of salamanders as predators of detrital food webs: effects on invertebrates, decomposition, and the carbon cycle. *Biodivers Conserv* 7:641–650.

Yan J, Barrett JN. 1998. Purification from bovine serum of a survival-promoting factor for cultured central neurons and its identification as selenoprotein P. *J Neurosci* 18:8682–8691.

Yan R, Gauthier D, Flamant G, Lu J, Zheng C. 2001. Fate of selenium in coal combustion: volatilization and speciation in the flue gas. *Environ Sci Technol* 35:1406–1410.

Yang GS, Wang RZ, Sun S. 1983. Endemic selenium intoxication of humans in China. *Am J Clin Nutrit* 37:872–881.

Yoneda S, Suzuki KT. 1997a. Equimolar Hg-Se complex binds to selenoprotein P. *Biochem Biophys Res Commun* 231:7–11.

Yoneda S, Suzuki KT. 1997b. Detoxification of mercury by selenium by binding of equimolar Hg-Se complex to a specific plasma protein. *Toxicol Appl Pharmacol* 143:274–280.

Yudovich YE, Ketris MP. 2006. Selenium in coal: a review. *Int J Coal Geol* 67:112–126.

Zhu J, Wang N, Li S, Li L, Su H, Liu C. 2008. Distribution and transport of selenium in Yutangba, China: impact of human activities. *Sci Tot Environ* 392:252–261.

4 Environmental Sources, Speciation, and Partitioning of Selenium

William Maher, Anthony Roach, Martina Doblin,
Teresa Fan, Simon Foster, Reid Garrett,
Gregory Möller, Libbie Oram, Dirk Wallschläger

CONTENTS

4.1   INTRODUCTION

Contamination by selenium (Se) is a significant global issue (Lemly 2002). It has occurred in many types of aquatic environments, including estuaries, wetlands, freshwater streams, and impoundments (Cutter 1989; Leighton 1989; Atalay 1990). Selenium is a naturally occurring element that is enriched in a range of geological sources such as crustal rock, black shale, phosphate rocks, and coal. While natural sources of Se make an important contribution to Se fluxes globally (Nriagu 1989), human activity is typically responsible for the greatest Se contamination on a regional scale. Many large-scale activities, including mining, agriculture, manufacturing, and petrochemical industries, use raw materials and disturb material containing Se (Skorupa 1998), which leads to its mobilization and contamination of the environment. Selenium may become an ecotoxicological problem under very different scenarios because of the diversity of its natural and anthropogenic sources. This multiplicity of natural and anthropogenic sources causes different forms (e.g., aqueous, particulate, and gaseous) and species of Se (e.g., selenate $[SeO_4^{-2}]$ and selenite $[SeO_3^{-2}]$) to enter the environment at different spatial and temporal scales, often making source identification, exposure pathway, and hazard assessment complex (Butler et al. 1996; Seiler et al. 2003).

Fundamental to understanding the environmental risks associated with any contaminant is predicting its exposure concentration, bioavailability, and toxicity to the organisms at risk. For Se, its bioavailability and toxicity are determined by its chemical form (speciation). In the natural environment, Se commonly exists as inorganic Se in 1 of 4 oxidation states (Se(0), Se(-II), Se(IV), and Se(VI)). Selenium, however, also readily forms volatile and methylated species and is incorporated within proteins and amino acids and other biochemical intermediates. Its ability to form this array of chemical and physical forms occurs because of its reactivity but also because it is an essential element necessary for the proper functioning of structural proteins and cellular processes (Hoffman 2002). Most organisms, such as micro-organisms, plants, and animals, therefore accumulate, metabolize, transform, and excrete Se, which results in its complex speciation. Different inorganic and organic Se species, however, partition differently in the environment, have differing bioavailability and toxicity, and as such predicting how Se will behave in an aquatic system requires an understanding of its speciation.

The sources, forms, and speciation of Se combine to determine how Se will be partitioned and transferred among the various environmental compartments (i.e., abiotic and biotic) in an aquatic system. The main abiotic compartments, sediment and water, can contain considerable amounts of Se in contaminated systems, but significantly, Se can readily enter the base of the food chain from these compartments (see Chapter 5) and be transferred among the other biotic and abiotic compartments. Bioaccumulation of Se may pose ecotoxicological risks to fish and birds, because elevated tissue concentrations can result in reproductive effects. Thus, an understanding of how Se is partitioned and transferred among the compartments in an aquatic ecosystem is important for assessing the exposure pathways and hence the ecological significance of this contaminant for any particular system.

This chapter describes

- the major natural and human-related sources of Se contamination;
- the Se species present in emission sources, water, air, sediment, and biota;
- the major abiotic and biotic environmental compartments where Se may partition; and
- the relative importance of these compartments and how their importance may vary among systems.

Where relevant, we provide appropriate methods for undertaking studies on Se and provide case studies to illustrate these issues.

A primary focus of this chapter was to develop 3 conceptual models summarizing Se sources, speciation, and environmental partitioning that could be used as a starting point for understanding the biogeochemical cycling of Se in the environment and its potential to cause adverse ecotoxicological effects. The importance of speciation measurements to understand processes and assess risk in ecosystems was also assessed.

4.2 SELENIUM SOURCES AND LOADS

Selenium is a naturally occurring element in the Earth's crust that is being redistributed in the environment through multiple natural and anthropogenic processes (Figure 4.1). The major geologic sources of Se include crustal rock such as black shale, phosphate rocks, and coal, while other formations such as igneous rock and limestone are minor sources (Guun et al. 1976). Natural processes redistributing Se include volcanic activity, terrestrial weathering of rocks and soils, wildfires, and volatilization from plants and water bodies. While making an important contribution to global Se fluxes (Mosher and Duce 1987; Nriagu 1989), on a regional basis, the largest Se sources are typically related to human activities not natural processes (Presser et al. 1990).

From these enriched geological sources, Se is redistributed into aquatic, sedimentary, and atmospheric as well as terrestrial compartments (see Section 4.4, Figure 4.6) (Nriagu 1991; Haygarth 1994). These processes not only transport Se but also transform it into different species (e.g., from selenate to selenite) and transfer them between phases (e.g., from liquid to solids; see Section 4.3, Figure 4.3). Importantly, the variation in Se source, phase, and speciation in different locations

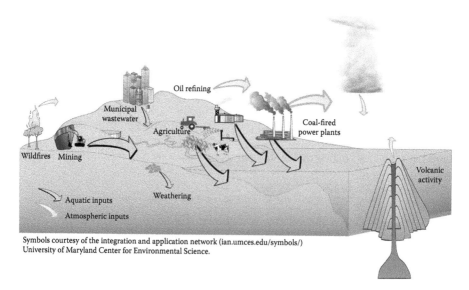

Symbols courtesy of the integration and application network (ian.umces.edu/symbols/)
University of Maryland Center for Environmental Science.

FIGURE 4.1   Major Se sources to aquatic ecosystems.

creates a region-specific risk profile, which should be recognized when managing Se contamination in the environment.

4.2.1   Sources of Se Contamination

Human activity is a major factor in the mobilization of Se. Fossil fuel combustion, oil refining, and fertilizer production are all anthropogenic sources of Se, but other activities involving land disturbance (e.g., agricultural irrigation and mining) can also be major contributors of Se to aquatic systems (Figure 4.1) (Presser et al. 1990; Muscatello and Janz 2009). Agricultural activities, including field application of bio-solids (from wastewater treatment facilities), manure, chemical fertilizers, and Se supplements, are relatively minor sources of Se to the environment (Bouwer 1989).

4.2.1.1   Oil Refining

Crude oil from certain geological formations such as marine shales can be rich in Se (Presser et al. 1990; Martens and Suarez 1997), making refinery effluents a major source of loading to aquatic environments, depending on their source of crude oil (e.g., San Francisco Bay [CA, USA]: Cutter 1989; Cutter and San Diego-McGlone 1990).

4.2.1.2   Fossil Fuel Combustion

Combustion of fossil fuels including coal and oil also generates the release of significant quantities of Se into the environment (Cumbie and Van Horn 1978; Woock et al. 1987). Historically, coal-fired plants emitted Se as SeO_2 (mostly in the vapor phase) to the atmosphere and as aqueous selenate. The primary route of Se into the aquatic environment from coal-fired power plants was from sluicing fly ash into settling ponds and subsequently discharging ash pond effluent. Several notable examples of the effects of fly ash disposal include Belews Lake and Hyco Reservoir in North Carolina and Martin Lake in Texas (Lemly 1985; Gillespie and Baumann 1986). At Hyco Reservoir, the fish population declined over several years and then collapsed shortly after the construction and commencement of operations of a fourth coal-fired unit in 1980 (Crutchfield 2000). Recently, however, power plants are adding wet flue gas scrubber systems to their process, changing the nature of discharges to include scrubber wastewaters (Electric Power Research Institute [EPRI] 2008).

4.2.1.3   Agriculture

Irrigation of saline and sometimes seleniferous soils, such as those found in parts of the Central Valley of California in the United States, has led to significant Se contamination in aquatic ecosystems from impoundment of irrigation drainage (e.g., Kesterson Wildlife Refuge in California: Ohlendorf 2003).

4.2.1.4   Mining

Coal, phosphate, and sulfidic ore mining produce overburden (waste rock), which is excavated during active mining to gain access to the ore body. Infiltration and runoff from waste rock piles, tailings impoundments, backfilled mining excavations, and reclaimed mined areas can release large amounts of Se which, in uncontrolled settings, can eventually make its way to aquatic ecosystems. In addition, mineral concentrate containing Se from ore mining is sometimes released to the environment through mishandling prior to smelting. Other Se releases can occur through the wastewater from mine tailings impoundments, which is typically treated and discharged through licensed (i.e., authorized) outfalls to receiving waters. Ore smelting and coal burning can also generate volatile and particulate forms of Se that escape to the atmosphere, which are then transported to other locations where deposition occurs.

4.2.2 ATMOSPHERIC TRANSPORT AND FLUXES OF SE

The atmosphere is the least characterized compartment in terms of Se fluxes. Selenium can be released to the atmosphere by geological processes, including terrestrial dust emissions, volcanic effluvia, wildfires, production of aerosols from natural sources such as sea spray, anthropogenic activity (Nriagu 1989; Wen and Carignan 2007), and microbial and plant biotransformation of Se to volatile products (Figure 4.1). The most recent estimates of Se emissions are (in tons per year) oceanic emissions, 35,000 (Amouroux et al. 2001); volcanic sources, 400 to 1200 (Mosher and Duce 1987); and anthropogenic activity, 15,600 (Nriagu 1989). There presently are no data to accurately estimate the magnitude of global atmospheric Se deposition (Wen and Carignan 2007). However, atmospheric transport processes such as particulate deposition and precipitation (wet deposition) can transport Se from relatively remote sources to aquatic ecosystems (Toy et al. 2002; Wen and Carignan 2007).

4.2.3 CHANGES IN SE SOURCES WITH TECHNOLOGICAL, ENVIRONMENTAL, AND SOCIAL CHANGE

Industrial development will continue to play a major role in determining human-related Se sources to the environment, particularly in light of projected global population growth. Climate change represents a major uncertainty in predicting future Se sources and will have effects on cellular- to global-scale processes such as temperature-regulated microbial production of dimethyl selenide [DMSe] (Guo et al. 2000) and atmospheric transport and circulation of Se aerosols. Fossil fuels currently supply 85% of global energy, but international agreements to decrease CO_2 effects to the climate system may change future use (Schwartz 2008). Hence, global responses to climate change will directly affect Se emissions from combustion sources.

Mobilization of Se in developed and developing countries may be abated by advancements in the treatment of waste and discharges that decrease or modify Se release. For example, installation of Flue Gas Desulfurization (FGD) technology (wet scrubbers) in coal-fired power plants can remove between 17% and 77% of atmospheric Se emissions by capturing the Se alongside sulfur (SO_2) in the scrubber wastewater (EPRI 2008). Selenium exists primarily as selenate and selenite in scrubber wastewater. However, in some scrubbers with strong oxidizing conditions and high temperatures such as those with forced oxidation systems, Se changes to mostly selenate before it is discharged (Akiho et al. 2008). Conventional treatment technologies, such as the use of adsorption on activated alumina, iron co-precipitation, reverse osmosis, and ion exchange, have proved ineffective or expensive for removing selenate from coal-fired power plant effluents (Merrill et al. 1986; Twidwell et al. 1999). Emerging anaerobic bioreactor technologies appear to be a cost-effective way to treat most forms of dissolved Se, including converting selenate to insoluble Se(0) (Lortie et al. 1992; Zhang and Frankenberger 2003; Pickett et al. 2006; Sonstegard et al. 2008).

4.2.4 Selenium Loadings to Aquatic Ecosystems

Selenium sources have variable Se concentration and speciation, which affects their solubility in water, volatility, sorption to particles, and partitioning into biota. Table 4.1 provides some indicative examples of the speciation of Se sources. The nature of the Se source, including its origin, its concentration, and speciation, and its mechanism and timing of delivery determine the Se loading into aquatic systems.

Selenium can be transported to aquatic environments in water or air, but water is the primary delivery mechanism for anthropogenic Se sources. Selenium can enter aquatic systems via point and non–point source discharges as either continuous (press) or episodic (pulse) inputs. Examples of point source discharges are streams, irrigation

TABLE 4.1

Selenium Speciation and Concentrations for Different Sources. Note: Values are Indicative Only, Due to Spatial and Temporal Variability

Source	Total Se	Se Speciation	Reference
Oil refineries San Francisco Bay, California, USA	48.4 μg/L (1987–1988) 16.4 μg/L (1999–2000)	Selenite = 64% Selenite = 14%	Cutter (1989) Cutter and Cutter (2004)
San Joaquin River, California, USA	1.2 ± 0.8 μg/L	Selenite = 71% Se(-II) = 23% Selenate = 6%	Cutter and Cutter (2004)
Upper Colorado River Basin, USA	>5 μg/L	Selenate = 95% Selenite = 0%–5%	Kharaka et al. (2001)
Groundwater, Nevada, Death Valley, California	0.5–10 μg/L		Johannesson et al. (1996)
Soil, Kesterson evaporation pond, California, USA	47.8 mg/kg dw	Se(0), Se(-II) = 86%	Martens and Suarez (1997)
Sediment, San Luis drain, California, USA	83.8 mg/kg dw	Se(0), Se(-II) = 91%	Martens and Suarez (1997)
Residential sludge (composted)		Selenite = 40% Selenate, Se(-II) = 60%	Cappon (1991)
Aqueous extracts of mine soil, Powder River basin, Wyoming, USA		Selenite = 2%–58% Selenate = 42%–98%	Sharmasarkar et al. (1998)
Rainwater, Delaware, USA	0.18 μg/L	Selenite = 24%–69%	Cutter and Church (1986)
North Atlantic Deep Water	0.11 mg/kg dw	Selenite = 43% Selenate = 57%	Cutter and Cutter (2001)

channels, and pipes, compared with non–point source discharges, such as atmospheric deposition or land runoff. Examples of continuous Se loads to aquatic ecosystems include managed discharges from coal fly ash ponds associated with power plants, whereas inputs of Se from erosion of agricultural lands tend to be pulse inputs occurring during high rainfall events. During rainfall events, Se can be discharged in the first flush of run-off with remobilization of particulate-bound Se as the flow increases. Of relevance to managers is that aquatic systems may have multiple Se sources whose Se concentration and delivery show climatic (i.e., weather dependent), seasonal, and long-term variability (Doblin et al. 2006) that need to be considered in assessing Se loads.

4.2.5 METHODS FOR ASSESSING LOADS TO SYSTEMS

The assessment of Se loads should begin by considering all potential Se sources to the system of interest, using the conceptual model outlined in Figure 4.1 as a guide. From there, an understanding of their concentration and supply is required, because

$$\text{Se concentration} \times \text{flow} = \text{Se load (mass / time)}.$$

In particular, information about the spatial and temporal nature of the inputs will determine whether a source is significant enough to require inclusion in the mass balance and, if so, what type of sampling regime needs to be used, including the location, frequency, and means of sample collection (e.g., possible use of automatic sampling devices). To avoid contamination and speciation changes, appropriate sample collection, preservation, and storage techniques need to be undertaken (Ralston et al. 2008).

When mass balance studies of Se sources are conducted via load-based sampling, sufficient samples need to be collected to accurately reflect the true flux of Se. For example, a first flush (high concentration in an initial discharge pulse) of a contaminant is common, as well as remobilization during higher flows. Accurate measures of discharge are required, such as the use of calibrated weirs, drainage pipes, etc.; there are well-established procedures to measure hydraulic discharge that can be employed in most situations (Lambie 1978; Ferguson 1987). In practice, discharge measurements contribute the highest degree of uncertainty in any load calculation rather than the determination of Se concentration. However, the determination of Se concentration by routine analytical laboratories can also be a significant source of error when Se is <10 µg/L or the water has a complex chemistry, like most industrial effluents, or both (Ralston et al. 2008).

Point source Se releases from pipes and drainage channels are the easiest loads to quantify, and in many developed countries, these are typically regulated through a governmental agency license to discharge (e.g., releases from coal fly ash settling ponds associated with power stations). Loadings from power stations depend on their size, operating run time, and the particular abatement technology applied to coal ash and smokestack emissions. Typically, water releases from coal power plant ash ponds are closely monitored and loads calculated using total Se estimates averaged from periodic water samples sent to certified analytical laboratories. In the United States, smokestack Se emissions are determined using data from the EPRI (1994), where measurements of particulate emission rates from smokestacks are modeled to estimate annual loads of Se and other elements to the atmosphere. This can be done for a variety of coal sources.

Emissions are variable between smokestacks, but typically scale is relative to power plant size (US Environmental Protection Agency [USEPA] 1999).

Cutter and San Diego-McGlone (1990) were able to use Se speciation signatures in San Francisco Bay to discriminate oil refinery from riverine sources, and in combination with refinery discharge and river flow estimates, calculate relative Se loads. Johnson et al. (1999) suggest that Se isotopes would also be effective source tracers, given that there is strong fractionation during reduction of selenate or selenite to Se(0) or Se(-II), but not during oxidation. The isotopic composition of Se in atmospheric aerosols should therefore reflect that of their source emission (Wen and Carignan 2007). There is variability in Se isotopic fractionation, with ratios changing under different environmental conditions (Johnson 2004). Clearly, further work to characterize the isotopic composition of different Se sources is necessary before this method can be used to reliably discriminate Se sources and relative loads to aquatic systems.

In situations where Se load assessment is not possible or incomplete, the risk of Se contamination is inevitably increased, because the magnitude of the problem cannot be quantified. This is particularly difficult in situations where the source is significant but unknown (e.g., undocumented groundwater inputs of relatively high Se concentration) or not easily managed (e.g., atmospheric deposition).

4.3   SELENIUM SPECIATION

4.3.1   INTRODUCTION

Selenium exists in a diverse array of chemical and physical forms, also called "species," which are linked by many biogeochemical transformation reactions. Major Se species found in the environment are mapped in Figure 4.2 while the cycling of major Se species in aquatic ecosystems is shown in Figure 4.3. These species can be grouped into the 4 major categories of

1) inorganic Se,
2) volatile and methylated Se,
3) protein and amino acid Se, and
4) non–protein amino acids and biochemical intermediates.

The chemical structures of the major species in these categories are shown in Attachment 1.

4.3.2   SELENIUM SPECIATION IN SOURCES

4.3.2.1   Lithosphere Mobilization

Selenium, an analogue of sulfur, is widely distributed in the lithosphere as a component of many minerals and soils. Selenium typically occurs in the environment in 4 oxidation states:

- Se(VI),
- Se(IV),

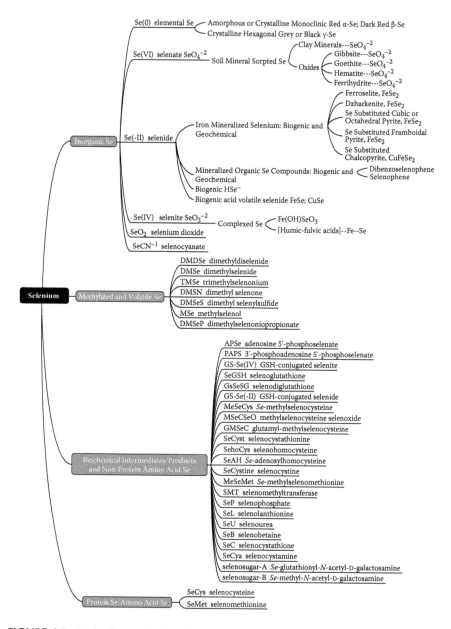

FIGURE 4.2   Major Se species found in the environment.

- Se(0), and
- Se(–II).

It is often found as the oxyanions selenate (SeO_4^{2-}) and selenite (SeO_3^{-2}) in oxidized systems, and as elemental Se [Se(0)] and selenides [Se(–II)] in anaerobic zones and unweathered mineral formations. Selenium geochemistry is fundamentally

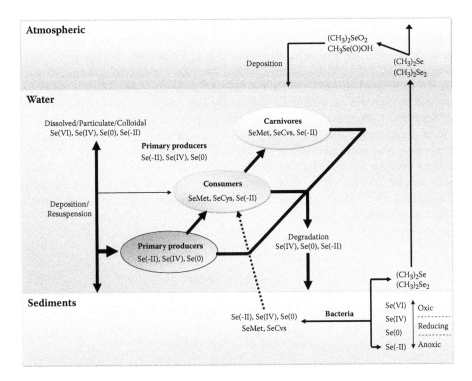

FIGURE 4.3   Cycling of major Se species in aquatic environments.

controlled by iron geochemistry (Howard 1977). Selenium is found in numerous minerals of sedimentary deposits, hydrothermal veins, and metamorphic rocks. Iron-Se minerals include ferroselite, dzharkenite, and Se-substituted pyrites of microbial origin. In aerobic systems, selenate and selenite can be adsorbed to iron oxide minerals such as gibbsite, goethite, hematite, and ferrihydrite as well as clay minerals. Distribution of selenite and selenate between the solid and solution phases is a function of pH and the mineral phases present (see Section 4.4).

Geological processes such as weathering and soil genesis mobilize mineralized and sequestered Se from the lithosphere to the atmosphere and the hydrosphere. Selenium released by reductive dissolution of minerals, microbial processes, or oxidation of reduced Se species, can be mobilized as selenate and selenite and some of these processes will yield Se isotopic fractionation (Johnson 2004). Reduced Se species such as Se(0), Se(-II), and strongly adsorbed Se species are insoluble and therefore more likely to be released as fine particulates to the atmosphere or as colloidal suspensions in surface waters.

4.3.2.2 Atmospheric Emission and Deposition

Selenium emitted to the atmosphere from natural and industrial combustion processes is generally assumed to be gaseous SeO_2, Se(0), and Se adsorbed to air-buoyant particles. The major anthropogenic atmospheric sources of Se are from

coal combustion and metal refining, accounting for 70% to 90% of the anthropogenic load (Wen and Carignan 2007). Coal combustion with efficient particulate
removal was initially estimated to release 93% of its Se load as vapor phase Se(0)
(Andren et al. 1975). More recently, cooled flue gases from coal combustion were
shown to be predominately Se(0) and SeO$_2$ (Yan et al. 2004). High temperature
combustion sources of volatile Se species are expected to be primarily composed
of Se(0) and SeO$_2$ that can be converted to Se(0) by co-emitted sulfur dioxide, an
effective reducing agent. Scavenging of aerosols during rainfall results in the wet
deposition of selenate and selenite (Ross 1984; Wen and Carignan 2007). The interactions between atmospheric sulphoxy radical anions and Se(0) transform Se(0) to
selenate and selenoselenate under oxidizing conditions (Bronikowski et al. 2000).
Theoretical predictions also suggest the following spontaneous reactions (Scheme 1)
(Monahan-Pendergast et al. 2007):

Scheme 1:

$$Se + OH \rightarrow SeOH$$

$$Se + H_2O \rightarrow SeO + OH$$

$$SeO + H_2O \rightarrow SeO_2 + OH$$

Dimethyl selenide [(CH$_3$)$_2$Se] and dimethyl diselenide [(CH$_3$)$_2$Se$_2$], as well as other
volatile organic Se compounds, are released to the atmosphere as the result of bio-
alkylation in soils, waters, and plants (Karlson and Frankenberger 1993; Terry et al.
2000; Amouroux et al. 2001; Guo et al. 2001). Once released to the atmosphere, it is
unlikely that significant re-deposition of these species occurs due to reactive oxidation processes. The half-life of (CH$_3$)$_2$Se is calculated to be less than 6 h (Atkinson
et al. 1990). A sequential oxidation pathway has been proposed (Scheme 2) (Musaev
et al. 2003; Wen and Carignan 2007).

Scheme 2:

$$(DMSe, DMDSe, MeSeH, DMSeS) + (OH, NO_3) \rightarrow CH_3 - Se - CH_2$$

$$CH_3 - Se - CH_2 \rightarrow CH_3 - Se - CH_2O$$

$$CH_3 - Se - CH_2O \rightarrow (Unknown\ oxidation\ pathway) \rightarrow Se^0, SeO_2, H_2SeO_3$$

Selenium speciation in atmospheric particulate matter is not well studied. Coal fly
ash has been found to contain selenite and, to a lesser extent, selenate (Niss et al.
1993). In other work, Se was identified as Se(IV) with some Se(0) with the observation that the Se(IV) fraction was insoluble on water extraction (Mattigod and Quinn
2003). Urban atmospheric particulate matter (PM10) samples from Spain have also
been shown to have selenite as the major Se species (Moscoso-Pérez et al. 2008).
Since coal combustion releases significant quantities of Se and global coal usage is

increasing, further research is needed to better understand the atmospheric reactions and cycling of Se from this source (Xie et al. 2006).

4.3.2.3 Stream Discharges

Dissolved or particulate adsorbed Se in aqueous discharge streams is generally a mixture of selenite and selenate. The relative abundance of these 2 species depends on the nature of the industrial process generating the release and any subsequent treatment prior to discharge into the environment. Some processes such as agricultural irrigation drainage in saline areas yield discharges that are predominantly selenate (Zhang et al. 1999; Gao et al. 2007), while other processes such as leaching of coal fly ash result in selenite as the primary released species (Wang et al. 2007). While some industrial processes initially contain significant amounts of other Se species in process waters (e.g., selenocyanate in petroleum refining) (Wallschläger and Bloom 2001), these waters may undergo extensive treatment prior to discharge, resulting in the formation of selenite and selenate (Sayre 1980). Municipal wastewater can have significant amounts of dissolved and suspended particulate Se; however, it appears that most of the Se is removed to biosolids via aerobic and/or anaerobic biological processing and other water treatment processes (Cappon 1991; Heninger et al. 1998).

Release of Se to ambient waters via industrial discharge streams is regulated in many developed countries, typically at a level that attempts to preserve the ecological integrity of the receiving waters. Treatment of selenate-containing waters usually requires an abiotic or biotic reduction step yielding selenite, Se(0), or other insoluble reduced Se species, which are then removed via filtration, flocculation, or settling (Murphy 1989; Manning and Burau 1995; Goodrich-Mahoney 1996; Meng et al. 2002; Lin and Terry 2003; Knotek-Smith et al. 2006; Zhang and Frankenberger 2006; Chen et al. 2008; Lenz et al. 2008a, 2008b). Biological treatment in reactors or constructed wetlands has the potential to create reduced soluble and volatile Se species, such as DMSe and DMDSe. Plants and algae have been used to remove Se from discharge waters (Gerhardt et al. 1991; Pilon-Smits et al. 1999; Carvalho and Martin 2001; LeDuc and Terry 2005), although production of volatile Se species and disposal of the concentrated Se biomass can be issues of concern.

4.3.3 Selenium Species Transformations

4.3.3.1 Water

Selenium species in water can exist in dissolved and suspended particulate forms. In the pH range of 6 to 8, only Se(0), selenite, $HSeO_3^{-1}$, and selenate are present in water (Milne 1998). Suspended particles containing Se can be an important part of the Se load in the water column. For example, selenite partitioning to amorphous iron oxyhydroxides and manganese dioxide on particles has been shown to be important in the sequestration of Se (Scott and Morgan 1996; Parida et al. 1997). Selenite and selenate are chemically stable in pure water, and do not interconvert to any appreciable extent (Lindemann et al. 2000). Dissolved oxygen is not an effective oxidant for selenite, because of its slow oxidation kinetics (Scott and Morgan 1996). Therefore, it would take either a stronger oxidant or a biologically mediated process to accomplish significant selenite oxidation. Consequently, selenite oxidation is enhanced by factors that increase the concentrations of strong oxidants in the water column such as UV radiation, redox-active transition metals, such as Fe, or simply by the presence of a high abundance of selenite-oxidizing bacteria. Thus, although selenite is thermodynamically unstable in oxic waters, it is frequently encountered because of its slow oxidation kinetics and the presence of reducing bacteria (Cutter and Bruland 1984). While selenite and selenate are normally stable in natural waters, they can be oxidized or reduced on mineral surfaces (Murphy 1989; Myneni et al. 1997; Nguyen et al. 2005). Selenate can be substantially and rapidly removed from the water column to sediment via reduction to Se(0) and binding to organic matter (Zhang and Moore 1997). Organo-Se compounds undergo photo-oxidation in water, and their mineralization eventually yields inorganic Se species (Chen et al. 2005). Volatile species of Se, H_2Se, CH_3SeH $(CH_3)_2Se$, and $(CH_3)_2Se_2$, are also known to be formed in Se-containing water bodies (Karlson and Frankenberger 1993; Masscheleyn and Patrick 1993; Karlson et al. 1994; Amouroux and Donard 1996; Zhang and Moore 1997; Amouroux et al. 2001; Lin and Terry 2003; Diaz et al. 2009).

4.3.3.2 Sediment

Aquatic sediments represent a dynamic, complex medium where Se speciation is controlled by micro- and macro-scale chemical and physical properties of sediments, as well as biotic factors. The digenesis of Se species in freshwater sediments has been extensively examined (Belzile et al. 2000). The observed geochemical pathways include

1) Se adsorption onto Fe–Mn oxyhydroxides at sediment surfaces,
2) release of adsorbed Se by the reduction of oxyhydroxides,
3) organic matter mineralization, and
4) Se removal from pore water as Se(0) and as seleno-pyrites.

Low redox conditions favor low Se solubility as iron selenide or Se(0) phases are formed (Masscheleyn et al. 1990; Tokunaga et al. 1997). Abiotic reduction of selenite to Se(0) in the presence of iron oxides has been shown to occur in lake sediments (Chen et al. 2008). Contact with FeS_2 and FeS reduces selenite to Se(0) and FeSe, respectively (Breynaert et al. 2008). Under oxidizing conditions, Se(0, II-) (rapidly) and selenite (slowly) will oxidize to selenate (Tokunaga et al. 1997).

Sediment-dwelling organisms, microorganisms, and aquatic plant rhizosphere exudates can enhance the formation of reduced inorganic and organic Se species, including Se-biomolecules and mineral selenides, and thus are an important part of the Se biogeochemical cycle in aquatic ecosystems (Zhang and Moore 1997; Peters et al. 1999a; Zhang et al. 1999; Lucas and Hollibaugh 2001; Stolz et al. 2002; Herbel et al. 2003; Siddique et al. 2006).

4.3.3.3 Organisms

Biological processing of Se is key to the formation of the diversity of Se species found in ecosystems (Figure 4.3). Selenium speciation in organisms of concern (e.g., predatory fish and water fowl; see Chapters 5 and 6) often is not characterized, and thus our understanding of Se biodynamics and potential for toxicity in an aquatic ecosystem is incomplete.

4.3.3.4   Microbial Se Species Transformations

The influence of microbial organisms on the formation and cycling of Se species (Figure 4.3) is significant (Stolz et al. 2006). Bacteria in sediments can remove Se from water via respiratory reduction (Losi and Frankenberger 1997), and some use selenite and selenate as a terminal electron receptor in respiration (Oremland et al. 1989; Stolz et al. 2006). Many microbial organisms also reduce selenate and selenite as a tolerance mechanism. Biological reduction by selenate-reducing microorganisms (usually sulfate-reducers) generally does not yield selenite, but more reduced Se species, such as Se(0), and those containing Se(-II), for example, $(CH_3)_2Se$ and selenoaminoacids (Long et al. 1990; Stolz et al. 2006). Challenger (1945) proposed a pathway to explain this process (Scheme 3) that involves the successive reduction and oxidative methylation of Se to dimethyl selenide in which methionine or S-adenosylmethionine provides the methyl groups (Cantoni 1952).

Scheme 3:

$$SeO_3^{2-} \longrightarrow Se^0 \longrightarrow HSeX \longrightarrow CH_3SeH \longrightarrow (CH_3)_2Se$$

Selenite   Elemental Se   Selenide   Methylselenol   Dimethylselenide

The number and diversity of selenate-reducing microbes increase in Se-enriched environments, but such organisms are also present in non-seleniferous environments (Ike et al. 2000). Selenium species in sediment pore waters undergo a rapid microbial transformation from selenate → selenite → Se(0). Microbes can also form a number of organic Se-containing metabolites, such as protein-bound selenocysteine (SeCys) and selenomethionine (SeMet) similar to animals (see Section 4.3.3.6). Elemental Se [Se(0)] in sediments can also be oxidized by microorganisms, largely to selenate (Dowdle and Oremland 1998; Zhang et al. 2004).

Although fungi also play a significant role in sediment Se transformations involving minerals and organic debris, their role in aquatic ecosystems with regard to Se speciation is relatively unknown (Gadd 2007).

4.3.3.5   Plants and Algae Se Species Transformations

Many reviews of the metabolism of Se in plants have been published (Terry et al. 2000; Ellis and Salt 2003; Dumont et al. 2006; White et al. 2007a). It is generally agreed that Se and sulfur follow the same metabolic pathways in plant species (Freeman et al. 2006; Sors et al. 2005; Terry et al. 2000; White et al. 2007a). After active absorption, selenate is transported to shoots or leafs, assimilated, and metabolized by S pathways and then is found in most S-containing biomolecules. The conversion of selenate to organo-Se compounds in plants is hypothesized to proceed through adenosine phosphoselenate (APSe), selenite, and reduced selenide, then to SeCys, via Cys synthase, from which SeMet is synthesized (Figure 4.4). However, many of

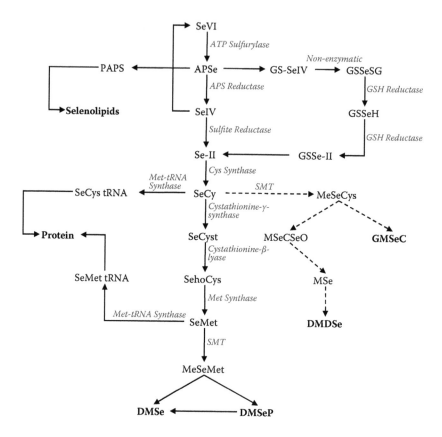

FIGURE 4.4   Transformations of Se in plants and animals.

these Se species have not yet been identified in plant tissues. After SeMet is synthesized, it can be incorporated into protein, methylated to Se-methylselenomethionine
(MeSeMet) and converted to dimethylselenide [(CH₃)₂Se] or dimethylselen](CH$_3$)$_2$Se] or dimethylselenioprop-
pionate (DMSeP). SeCys can also be incorporated into proteins, methylated to
Se-methylselenocysteine (MeSeCys), or transformed to SeMet through selenocystathionine (SeCyst) and selenohomocysteine (SehoCys) (Figure 4.4). Selenium accumulation in plants occurs when the final metabolites are not excreted from the plant
(White et al. 2007a). Alternatively, Se can be directed toward pathways that lead to
the formation of volatile Se species, thereby avoiding the accumulation of excess Se
(Terry et al. 2000).

4.3.3.6 Animal Se Species Transformations

The bioavailability and metabolism of Se and its distribution within an organism
depend on the Se species, the route of entry (dissolved or food), and the organism's
physiology. The major pathway of Se metabolism in animals is via the reduction of
organic and inorganic sources of Se and subsequent incorporation into proteins or
excretion as methylated species (Figure 4.5). It should be noted that several of the
intermediates in this hypothesized pathway have not been verified experimentally,

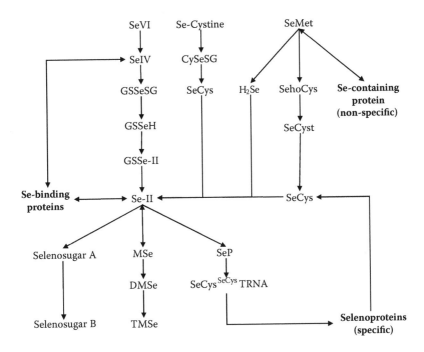

FIGURE 4.5 Transformations of Se in animals.

particularly the less stable intermediates, such as hydrogen selenide, which represents a major knowledge gap in Se speciation and reactive pathways.

Research on the fate of ingested Se(0) and selenides is limited, but Se(0) is probably transformed into methylated metabolites prior to being excreted or exhaled (Agency for Toxic Substances and Disease Registry [ATSDR] 2003; Suzuki et al. 2005).

Ingested selenite, selenate, SeCyst, and SeMet are readily absorbed from the intestine and then stored as SeMet or SeCys in proteins (Figure 4.5) or as selenite complexed to proteins (Lobinski et al. 2000; Nakamuro et al. 2000; Commandeur et al. 2001; ATSDR 2003; Dumont et al. 2006). There is no indication that SeCys, selenate, and selenite are incorporated non-specifically into proteins, indicating that these forms are metabolized by specific Se metabolic processes. SeCys is specifically incorporated into selenoproteins, not into other proteins in place of cysteine. In contrast, SeMet can be incorporated non-specifically into proteins (ATSDR 2003). SeMet incorporation into yolk proteins of fish and bird eggs via maternal transfer is an important pathway leading to embryonic and developmental toxicity (Fan et al. 2002; Palace et al. 2004; Freeman et al. 2006). Because many of the proteins containing Se have not been identified, nor their functions determined, many of the mechanisms of metabolism and toxicity remain unknown (Lobinski et al. 2000; Behne and Kyriakopoulos 2001; Gladyshev 2001; ATSDR 2003).

Methylation of selenide and other Se species (e.g., Se(0)) and excretion in urine may be a major detoxification pathway for Se (ATSDR 2003; Nakamuro et al. 2000). The major forms of Se in urine are now thought to be selenosugars (Suzuki et al. 2005).

4.3.4 Use of Se Speciation Information for Risk Assessments

Selenium species control the movement and environmental partitioning of Se in aquatic ecosystems. For example, 2 marine algae species were shown to rapidly accumulate selenite but not selenate (Hu et al. 1997), so Se accumulation in this type of food chain would be greater if the source of Se is selenite. Microbial conversion can shift the dominant waterborne Se form from selenate to selenite. It is also well established that the toxicity of Se is dependent on its chemical species (ATSDR 2003). For example, in aqueous solution, selenate has been found to be more toxic than selenite to fish (Hodson 1988), while in bird eggs SeMet enrichment is related to deformities and other toxic effects (Heinz et al. 1989). A mechanistic understanding of the toxic species of Se has the promise of providing a "universal" marker for endpoint assessment of Se toxicity or remediation that transcends biological or site specificity. Thus, in risk assessment, when the knowledge of where Se accumulates and its potential toxicity is required, it is critical to have information on the Se species present.

4.3.5 Methods for Assessing Se Speciation in Systems

Several recent comprehensive reviews on measuring Se species in environmental matrices are available (B'hymer and Caruso 2006; Polatajko et al. 2006; Pedrero and Madrid 2008). General comments on the measurement of important Se species are provided below.

4.3.5.1 Selenate and Selenite in Waters

Inorganic Se species are relatively easy to measure in waters by selective hydride generation or ion-chromatography coupled to either atomic fluorescence spectrometry or inductively coupled plasma-mass spectrometry (ICP-MS). Methods often depend on the reduction of selenate to selenite by heating with HCl (Cutter 1978) or HBr–KBr (Raptis et al. 1983), which may result in losses of Se. Selenium species are chemically stable, but sample preservation may be required to prevent precipitation reactions and microbial conversion of species within samples (Ralston et al. 2008).

4.3.5.2 $(CH_3)_2Se$ and $(CH_3)_2Se_2$ in Waters

A variety of analytical procedures have been published on the measurement of volatile Se species in aquatic ecosystems (Amouroux and Donard 1996; Gomez-Ariza et al. 1998; Pyrzynska 1998, 2002; Carvalho and Martin 2001; Diaz et al. 2009). These species can be purged relatively quickly from waters into cryogenically cooled traps, transported to the laboratory and measured by gas chromatography(GC)-ICP-MS (Amouroux et al. 1998) or GC-MS (Dauchy et al. 1994). Care must be taken with elution of Se species from traps as other volatile Se-S species may also be present (Amouroux et al. 1998).

4.3.5.3 Selenium Speciation in Sediments

At present, the main analytical approaches available for understanding Se speciation in sediments are selective extraction procedures (SEP) and X-ray absorption spectrometry (XAS). SEP normally employs hydride generation with atomic absorption spectrometry, atomic fluorescence spectrometry or optical emission spectrometry or ICP-MS for quantitative analysis of Se in the extraction solutions (Martens and Suarez 1997; Kulp and Pratt 2004; Zhang and Frankenberger 2003). Selective chemical extraction of trace element species in soils and sediments has been reviewed and limitations noted (Gleyzes et al. 2002; Wright et al. 2003; Bacon and Davidson 2008). Operationally defined selective extractions are susceptible to Se redistribution during extraction and inaccurate categorization of Se species (e.g., categorization of organo-Se fraction including true C-Se covalently bonded compounds as well as Se complexed to organic matter), validation concerns regarding the difference in extraction behavior of added and native Se species, and alteration of Se species during the fractionation steps (Gruebel et al. 1988; Petrovic and Kastelan-Macan 1996; Martens and Suarez 1997; Jackson and Miller 2000; Shaw et al. 2000; Kulp and Pratt 2004; Bacon and Davidson 2008). Consistency between SEP and synchrotron-based micro-X-ray fluorescence (μ-SXRF) measurements has been demonstrated for freshwater sediments, suggesting that SEP approaches may have the capability to reliably determine Se speciation in natural sediments (Oram et al. 2008). Although this work is not conclusive for all sediment types, SEP holds promise for aquatic ecosystem Se species assessment in routine practice.

Synchrotron-based XAS techniques (Manceau et al. 2002) allow direct determination of Se speciation in solids at much lower concentration levels than most other solid-state spectroscopic techniques. The most popular XAS techniques are the near-edge region of the XAS spectrum (XANES) and extended X-ray absorption fine-structure (EXAFS) (Pickering et al. 1995; Tokunaga et al. 1998; Ryser et al. 2006), and μ-SXRF (Sutton et al. 1995; Oram et al. 2008).

Collection of XANES speciation spectra requires extended X-ray beam exposure that can result in radiation-induced changes in oxidation states (Ryser et al. 2006). Cryostats and controlled atmosphere chambers in XANES analysis are essential to prevent oxidation changes. An additional limitation of EXAFS is that it cannot differentiate atoms that have similar radii, and resolving structural differences of similar molecules is not possible. Micro-SXRF has been successfully used to map Se species in sediment samples, although the precise analysis of specific Se species requires large amounts of beam-time when applied to sediment with low Se concentrations (Oram et al. 2008). Application of these approaches for a variety of freshwater and marine sediments indicates the presence of Se(VI), Se(IV), as well as Se(0 or -II) that cannot be distinguished. In general, synchrotron beam-time access and expense, along with the technical expertise required, limit the routine application of these techniques for identifying Se species in sediments. Perhaps the best utility of X-ray molecular spectroscopy at present is in providing a feedback loop in the development, validation, and application of other sediment speciation approaches such as selective extraction procedures.

4.3.5.4 Selenium Speciation in Biota

Complete and accurate determination of all Se species in tissues is a challenging task and not routine (Polatajko et al. 2006). A variety of approaches to analyze protein-bound seleno amino acids in biological tissue have been reported (Gilon et al.

1995; Hammel et al. 1997; Larsen et al. 2001; Zheng et al. 2003; Encinar et al. 2004; Liang et al. 2006; Su et al. 2008). Many studies have reported SeCys and SeMet concentrations in biota, but the reliability of these measurements is questionable. Most procedures rely on the extraction of a protein fraction, hydrolysis of proteins with acid or enzymes, and subsequent quantification by GC-MS or ion chromatography-ICP-MS. SeMet is relatively stable during extraction and hydrolysis but can oxidize to SeMet-selenoxide in extracts (Ayouni et al. 2006). SeCys is easily oxidized to its dimer selenocystine and Se(0) during hydrolysis and needs to be protected by derivitization before extraction-hydrolysis for reliable quantification (Lobinski et al. 2000).

4.3.6   Case Studies that Document Se Transformations in Systems

4.3.6.1   Lake Macquarie, New South Wales, Australia

Lake Macquarie is a large estuarine barrier lake near the city of Newcastle, New South Wales (NSW). The lake extends approximately 22 km in a north–south direction, from Cockle Bay to Mannering Bay. Lake Macquarie has a maximum width of about 10 km, a maximum depth of approximately 11 m, and an average depth of 8 m (Maunsell and Partners 1974). The lake catchment occupies an area of approximately 622 km². The lake is separated from the ocean by a narrow entrance channel and sandbars. The tidal range in Lake Macquarie is small, with the spring tidal range being estimated at 0.15 m at the western end of the tidal channel (Stone 1964), decreasing with distance from the entrance.

Peters et al. (1999b) measured Se concentrations in surficial sediments ranging from 0.4 to 0.8 mg/kg dw at Kilaben Bay (northeast Lake Macquarie) and up to 9 ± 2 mg/kg dw from Mannerin Bay adjacent to a coal-fired power station at Vales Point. Sediment cores indicated Se concentrations up to 20 mg/kg dw at 50 mm depths (Peters et al. 1999a). Concern about Se contamination of Lake Macquarie became public after the release of reports (Roberts 1994; Kirby 2001; Kirby et al. 2001b) of Se concentrations in mullet (*Mugil cephalus*) and silverbiddy (*Gerres subfasciatus*) that were up to 12 times the acceptable limit for human consumption (1 mg/kg ww). The fish analyzed in these studies were taken from areas close to coal-fired power stations.

Speciation studies have been conducted to understand the partitioning of Se and fluxes within the lake. Selenium enters the lake as selenite and is precipitated with iron oxyhydroxide into sediments, explaining why Se concentrations in lake waters are low (<0.1 μg/L). Measurements of volatile Se species [$(CH_3)_2Se$ and $(CH_3)_2Se_2$] in the water column revealed that a considerable mass of Se is lost by volatilization from the Lake (Kirby and Maher, U of Canberra [Australia], unpublished results), thus with improved fly ash handling procedures, the Se content of sediments was predicted to dramatically decrease, as should the content of fish foraging on sediments. Subsequent measurements of Se concentrations in sediments and fish after ash-handling procedures were changed (Kirby et al. 2001a) revealed this to be the case. Bacteria in the sediment also convert selenite to selenate, Se(0), Se (-II), and Se proteins controlling the remobilization of Se from sediments (Peters et al. 1999b). Selenium in biota was converted to protein-bound SeMet and possibly SeCys; a considerable percentage of Se in biota was uncharacterized, but no inorganic Se species were present. The presence of Se entirely as organo-Se species, which are known to be highly bioavailable, accounts for Se biomagnification in this ecosystem (Barwick and Maher 2003).

4.3.6.2 Saline Agricultural Drainage Systems of the San Joaquin Valley, California, USA

Selenium transformations in biota have been demonstrated both in laboratory and field studies related to the agricultural drainage systems of the San Joaquin Valley, California, USA.

4.3.6.2.1 Laboratory Studies

Fan et al. (1997) investigated Se volatilization from selenite by a euryhaline phytoplankton (*Chlorella* sp.) isolated from Se-laden drainage pond waters. The dominant volatile Se species observed was $(CH_3)_2Se$, while the contribution of $(CH_3)_2Se_2$ and $(CH_3)_2SeS$ was minor. The metabolic precursor to $(CH_3)_2Se$ production was identified as dimethylselenonium propionate (DMSeP). The Se analogue of dimethylsulfonium propionate is known to be an important osmolyte for certain macrophytes and microalgae in saline waters (Kiene 1991; Colmer et al. 1996).

The discovery of DMSeP in the halophytic *Chlorella* implicates its occurrence in other halophytes that synthesize DMSP for osmoregulation. DMSeP as a major sink of biotransformed Se could dramatically alter the fate of transformed Se in biota. Unlike selenoamino acids (e.g., SeMet), DMSeP is not incorporated into proteins and readily dissipated via $(CH_3)_2Se$ volatilization, which implies its resistance to transfer through the food web. The synthesis of DMSeP may also divert selenite from transformation into selenomethionine (SeMet), which is known to be toxic to aquatic predators (Hoffman and Heinz 1988; Hoffman et al. 1996; Finley 2001; Palace et al. 2004). This is consistent with the trace concentrations of SeMet found in *Chlorella*.

4.3.6.2.2 Field Observations

In a separate study, Fan et al. (2002) surveyed food web organisms collected from the agricultural drainage systems of the San Joaquin Valley (CA, USA), including microphytes, macroinvertebrates, fish, and bird embryos. Selenium distribution in biomass and its water-soluble and protein fractions were analyzed. Water-soluble Se constituted a major fraction of the Se biomass, while the protein-bound Se contributed significantly to total Se in the biomass and water-soluble fractions for all food web organisms. Selenomethionine was present in the protein fraction of lower trophic organisms, while it dominated in the embryos of several top trophic bird species. The distribution of protein-bound SeMet in microphytes was negatively related to the salinity of the site waters. This phenomenon could be linked to an increasing Se volatilization capacity with salinity observed for the agricultural drainage systems (Fan and Higashi 2001, 2005), which is also consistent with the laboratory findings described above. A more consistent biomagnification of protein-bound Se and SeMet than biomass Se was evident in the drainage food web, indicating the propensity for trophic transfer of SeMet (Fan et al. 2002).

Among the fish species analyzed, protein-bound SeMet concentrations were generally higher in liver and ovary than in muscle tissues. The reason for the preferential accumulation of SeMet into ovarian and liver proteins is unclear. However, it is reasonable to conjecture that ovarian and liver proteins may be higher in methionine content than muscle proteins, thereby enabling an enhanced SeMet incorporation by replacing the methionine residue. The increase in SeMet deposition into hepatic proteins could be correlated with ovarian lesions observed for the omnivorous carp (*Cyprinus carpio*) (Fan et al. 2002), which is consistent with the toxic effect of SeMet on fish embryos observed in the laboratory (Palace et al. 2004).

4.4 ENVIRONMENTAL PARTITIONING

The multiplicity of Se species in the environment and the importance of Se as an essential trace element for algae, animals, and higher plants, facilitate its partitioning into abiotic and biotic compartments in aquatic environments (Figure 4.6).

Sediment and water often contain most of the Se in many contaminated systems, but Se can readily enter food chains. As such, an understanding of how Se is partitioned and transferred among compartments in an aquatic ecosystem is important when assessing the ecological risk of this potential toxicant.

4.4.1 Aqueous–Solid Phases

4.4.1.1 Adsorption

The most important process that influences the distribution of Se in aquatic ecosystems is the partitioning between the aqueous and solid phases. Under normal conditions, Se in waters is present as selenite or selenate. Both of these species, however, have different adsorption behaviors that result in different partitioning

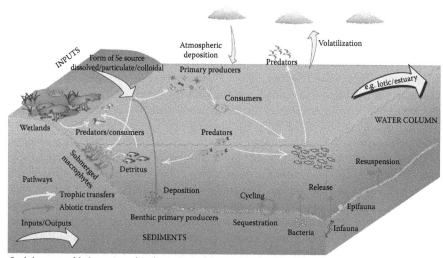

Symbols courtesy of the Integration and Application Network (ian.umces.edu/symbols/)
University of Maryland Center for Environmental Science.

FIGURE 4.6   Partitioning of Se in aquatic environments.

in aquatic systems. Selenite partitions strongly to the oxy-hydroxide minerals of Fe and Mn, whereas selenate exhibits only partial partitioning to Fe compounds at sub-neutral pH and does not adsorb to manganese dioxide (Balistrieri and Chao 1987a, 1990; Parida et al. 1997; Foster et al. 2003; Martínez et al. 2006; Fukushi and Sverjensky 2007; Rovira et al. 2008). In the lithosphere, Fe accounts for 3.92% of total mass and Mn accounts for 0.06%; in soil, Fe is 3.8% and Mn is 0.085% of total mass (Chesworth 2008). Thus, most sediments will have much higher Fe than Mn concentrations; therefore, Fe oxy-hydroxides play an important role in selenite partitioning and, to a lesser degree, that of selenate, depending on local geochemistry. Only in the absence of Fe and Mn oxy-hydroxide minerals do other types of minerals (e.g., Al and Si oxy-hydroxide minerals, including clay minerals and Fe-sulfide minerals) contribute significantly to Se partitioning to sediments. Adsorption of selenite and selenate by these minerals is similar, and they show comparable response to changes in pH and redox as discussed below for Fe and Mn oxy-hydroxide minerals. The main factors that influence the adsorption of selenate and selenite to Fe and Mn oxy-hydroxide minerals are pH, competing anions, and temperature.

4.4.1.2 pH

The effect of pH on Se partitoning to sediments is complex. Increasing pH decreases the binding of both selenite and selenate because it changes the surface charge of the minerals and the hydroxide ions compete with Se for binding sites. The point of zero charge (i.e., the pH above which minerals change from net positive to negative surface charge) for Fe and Mn oxy-hydroxide minerals lies around pH 6 (Glasby and Schulz 1999), but the zero point charge of minerals can vary by several pH units (Brookins 1988; Glasby and Schulz 1999). The sorption of selenite and selenate on hematite, goethite, and magnetite decreases at alkaline pH (Martínez et al. 2006; Rovira et al. 2008). Selenium adsorption increases under acidic conditions; however, decreasing the pH leads to the dissolution of Fe and Mn hydroxide minerals (Davis et al. 1978). Consequently, an optimum pH for adsorption of selenite from solution lies around pH 7 (Davis et al. 1978; Saeki et al. 1995).

4.4.1.3 Redox

Redox potential influences partitioning of Se to sediments. The reductions Mn(IV) $\rightarrow$ Mn(II) and Fe(III) $\rightarrow$ Fe(II) occur under very different redox conditions: while the reduction of Mn(IV) occurs at redox potentials near +600 mV, the reduction of Fe(III) occurs near ± 0 mV (Lovley 1991). By comparison, the reduction cascade Se(VI) $\rightarrow$ Se(IV) $\rightarrow$ Se(0) occurs near +500 mV and +100 mV, respectively (Masscheleyn et al. 1991). Consequently, the reduction of Se(IV) to Se(0) leads to its removal from aqueous phases due to reduced solubility. However, this occurs at the same redox conditions as the reductive dissolution of Fe oxy-hydroxide minerals. Therefore, the optimal redox conditions for adsorptive Se removal from waters by Fe oxy-hydroxide minerals lies in the range of +400 to +100 mV. Low redox conditions found in anoxic sediments can also influence Se adsorption and reduction. This is illustrated by the reduction of selenate by the Fe (II, III) oxide called "green rust" (Myneni et al. 1997) and the sorption–reduction of Se by iron pyrite (Naveau et al. 2007).

4.4.1.4 Temperature

Adsorption processes of Se are decreased at elevated temperatures because adsorption is an exothermic process (Balistrieri and Chao 1987b). Hence, if all other things remained equal, partitioning of Se to sediments is greater at higher latitudes and during seasonal cooling of aquatic systems.

4.4.1.5 Competitive Ions

The main competing anions are carbonate, sulfate from acid rain and many Se emission sources, and phosphate from agricultural use. Increasing anion concentrations decrease the binding of both selenite and selenate to mineral surfaces (Dhillon and Dhillon 2003). For selenate, this competition leads to the absence of any significant adsorption in most ambient waters. Since sea waters generally contain more sulfate and carbonate than freshwaters, adsorption of selenite to sediments or suspended particles should be lower than in fresh waters. In specific cases, other anionic trace element species in excess such as arsenate and chromate, may also cause a decrease in selenite adsorption by competition (Goh and Lim 2004).

4.4.1.6 Precipitation

Another mechanism that transfers Se from aqueous to the solid phases is precipitation (Myneni et al. 1997; Duc et al. 2003). Precipitation is defined as the formation of compounds that exceed their solubility in water, and it needs to be distinguished from adsorption, because both processes have different mechanisms and influencing factors. While the mechanisms of precipitation are specific to the individual Se species (Selivanova et al. 1959; Sharmasarkar et al. 1996), there are some common factors influencing the rate of precipitation, including low temperatures and the availability of nuclei on which the precipitate can begin to form. Precipitation reactions in the environment usually initially produce suspended colloidal solids (defined as particles <0.22 or 0.45 μm diameter) (Gustafsson and Gschwend 1997), followed by the precipitation of amorphous products, which only convert to the thermodynamically more stable crystalline forms over very long time periods (Ihnat 1989). Because of the relatively low Se concentrations in ambient waters, it is uncommon to observe the in situ formation of discrete Se minerals as the result of precipitation reactions, even in the presence of large Se emissions. More commonly, one observes co-precipitation of Se species with major minerals, or adsorption of Se species to pre-existing major minerals.

Selenate generally forms fairly soluble minerals (Ihnat 1989), but by analogy to sulfate, it is possible that calcium selenate could precipitate from ambient waters, especially as crystalline matrix inclusions in gypsum crystals in highly mineralized sulfate brines. Selenite salts tend to be even more soluble than selenate salts (Séby et al. 2001); calcium selenite will precipitate from waters only under exceptional circumstances. Elemental Se(0) is insoluble in water and will precipitate from solution immediately after formation; however, relatively stable colloids may be formed, which resist further aggregation and subsequent precipitation (Mees et al. 1995) but may be available for microbial oxidation (Dowdle and Oremland 1998). In anoxic environments, oxidized Se can be reduced to selenide, forming insoluble salts with transition metals. Work with Se-respiring bacteria in anoxic sediments has shown a microbial pathway for the formation of metal selenides (Herbel et al. 2003). Anaerobic bioreactor biofilms, which contain elevated concentrations of Zn, Cu, Fe, and Na selenide products, provide additional evidence of a microbial pathway of metal selenide formation (Lenz et al. 2008a). Metallic sulfides, such as pyrite, have the capability to adsorb and reduce oxidized Se species (Naveau et al. 2007). The formation and interchange of selenide minerals in sediments is likely complex and dynamic, similar to the sediment biogeochemistry of sulfur (Rickard and Morse 2005).

Because Se(0) is insoluble, microbial reduction of selenite and selenate in sediment pore waters is an important removal process from overlying waters to solid phases. Redox processes are driven directly or indirectly by microbial activity (Schink 1997), and thus formation of Se(0) is consequently influenced by temperature and the availability of the required microbial substrates (Stolz et al. 2006).

4.4.2 AQUEOUS–GAS PHASES

A common mechanism, through which Se is volatilized to the atmosphere and thereby lost to aquatic ecosystems, is biotransformation leading to the production of volatile Se species, principally $(CH_3)_2Se$ and $(CH_3)_2Se_2$ (see Figure 4.3) (Besser et al. 1989; Thompson-Eagle et al. 1989; Ansede and Yoch 1997; Amouroux et al. 1998; Fan and Higashi 1998; Fujita et al. 2005). The extent of Se partitioning into volatile fractions is much less understood than water–sediment partitioning. However, a few studies have demonstrated the potential significance of volatilization (Thompson-Eagle et al. 1989; Fan and Higashi 1998; Fan et al. 1998; Hansen et al. 1998; Fujita et al. 2005).

4.4.3 SELENIUM PARTITIONING INTO BIOTA

The role of Se speciation in partitioning into biota has been reviewed in Section 4.3. Other aspects that control partitioning are discussed below.

4.4.3.1 Phytoplankton and Bacteria

Uptake into primary producers from water represents the largest bioaccumulation step in the aquatic food web (see Chapter 5). Laboratory experiments show phytoplankton cultures are considerably more enriched in Se than estuarine seston or sediments (Doblin et al. 2006), suggesting that a relatively small contribution of phytoplankton particles would be effective in increasing the suspended particulate Se concentration and dramatically increasing exposure of the food web to Se. Partitioning of Se into microalgae (e.g., suspended phytoplankton, attached microphytobenthos or epiphytes) and bacteria is dependent on Se species and concentration, as well as conditions that control growth and biomass, such as temperature, salinity, oxygen, light availability, dissolved organic matter, and cellular nutrient status (Baines et al. 2001). Phytoplankton biomass and hence partitioning will also be dependent on grazing and physical loss processes. Selenite and dissolved organic selenides are taken up at similar rates by phytoplankton (Baines et al. 2001), but selenate has a relatively low bioavailability to microalgae. A detailed discussion of Se uptake mechanisms into phytoplankton is found in Chapter 5. Once incorporated into primary producers, selenite and selenate are reduced to organic selenides (Besser et al. 1994); intracellular Se is mainly seleno-amino acids in peptides and proteins (Wrench 1978; Fan et al. 1998; Vandermeulen and Foda 1988), but it is also found in lipid, soluble carbohydrate, and polysaccharide fractions (Bottino et al. 1984). Euryhaline primary producers can accumulate a significant amount of Se as DMSeP, the Se analogue of the sulfur osmolyte (Fan et al. 1997; Ansede et al. 1999). Thus, from a trophic transfer perspective, Se in biogenic particles is principally comprised of organic selenide (Cutter and Bruland 1984) and highly bioavaliable.

4.4.3.2 Plants

The uptake of Se into plants is primarily through roots, influenced by the Se species and concentration in water and sediment, sediment redox conditions, pH, and the presence of competing anions (Gissel-Nielsen 1971; Haygarth 1994). Selenate is actively taken up by a sulfate transport protein against the electrochemical gradient (Ferrari and Renosto 1972). Selenite is accumulated in plants through passive diffusion and can be inhibited by phosphate (Wells and Richardson 1985; Sors et al. 2005). The trend is that non-Se accumulators discriminate against selenate uptake relative to sulfate, whereas Se accumulators preferentially absorb Se over sulfur, but the mechanisms for the increased preferential uptake and sequestration of Se remain largely unknown (Sors et al. 2005; White et al. 2007b). In terms of Se distribution, selenate translocation parallels sulfate, and Se can accumulate in roots, seeds, stems, leaves, and shoots (Sors et al. 2005). The distribution of different Se species is dependent on development, with selenate concentrating generally in older leaves and organic Se species in younger tissues (Pickering et al. 2000, 2003). Young leaves are a site for selenate reduction and for incorporation of Se into organic forms (Freeman et al. 2006). Selenium accumulation ultimately appears to be determined by Se metabolism (White et al. 2007b). Loss of Se from plants occurs via volatilization, normally when Se is supplied to plants in excess of their Se requirement (Terry et al. 2000). Most of the Se produced by plants is volatilized from roots (Terry and Zayed 1994). In addition, experiments have shown that roots supplied with selenite can release 10-fold more volatile Se species than the leaves of plants supplied with selenate (Terry and Zayed 1994). As for phytoplankton, plant biomass and hence partitioning ultimately will be dependent on grazing and other loss processes, including senescence.

4.4.3.3 Animals

Selenium enters animals primarily via food, thus animal diets and their Se content will determine the partitioning of Se to higher trophic levels (Mayland 1994). The portion assimilated, which is a function of the Se species in the food and an animal's physiology, will determine bioaccumulation within an animal (Mayland 1994). After entering an organism, the quantity of Se absorbed is subject to homeostasis, a process by which Se in metabolic excess is eliminated by urinary excretion (Windisch 2002). Selenium can also be dermally absorbed or inhaled; however, this route is considered a less important source. Although published research on the metabolic pathways in aquatic organisms following absorption of different Se species is sparse, it appears that ingested selenite, selenate, and SeMet can be readily absorbed from the intestine, often to >80% of the administered dose (ATSDR 2003; Dumont et al. 2006). Elemental Se is generally considered to be less available, although Se(0) can be mobilized when reduced by gut bacteria of benthic invertebrates (Reinfelder et al. 1998; Stewart et al. 2004). Selenium is distributed and accumulated in all tissues, and stored primarily as SeMet and SeCys in proteins (Lobinski et al. 2000; Nakamuro et al. 2000; Commandeur et al. 2001; ATSDR 2003; Dumont et al. 2006); thus, it is highly bioavailable and potentially toxic to higher trophic organisms.

4.4.4 Environmental Partitioning: Contrasting Case Studies

To illustrate how environmental partitioning of Se can vary between aquatic ecosystems and influence the accumulation of Se by biota, examples of a benthic detritus–based system (Lake Macquarie) and pelagic phytoplankton–based ecosystem (San Francisco Estuary) are shown in Figure 4.7.

4.4.4.1 Lake Macquarie, New South Wales, Australia

The Lake Macquarie system and study context have been described in Section 4.3.6.1. Selenium enters the lake from the overflow of a power station ash dam and is precipitated with iron hydroxides as freshwater inputs mix with the high-salinity estuarine water, resulting in considerable partitioning to the sediments. Selenium concentrations in the lake water are generally below 0.05 µg/L, and the sediments range between 0.4 and 9 mg/kg dw but can exceed 20 mg/kg dw at depth (<63 um fraction) (Peters et al. 1999a,b). Selenium enters the food web through accumulation by the benthos (e.g., bacteria, benthic algae, and bivalves) (Peters et al. 1999a, 1999b). Fish such as mullet (*Mugil cephalus*) foraging in sediment consume seagrass detritus and benthic biota (Kirby et al. 2001a,b). Selenium biomagnification has been shown to occur through successive trophic levels (Barwick and Maher 2003), but the response is variable (Roach et al. 2008). Phytoplankton biomass is low and a relatively unimportant vector for the accumulation of Se in the food web.

4.4.4.2 San Fransico Bay, California, USA

Selenium enters the San Francisco Bay in dissolved forms via the Sacramento and San Joaquin Rivers, as well as oil refinery and municipal waste discharges (Cutter and Cutter 2004). Selenium in the water column ranges between 0.08 to 0.24 µg/L (Cutter and Cutter 2004), while suspended particulate Se concentrations range from 0.03 to 1.66 mg/kg dw comprising approximately 10% of the total water column inventory (Doblin et al. 2006). Sediments are less enriched with Se (0.16 to 0.65 mg/kg dw) (Meseck 2002). In contrast to Lake Macquarie, phytoplankton are the major source of food for invertebrate biota, and benthic filter-feeding clams are the primary point of Se entry into higher trophic levels (Stewart et al. 2004).

4.4.5 Methods for Assessing Biological Partitioning of Se (Exposure-Dose and Trophic Transfer) in Systems

Identifying the trophic compartments and linkages among species in aquatic ecosystems is complex but essential for understanding the biological partitioning

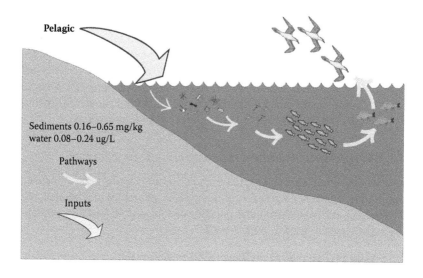

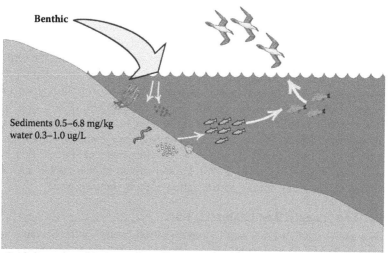

Symbols courtesy of the integration and application network (ian.umces.edu/symbols/)
University of Maryland Center for Environmental Science.

FIGURE 4.7 Comparison of the partitioning of Se in pelagic phytoplankton and benthic detritus–based ecosystems.

and trophic transfer of Se in the food web. Assessment of partitioning and rates of trophic transfer among environmental compartments, primarily between sediment/water to biota, can be undertaken by using several approaches as outlined below:

4.4.5.1   Laboratory Microcosms and Mesocosms

In these studies, animals (typically bivalves or fish) are placed in vessels containing water or water and sediment (either Se-spiked or naturally contaminated) mixtures. Standard protocols (e.g., ASTM 2008) are available to provide guidance on the selection of appropriately sized containers, ratios of water and sediment volume to organism size to conduct experiments. Guidance on other issues such as aeration, feeding, and assessment of organism condition is also available (e.g., Salazar and Salazar 2005). An important consideration when determining sediment to biota transfer, if spiked sediments are used, is ensuring that the spiking of sediments is performed rigorously to ensure Se is on particles and not in pore waters. Typically, sediments can require equilibration times on the order of weeks before Se partitions to particles (Simpson et al. 2004). Adjustment of sediment pH may also be required, and pore water concentrations must be measured to ensure that Se has been adsorbed.

4.4.5.2   Transplantation Experiments

Organisms can also be transplanted into the environment. Typically, filter-feeding organisms such as oysters are suspended in bags in the water column (Spooner et al. 2003), while sediment-dwelling organisms such as clams or cockles are placed in mesh cages that allow intimate contact of organisms with the sediment (Burt et al. 2007). For guidance with designing and conducting these experiments, publications such of those of Salazar and Salazar (2005) should be consulted. The main drawback of transplantation experiments is vandalism or natural predation (due to poor design of sample bag or cage), so redundancy should be incorporated into experimental designs to cover the likelihood of missing samples.

4.4.5.3   In Situ Measurements

The third approach is to measure trophic compartments in situ. Gut or stomach dietary analysis is the traditional method for determining food web relationships (Hyslop 1980; West et al. 2003). This approach is often difficult because food webs are complex and spatial or temporal variations in food availability occur. It is also labor intensive and difficult to delineate what was actually digested and assimilated by consumers. The use of C and N stable isotopes provides an alternative approach because the assimilation of energy or mass flow through all of the different trophic linkages leading to an organism is measured (Post 2002). This analysis measures the enrichment of $\delta15N$ in each species relative to the base level in the food chain (e.g., phytoplankton) or primary herbivore (e.g., phytoplankton grazer) to identify trophic level. It is based on the selective metabolic partitioning of nitrogen, which leads to a preferential elimination of lighter isotopes during respiration, metabolism, and excretion (DeNiro and Epstein 1978) and thus the enrichment of $\delta15N$ in higher trophic-level organisms. By comparison, $\delta13C$ changes little as carbon moves through food webs and can be used to identify organisms consuming different carbon sources in their diet such as C3 or C4 plants (Post 2002). In combination, $\delta15N$ and $\delta13C$ provide a measure of the relative trophic positions of organisms in a food web. Stable isotope analysis has been used widely in contaminant studies (Fisk et al. 2001; Hop et al. 2002), including Se (Stewart et al. 2004).

4.5 CONCLUSIONS

4.5.1 SOURCES

A wide range of sources span natural (e.g., terrestrial geochemical weathering and mobilization, wildfires, volcanic activity) and anthropogenic activities (e.g., agriculture, mining, coal-fired power plants, and municipal wastewater treatment). The relative magnitude of these sources varies spatially and temporally, as do their delivery to aquatic systems (e.g., episodic, continual).

Major findings are as follows:

- Selenium is being redistributed in the environment by human activities.
- Stochastic processes of volcanic activity and wildfires can periodically be regionally important Se sources.
- Selenium sources, species, and loads are spatially and temporally variable.
- There are multiple Se sources and sinks in the environment.
- Water is the most important vector for delivering Se to aquatic systems.

4.5.2 SPECIATION

Knowledge of Se speciation is essential to understanding its mobility, transformation, and partitioning in the environment and potential risk to aquatic ecosystems. There are multiple Se species associated with 3 major processes in aquatic systems:

1) deposition or resuspension (selenate, selenite, elemental Se, and Se(-II));
2) trophic transfer involving algae, plants, and animals (selenomethionine, selenocysteine, Se(-II)); and
3) microbial processes (selenate, selenite, elemental Se, Se(-II), and gaseous forms dimethylselenide and dimethyldiselenide).

Major findings are as follows:

- Knowledge of speciation is important to understand Se transport, partitioning, and effects.
- Major Se species in biota are often mischaracterized.
- In certain cases, Se speciation can be used to ascribe Se to a particular source.
- Development and application of Se speciation methods and models should be driven by the need to understand effects.
- Our understanding of Se metabolism in aquatic biota stems mainly from studies of terrestrial plant and animal systems.

4.5.3 ENVIRONMENTAL PARTITIONING

The major redistribution of Se between compartments can occur immediately on delivery to the aquatic system (e.g., adsorption of Se on hydrated iron oxides, release of Se from particles). Selenium redistribution within the system is then dependent on the structure of the aquatic food web (e.g., detrital vs. phytoplankton-based food webs) and the hydraulic residence time.

Major findings are as follows:

- to determine Se partitioning, it is necessary to understand the specific aquatic ecosystem under study (e.g., food web structure, hydrology, residence time).
- being a required element in biota, Se uptake is facilitated across most biological membranes, making its partitioning in the environment unique among metalloid contaminants.

4.5.4 KNOWLEDGE GAPS

The following knowledge gaps were identified:

- The atmosphere is the least understood of all the Se compartments. Extensive knowledge is available on water, sediment, and biota, but there are few data for the atmosphere.
- A major uncertainty in calculating Se budgets is Se volatilization. Few data are available for this major loss and mobilization process.
- Data are lacking on how future approaches and technological advances in energy production will affect Se loading to the environment.
- Se atmospheric re-deposition processes—global distillation or regional deposition adjacent to source—are not sufficiently understood in the context of what is known about other contaminants such as mercury where diffuse global transport is significant.
- Microbial organisms are important in transforming Se in the environment; however, there are few flux estimates in contaminated systems.
- Because of the lack of identifiable quality control and quality assurance procedures in published data, the reliability of some total Se and Se species measurements is of concern.
- There is variability in the measurement of Se concentration and loads with respect to environmental dynamics such as diurnal, seasonal, and climatic processes.

4.5.5 RECOMMENDATIONS

The following recommendations are provided:

- Investigations of the fate of Se and partitioning in aquatic ecosystems should include estimates of Se volatilization and aerosol deposition.
- Technological advances in managing Se loads from industries such as power generation, mining, and agriculture and their potential effect on Se loads should be evaluated and included in risk assessment.
- Development of reliable methods for measuring Se species in all environmental compartments is required.

- Establishment of interlaboratory comparison studies and use of suitable certified reference materials is required to ensure high-quality analytical data.
- Publication of Se data should contain reference to quality control and quality assurance procedures.
- To understand Se partitioning, there is a need to calculate accurate mass balances and thus loads, using knowledge of source discharges (pulse, press, non-point, and point) as well as the correct interpretation of the hydrologic discharge profiles.
- Loads should be modeled with Se speciation, form, and phase to understand accumulation in different environmental compartments.

In summary, Se remains a challenge for the future because global release of Se through anthropogenic activities will continue to load aquatic ecosystems, thus increasing potential for risk. The complex interplay of Se species with the environment demands a thorough understanding of release, transport, and effects processes based on known-quality data collected through quality-assured processes.

4.6 REFERENCES

Akiho H, Ito S, Matsuda H. 2008. Effect of oxidation agents on Se^{6+} formation in a wet FGD. Power Plant Air Pollution Control "Mega" Symposium; 2008 Aug 25–28; Baltimore, MD, USA. Paper No. 20.

Amouroux D, Donard OFX. 1996. Maritime emission of selenium to the atmosphere in eastern Mediterranean seas. *Geophys Res Lett* 23:1777–1780.

Amouroux D, Tessier E, Pecheyran C, Donard OFX. 1998. Sampling and probing volatile metal (loid) species in natural waters by in-situ purge and cryogenic trapping followed by gas chromatography and inductively coupled plasma mass spectrometry (P-CT–GC–ICP/MS). *Anal Chim Acta* 377:241–254.

Amouroux D, Liss PS, Tessier E, Hamren-Larsson M, Donard OFX. 2001. Role of oceans as biogenic sources of selenium. *Earth Planet Sci Lett* 189:277–283.

Andren AW, Klein DH, Talmi Y. 1975. Selenium in coal-fired steam plant emissions. *Environ Sci Technol* 9:856–858.

Ansede JH, Yoch DC. 1997. Comparison of selenium and sulfur volatilization by dimethylsulfoniopropionate lyase (DMSP) in two marine bacteria and estuarine sediments. *FEMS Microbiol Ecol* 23:315–324.

Ansede JH, Pellechia PJ, Yoch DC. 1999. Selenium biotransformation by the salt marsh cordgrass *Spartina alterniflora*: evidence for dimethylselenonioproprionate formation. *Environ Sci Technol* 33:2064–2069.

[ASTM] American Society for Testing and Materials International. 2008. Standard guide for collection, storage, characterization, and manipulation of sediments for toxicological testing and for selection of samplers used to collect benthic invertebrates. West Conshohocken (PA, USA): ASTM. E1391-03.

Atalay A. 1990. Environmental impact assessment of selenium from coal mine spoils. Norman (OK, USA): School of Civil Engineering and Environmental Science, Oklahoma Univ. DOE/PC/89782-T4.

Atkinson R, Aschmann SM, Hasegawa D, Thompson-Eagle ET, Frankenberger WT Jr. 1990. Kinetics of the atmospherically important reactions of dimethyl selenide. *Environ Sci Technol* 24:1326–1332.

[ATSDR] Agency for Toxic Substances and Disease Registry. 2003. Toxicological profile for selenium. Atlanta (GA, USA): ATSDR, US Department of Health and Human Services, Public Health Service.

Ayouni L, Barbier F, Imbert J, Gauvrit J, Lantéri P, Grenier-Loustalot M. 2006. New separation method for organic and inorganic selenium compounds based on anion exchange chromatography followed by inductively coupled plasma mass spectrometry. *Anal Bioanal Chem* 385:1504–1512.

B'hymer C, Caruso JA. 2006. Selenium speciation analysis using inductively coupled plasma-mass spectrometry. *J Chromatogr A* 1114:1–20.

Bacon JR, Davidson CM. 2008. Is there a future for sequential chemical extraction? *Analyst* 133:25–46.

Baines SB, Fisher NS, Doblin MA, Cutter GA. 2001. Accumulation of dissolved organic selenides by marine phytoplankton. *Limnol Oceanogr* 46:1936–1944.

Balistrieri LS, Chao TT. 1987a. Adsorption of selenium by iron and manganese oxides: environmental implications. Proceedings of the Third Technical Meeting, U.S. Geological Survey Program on Toxic Waste-Ground-Water Contamination. Pensacola (FL, USA). p E19–E20.

Balistrieri LS, Chao TT. 1987b. Selenium adsorption by goethite. *Soil Sci Soc Am J* 51:1145–1151.

Balistrieri LS, Chao TT. 1990. Adsorption of selenium by amorphous iron oxyhydroxide and manganese dioxide. *Geochim Cosmochim Acta* 54:739–751.

Barwick M, Maher W. 2003. Biotransference and biomagnification of selenium, copper, cadmium, zinc, arsenic and lead in a temperate seagrass ecosystem from Lake Macquarie Estuary, NSW, Australia. *Mar Environ Res* 56:471–502.

Behne D, Kyriakopoulos A. 2001. Mammalian selenium-containing proteins. *Annu Rev Nutr* 21:453–473.

Belzile N, Chen YW, Xu R. 2000. Early diagenetic behaviour of selenium in freshwater sediments. *Appl Geochem* 15:1439–1454.

Besser JM, Huckins JN, Little EE, La Point TW. 1989. Distribution and bioaccumulation of selenium in aquatic microcosms. *Environ Pollut* 62:1–12.

Besser JM, Huckins JN, Clark RC. 1994. Separation of selenium species released from Se-exposed algae. *Chemosphere* 29:771–780.

Bottino NR, Banks CH, Irgolic KJ, Micks P, Wheeler AE, Zingaro RA. 1984. Selenium containing amino acids and proteins in marine algae. *Phytochemistry* 23:2445–2452.

Bouwer H. 1989. Agricultural contamination: problems and solutions. *Water Environ Technol* 1:292–297.

Breynaert E, Bruggeman C, Maes A. 2008. XANES- EXAFS analysis of Se solid-phase reaction products formed upon contacting Se (IV) with FeS_2 and FeS. *Environ Sci Technol* 42:3595–3601.

Bronikowski T, Pasiuk-Bronikowska W, Ulejczyk M, Nowakowski R. 2000. Interactions between environmental selenium and sulphoxy radicals. *J Atmos Chem* 35:19–31.

Brookins DG. 1988. Eh-pH diagrams for geochemistry. New York (NY, USA): Springer-Verlag. 176 p.

Burt A, Maher W, Roach A, Krikowa F, Honkoop P, Bayne B. 2007. The accumulation of Zn, Se, Cd, and Pb and physiological condition of *Anadara trapezia* transplanted to a contamination gradient in Lake Macquarie, New South Wales, Australia. *Mar Environ Res* 64:54–78.

Butler DL, Wright WG, Stewart KC, Osmundson BC, Krueger RP, Crabtree DW. 1996. Detailed study of selenium and other constituents in water, bottom sediment, soil, alfalfa, and biota associated with irrigation drainage in the Uncompahgre Project area and in the Grand Valley, west-central Colorado, 1991–93. Menlo Park (CA, USA): USGS Water-Resources Investigations Report No. 96-4138. 136 p.

Cantoni G. 1952. The nature of the active methyl donor formed enzymatically from L-methionine and adenosinetriphosphate. *J Am Chem Soc* 74:2942–2943.

Cappon CJ. 1991. Sewage sludge as a source of environmental selenium. *Sci Tot Environ* 100:177–205.

Carvalho KM, Martin DF. 2001. Removal of aqueous selenium by four aquatic plants. *J Aquat Plant Manage* 39:33–36.

Challenger F. 1945. Biological methylation. *Chem Rev* 36:315–361.

Chen YW, Zhou MD, Tong J, Belzile N. 2005. Application of photochemical reactions of Se in natural waters by hydride generation atomic fluorescence spectrometry. *Anal Chim Acta* 545:142–148.

Chen YW, Truong HYT, Belzile N. 2008. Abiotic formation of elemental selenium and role of iron oxide surfaces. *Chemosphere* 74:1079–1084.

Chesworth W. 2008. Encyclopedia of soil science. London (UK): Kluwer Academic. 902 p.

Colmer TD, Teresa WF, Läuchli A, Higashi RM. 1996. Interactive effects of salinity, nitrogen and sulphur on the organic solutes in *Spartina alterniflora* leaf blades. *J Exp Bot* 47:369–375.

Commandeur JN, Rooseboom M, Vermeulen NP. 2001. Chemistry and biological activity of novel selenium-containing compounds. *Adv Exp Med Biol* 500:105–112.

Crutchfield JU. 2000. Recovery of a power plant cooling reservoir ecosystem from selenium bioaccumulation. *Environ Sci Policy* 3:145–163.

Cumbie PM, Van Horn SL. 1978. Selenium accumulation associated with fish mortality and reproductive failure. *Proc Ann Conf Southeastern Assoc Fish Wildl Agencies* 32:612–624.

Cutter GA. 1978. Species determination of selenium in natural waters. *Anal Chim Acta* 98:59–66.

Cutter GA. 1989. The estuarine behaviour of selenium in San Francisco Bay. *Estuar Coastal Shelf Sci* 28:13–34.

Cutter GA, Bruland KW. 1984. The marine biogeochemistry of selenium: a re-evaluation. *Limnol Oceanogr* 29:1179–1192.

Cutter GA, Church TM. 1986. Selenium in western Atlantic precipitation. *Nature* 322:720–722.

Cutter GA, San Diego-McGlone MLC. 1990. Temporal variability of selenium fluxes in San Francisco Bay. *Sci Total Environ* 97:235–250.

Cutter GA, Cutter LS. 2001. Sources and cycling of selenium in the western and equatorial Atlantic Ocean. *Deep Sea Res Part II* 48:2917–2931.

Cutter GA, Cutter LS. 2004. Selenium biogeochemistry in the San Francisco Bay estuary: changes in water column behavior. *Estuar Coastal Shelf Sci* 61:463–476.

Dauchy X, Potin-Gautier M, Astruc A, Astruc M. 1994. Analytical methods for the speciation of selenium compounds: a review. *Anal Bioanal Chem* 348:792–805.

Davis JA, James RO, Leckie JO. 1978. Surface ionization and complexation at the oxide/water interface: I. Computation of electrical double layer properties in simple electrolytes. *J Colloid Interface Sci* 63:480–499.

DeNiro MJ, Epstein S. 1978. Influence of diet on the distribution of carbon isotopes in animals. *Geochim Cosmochim Acta* 42:495–506.

Dhillon KS, Dhillon SK. 2003. Distribution and management of seleniferous soils. *Adv Agron* 79:119–184.

Diaz X, Johnson WP, Oliver WA, Naftz DL. 2009. Volatile selenium flux from the Great Salt Lake, Utah. *Environ Sci Technol* 43:53–59.

Doblin MA, Baines SB, Cutter LS, Cutter GA. 2006. Sources and biogeochemical cycling of particulate selenium in the San Francisco Bay estuary. *Estuar Coastal Shelf Sci* 67:681–694.

Dowdle PR, Oremland RS. 1998. Microbial oxidation of elemental selenium in soil slurries and bacterial cultures. *Environ Sci Technol* 32:3749–3755.

Duc M, Lefevre G, Fedoroff M, Jeanjean J, Rouchaud JC, Monteil-Rivera F, Dumonceau J, Milonjic S. 2003. Sorption of selenium anionic species on apatites and iron oxides from aqueous solutions. *J Environ Radioact* 70:61–72.

Dumont E, Vanhaecke F, Cornelis R. 2006. Selenium speciation from food source to metabolites: a critical review. *Anal Bioanal Chem* 385:1304–1323.

Ellis DR, Salt DE. 2003. Plants, selenium and human health. *Curr Opin Plant Biol* 6:273–279.

Encinar JR, Schaumloffel D, Ogra Y, Lobinski R. 2004. Determination of selenomethionine and selenocysteine in human serum using speciated isotope dilution-capillary HPLC — inductively coupled plasma collision cell mass spectrometry. *Anal Chem* 76:6635–6642.

[EPRI] Electric Power Research Institute. 1994. Electric utility trace substances synthesis report, Volume l. Palo Alto (CA, USA): EPRI. Report TR-104614.

[EPRI] Electric Power Research Institute. 2008. Multimedia fate of selenium and boron at coal-fired power plants equipped with particulate and wet FGD controls. Palo Alto (CA, USA): EPRI. Report 1015615.

Fan TWM, Higashi RM. 1998. Biochemical fate of selenium in microphytes: natural bioremediation by volatilization and sedimentation in aquatic environments. In: Frankenberger WT Jr, Engberg RA, editors. Environmental chemistry of selenium. Boca Raton (FL, USA): CRC Pr. p 545–563.

Fan TWM, Higashi RM. 2001. Mitigating selenium ecotoxic risk by combining foodchain breakage with natural remediation. 2000–2001 Technical Progress Reports: Univ of California Salinity/Drainage Program Annual Report (CA, USA). p 1–17.

Fan TWM, Higashi RM. 2005. Interaction of Se biogeochemistry with foodchain disruption in full-scale evaporation basins and pilot-scale drain water systems. University of California Salinity/Drainage Annual Report. Riverside (CA, USA): Univ of California.

Fan TWM, Lane AN, Higashi RM. 1997. Selenium biotransformations by a euryhaline microalga isolated from a saline evaporation pond. *Environ Sci Technol* 31:569–576.

Fan TWM, Higashi RM, Lane AN. 1998. Biotransformations of selenium oxyanion by filamentous cyanophyte-dominated mat cultured from agricultural drainage waters. *Environ Sci Technol* 32:3185–3193.

Fan TWM, Teh SJ, Hinton DE, Higashi RM. 2002. Selenium biotransformations into proteinaceous forms by foodweb organisms of selenium-laden drainage waters in California. *Aquat Toxicol* 57:65–84.

Ferguson RI. 1987. Accuracy and precision of methods for estimating river loads. *Earth Surf Processes Landforms* 12:59–104.

Ferrari G, Renosto F. 1972. Regulation of sulfate uptake by excised barley roots in the presence of selenate. *Plant Physiol* 49:114–116.

Finley JW. 2001. Selenium (Se) from high-selenium broccoli is utilized differently than selenite, selenate and selenomethionine, but is more effective in inhibiting colon carcinogenesis. *Biofactors* 14:191–196.

Fisk AT, Hobson KA, Norstrom RJ. 2001. Influence of chemical and biological factors on trophic transfer of persistent organic pollutants in the Northwater Polynya marine food web. *Environ Sci Technol* 35:732–738.

Foster AL, Brown GE, Parks GA. 2003. X-ray absorption fine structure study of As (V) and Se (IV) sorption complexes on hydrous Mn oxides. *Geochim Cosmochim Acta* 67:1937–1953.

Freeman JL, Zhang LH, Marcus MA, Fakra S, McGrath SP, Pilon-Smits EAH. 2006. Spatial imaging, speciation, and quantification of selenium in the hyperaccumulator plants *Astragalus bisulcatus* and *Stanleya pinnata*. *Plant Physiol* 142:124–134.

Fujita M, Ike M, Hashimoto R, Nakagawa T, Yamaguchi K, Soda SO. 2005. Characterizing kinetics of transport and transformation of selenium in water–sediment microcosm free from selenium contamination using a simple mathematical model. *Chemosphere* 58:705–714.

Fukushi K, Sverjensky DA. 2007. A surface complexation model for sulfate and selenate on iron oxides consistent with spectroscopic and theoretical molecular evidence. *Geochim Cosmochim Acta* 71:1–24.

Gadd GM. 2007. Bacterial and fungal transformations of minerals, metals and metalloids. *Geophys Res Abstr* 9:179.

Gao S, Tanji KK, Dahlgren RA, Ryu J, Herbel MJ, Higashi RM. 2007. Chemical status of selenium in evaporation basins for disposal of agricultural drainage. *Chemosphere* 69:585–594.

Gerhardt MB, Green FB, Newman RD, Lundquist TJ, Tresan RB, Oswald WJ. 1991. Removal of selenium using a novel algal-bacterial process. *J Water Poll Contr Fed* 63:799–805.

Gillespie RB, Baumann PC. 1986. Effects of high tissue concentrations of selenium on reproduction by bluegills. *Trans Am Fish Soc* 115:208–213.

Gilon N, Astruc A, Astruc M, Potin-Gautier M. 1995. Selenoamino acid speciation using HPLC-ETAAS following an enzymic hydrolysis of selenoprotein. *Appl Organomet Chem* 9:623–628.

Gissel-Nielsen G. 1971. Influence of pH and texture of the soil on plant uptake of added selenium. *J Agric Food Chem* 19:1165–1167.

Gladyshev VN. 2001. Identity, evolution and function of selenoproteins and selenoprotein genes. In: Hatfield DL, editor. Selenium: its molecular biology and role in human health. Amsterdam (NL): Kluwer Academic. p 99–114.

Glasby GP, Schulz HD. 1999. Eh-pH diagrams for Mn, Fe, Co, Ni, Cu and As under seawater conditions: Application of two new types of Eh-pH diagrams to the study of specific problems in marine geochemistry. *Aquat Geochem* 5:227–248.

Gleyzes C, Tellier S, Astruc M. 2002. Fractionation studies of trace elements in contaminated soils and sediments: a review of sequential extraction procedures. *Trends Anal Chem* 21:451–467.

Goh KH, Lim TT. 2004. Geochemistry of inorganic arsenic and selenium in a tropical soil: effect of reaction time, pH, and competitive anions on arsenic and selenium adsorption. *Chemosphere* 55:849–859.

Gomez-Ariza JL, Pozas JA, Giraldez I, Morales E. 1998. Speciation of volatile forms of selenium and inorganic selenium in sediments by gas chromatography–mass spectrometry. *J Chromatogr A* 823:259–277.

Goodrich-Mahoney JW. 1996. Constructed wetland treatment systems applied research program at the Electric Power Research Institute. *Water Air Soil Pollut* 90:205–217.

Gruebel KA, Leckie JO, Davis JA. 1988. The feasibility of using sequential extraction techniques for arsenic and selenium in soils and sediments. *Soil Sci Soc Am J* 52:390–397.

Guo L, Jury WA, Frankenberger WT Jr. 2000. Measurement of the Henry's constant of dimethyl selenide as a function of temperature. *J Environ Qual* 29:1715–1717.

Guo L, Jury WA, Frankenberger WT Jr. 2001. Coupled production and transport of selenium vapor in unsaturated soil: evaluation by experiments and numerical simulation. *J Contam Hydrol* 49:67–85.

Gustafsson O, Gschwend PM. 1997. Aquatic colloids: concepts, definitions, and current challenges. *Limnol Oceanogr* 43:519–528.

Guun SA, Harr JA, Olson OS, Schroder HJ, Allaway WH, Lakin THW, Boaz JTD. 1976. Medical and biological effects of environmental pollutants. Selenium. Washington (DC, USA): National Academy of Sciences. 203 p.

Hammel C, Kyriakopoulos A, Rösick U, Behne D. 1997. Identification of selenocysteine and selenomethionine in protein hydrolysates by high-performance liquid chromatography of their o-phthaldialdehyde derivatives. *Analyst* 122:1359–1364.

Hansen D, Duda PJ, Zayed A, Terry N. 1998. Selenium removal by constructed wetlands: role of biological volatilization. *Environ Sci Technol* 32:591–597.

Haygarth PM. 1994. Global importance and global cycling of selenium. In: Frankenberger WT Jr, Benson S, editors. Selenium in the environment. New York (NY, USA): Marcel Dekker. p 1–27.

Heinz GH, Hoffman DJ, Gold LG. 1989. Impaired reproduction of mallards fed an organic form of selenium. *J Wildlife Manag* 53:418–428.

Heninger I, Potin-Gautier M, Astruc M, Galvez L, Vignier V. 1998. Speciation of selenium and organotin compounds in sewage sludge applied to land. *Chem Speciat Bioavail* 10:1–10.

Herbel MJ, Blum JS, Oremland RS, Borglin SE. 2003. Reduction of elemental selenium to selenide: experiments with anoxic sediments and bacteria that respire Se-oxyanions. *Geomicrobiol J* 20:587–602.

Hodson PV. 1988. Effect of metal metabolism on uptake, disposition, and toxicity in fish. *Aquat Toxicol* 11:3–18.

Hoffman DJ. 2002. Role of selenium toxicity and oxidative stress in aquatic birds. *Aquat Toxicol* 57:11–26.

Hoffman DJ, Heinz GH. 1988. Embryotoxic and teratogenic effects of selenium in the diet of mallards. *J Toxicol Environ Health* 24:477–490.

Hoffman DJ, Heinz GH, LeCaptain LJ, Eisemann JD, Pendleton GW. 1996. Toxicity and oxidative stress of different forms of organic selenium and dietary protein in mallard ducklings. *Arch Environ Contam Toxicol* 31:120–127.

Hop H, Borga K, Gabrielsen GW, Kleivane L, Skaares JU. 2002. Food web magnification of persistent organic pollutants in poikilotherms and homeotherms from the Barents Sea. *Environ Sci Technol* 36:2589–2597.

Howard JH. 1977. Geochemistry of selenium: formation of ferroselite and selenium behavior in the vicinity of oxidizing sulfide and uranium deposits. *Geochim Cosmochim Acta* 41:1665–1678.

Hu M, Yang Y, Martin JM, Yin K, Harrison PJ. 1997. Preferential uptake of Se (IV) over Se (VI) and the production of dissolved organic Se by marine phytoplankton. *Mar Environ Res* 44:225–231.

Hyslop EJ. 1980. Stomach contents analysis — a review of methods and their application. *J Fish Biol* 17:411–429.

Ihnat M. 1989. Occurrence and distribution of selenium. Boca Raton (FL, USA): CRC Pr. 347 p.

Ike M, Takahashi K, Fujita T, Kashiwa M, Fujita M. 2000. Selenate reduction by bacteria isolated from aquatic environment free from selenium contamination. *Wat Res* 34:3019–3025.

Jackson BP, Miller WP. 2000. Effectiveness of phosphate and hydroxide for desorption of arsenic and selenium species from iron oxides. *Soil Sci Soc Am J* 64:1616–1622.

Johannesson KH, Stetzenbach KJ, Kreamer DK, Hodge VF. 1996. Multivariate statistical analysis of arsenic and selenium concentrations in groundwaters from south-central Nevada and Death Valley, California. *J Hydrol* 178:181–204.

Johnson TM. 2004. A review of mass-dependent fractionation of selenium isotopes and implications for other heavy stable isotopes. *Chem Geol* 204:201–214.

Johnson TM, Herbel MJ, Bullen TD, Zawislanski PT. 1999. Selenium isotope ratios as indicators of selenium sources and oxyanion reduction. *Geochim Cosmochim Acta* 63:2775–2783.

Karlson U, Frankenberger WT Jr. 1993. Biological alkylation of selenium and tellurium. *Met Ions Biol Syst* 29:185–227.

Karlson U, Frankenberger WT Jr, Spencer WF. 1994. Physicochemical properties of dimethyl selenide and dimethyl diselenide. *J Chem Eng Data* 39:608–610.

Kharaka YK, Kakouros EG, Miller JB. 2001. Natural and anthropogenic loading of dissolved selenium in Colorado River Basin. In: Cidu R, editor. Proceedings of the Tenth International Symposium WRI-10; 2001 Jun 10–15; Villasimius, Italy: A.A. Balkema, volume 2, p 1107–1110.

Kiene RP. 1991. Decomposition of dissolved DMSP and DMS in estuarine waters: dependence on temperature and substrate concentration. *Mar Ecol Prog Ser* 76:1–11.

Kirby J, Maher W, Harasti D. 2001a. Changes in selenium, copper, cadmium, and zinc concentrations in mullet (*Mugil cephalus*) from the southern basin of Lake Macquarie, Australia, in response to alteration of coal-fired power station fly ash handling procedures. *Arch Environ Contam Toxicol* 41:171–181.

Kirby J, Maher W, Krikowa F. 2001b. Selenium, cadmium, copper, and zinc concentrations in sediments and mullet (*Mugil cephalus*) from the southern basin of Lake Macquarie, NSW, Australia. *Arch Environ Contam Toxicol* 40:246–256.

Knotek-Smith HM, Crawford DL, Möller G, Henson RA. 2006. Microbial studies of a selenium-contaminated mine site and potential for on-site remediation. *J Ind Microbiol Biotechnol* 33:897–913.

Kulp TR, Pratt LM. 2004. Speciation and weathering of selenium in Upper Cretaceous chalk and shale from South Dakota and Wyoming, USA. *Geochim Cosmochim Acta* 68:3687–3701.

Lambie JC. 1978. Measurement of flow: velocity-area methods. In: Herschy RW, editor. Hydrometry: principles and practices. Chichester (UK): J Wiley. p 1–52.

Larsen EH, Hansen M, Fan T, Vahl M. 2001. Speciation of selenoamino acids, selenonium ions and inorganic selenium by ion exchange HPLC with mass spectrometric detection and its application to yeast and algae. *J Anal At Spectrom* 16:1403–1408.

LeDuc DL, Terry N. 2005. Phytoremediation of toxic trace elements in soil and water. *J Ind Microbiol Biotechnol* 32:514–520.

Leighton FAT. 1989. Pollution and wild birds: North America in the 1980's. *Can Vet J* 30:783–785.

Lemly AD. 1985. Ecological basis for regulating aquatic emissions from the power industry: the case with selenium. *Regul Toxicol Pharm* 5:465–486.

Lemly AD. 2002. Selenium assessment in aquatic ecosystems: a guide for hazard evaluation and water quality criteria. Springer (NY, USA). p 169.

Lenz M, Hullebusch EDV, Hommes G, Corvini PFX, Lens PNL. 2008a. Selenate removal in methanogenic and sulfate-reducing upflow anaerobic sludge bed reactors. *Wat Res* 42:2184–2194.

Lenz M, Smit M, Binder P, Van Aelst AC, Lens PNL. 2008b. Biological alkylation and colloid formation of selenium in methanogenic UASB reactors. *J Environ Qual* 37:1691–1700.

Liang L, Mo S, Zhang P, Cai Y, Mou S, Jiang G, Wen M. 2006. Selenium speciation by high-performance anion-exchange chromatography–post-column UV irradiation coupled with atomic fluorescence spectrometry. *J Chromatogr A* 1118:139–143.

Lin ZQ, Terry N. 2003. Selenium removal by constructed wetlands: quantitative importance of biological volatilization in the treatment of selenium-laden agricultural drainage water. *Environ Sci Technol* 37:606–615.

Lindemann T, Prange A, Dannecker W, Neidhart B. 2000. Stability studies of arsenic, selenium, antimony and tellurium species in water, urine, fish and soil extracts using HPLC/ICP-MS. *Fresenius J Anal Chem* 368:214–220.

Lobinski R, Edmonds JS, Suzuki KT, Uden PC. 2000. Species-selective determination of selenium compounds in biological materials. *Pure Appl Chem* 72:447–462.

Long RHB, Benson SM, Tokunaga TK, Yee A. 1990. Selenium immobilization in a pond sediment at Kesterson Reservoir. *J Environ Qual* 19:302–311.

Lortie L, Gould WD, Rajan S, McCready RGL, Cheng KJ. 1992. Reduction of selenate and selenite to elemental selenium by a *Pseudomonas stutzeri* isolate. *Appl Environ Microbiol* 58:4042–4044.

Losi ME, Frankenberger WT Jr. 1997. Reduction of selenium oxyanions by *Enterobacter cloacae* SLD1a-1: isolation and growth of the bacterium and its expulsion of selenium particles. *Appl Environ Microbiol* 63:3079–3084.

Lovley DR. 1991. Dissimilatory Fe (III) and Mn (IV) reduction. *Microbiol Mol Biol Rev* 55:259–287.

Lucas FS, Hollibaugh JT. 2001. Response of sediment bacterial assemblages to selenate and acetate amendments. *Environ Sci Technol* 35:528–534.

Manceau A, Marcus MA, Tamura N. 2002. Quantitative speciation of heavy metals in soils and sediments by synchrotron X-ray techniques. *Rev Mineral Geochem* 49:341–428.

Manning BA, Burau RG. 1995. Selenium immobilization in evaporation pond sediments by in situ precipitation of ferric oxyhydroxide. *Environ Sci Technol* 29:2639–2646.

Martens DA, Suarez DL. 1997. Selenium speciation of marine shales, alluvial soils, and evaporation basin soils of California. *J Environ Qual* 26:424–432.

Martínez M, Giménez J, de Pablo J, Rovira M, Duro L. 2006. Sorption of selenium (IV) and selenium (VI) onto magnetite. *Appl Surf Sci* 252:3767–3773.

Mascheleyn PH, Patrick WH Jr. 1993. Biogeochemical processes affecting selenium cycling in wetlands. *Environ Toxicol Chem* 12:2235–2243.

Mascheleyn PH, Delaune RD, Patrick WH Jr. 1990. Transformations of selenium as affected by sediment oxidation-reduction potential and pH. *Environ Sci Technol* 24:91–96.

Mascheleyn PH, Delaune RD, Patrick WH Jr. 1991. Arsenic and selenium chemistry as affected by sediment redox potential and pH. *J Environ Qual* 20:522–527.

Mattigod SV, Quinn TR. 2003. Selenium content and oxidation states in fly ash from Western US coals. In: Sajwan KS, Alva AK, Keeffe RF, editors. Chemistry of trace elements in fly ash. New York (NY, USA): Kluwer Academic/Plenum. p 143–153.

Maunsell and Partners. 1974. Lake Macquarie foreshore study. Sydney (NSW, Australia): New South Wales Department of Public Works. 39 p.

Mayland HF. 1994. Selenium in plant and animal nutrition. In: Frankenberger WT Jr, Benson S, editors. Selenium in the environment. New York (NY, USA): Marcel Dekker. p 29–46.

Mees DR, Pysto W, Tarcha PJ. 1995. Formation of selenium colloids using sodium ascorbate as the reducing agent. *J. Colloid Interface Sci* 170:254–260.

Meng X, Bang S, Korfiatis GP. 2002. Removal of selenocyanate from water using elemental iron. *Wat Res* 36:3867–3873.

Merrill DT, Manzione MA, Peterson JJ, Parker DS, Chow W, Hobbs AO. 1986. Field evaluation of arsenic and selenium removal by iron coprecipitation. *J Water Poll Contr Fed* 58:18–26.

Meseck SL. 2002. Modeling the biogeochemical cycle of selenium in the San Francisco Bay. PhD Dissertation. Norfolk (VA, USA): Old Dominion Univ. 249 p.

Milne JB. 1998. The uptake and metabolism of inorganic selenium species. In: Frankenberger WT Jr, Engberg RA, editors. Environmental chemistry of selenium. Boca Raton (FL, USA): CRC Pr. p 459–476.

Monahan-Pendergast MT, Przybylek M, Lindblad M, Wilcox J. 2007. Theoretical predictions of arsenic and selenium species under atmospheric conditions. *Atmos Environ* 42:2349–2357.

Moscoso-Pérez C, Moreda-Piñeiro J, López-Mahía P, Muniategui-Lorenzo S, Fernández-Fernández E, Prada-Rodríguez D. 2008. Pressurized liquid extraction followed by high performance liquid chromatography coupled to hydride generation atomic fluorescence spectrometry for arsenic and selenium speciation in atmospheric particulate matter. *J Chromatogr A* 1215:15–20.

Mosher B, Duce R. 1987. A global atmospheric selenium budget. *J Geophys Res* 92:13289–13298.

Murphy AP. 1989. A water treatment process for selenium removal. *J Water Poll Contr Fed* 61:361–362.

Musaev DG, Geletii YV, Hill CL. 2003. Theoretical studies of the reaction mechanisms of dimethylsulfide and dimethylselenide with peroxynitrite. *J Phys Chem A* 107:5862–5873.

Muscatello JR, Janz DM. 2009. Assessment of larval deformities and selenium accumulation in Northern Pike (*Esox lucius*) and White Sucker (*Catostomus commersoni*) exposed to metal mining effluent. *Environ Toxicol Chem* 28:609–618.

Myneni SC, Tokunaga TK, Brown GE Jr. 1997. Abiotic selenium redox transformations in the presence of Fe (II, III) oxides. *Science* 278:1106–1109.

Nakamuro K, Okuno T, Hasegawa T. 2000. Metabolism of selenoamino acids and contribution of selenium methylation to their toxicity. *J Health Sci* 46:418–421.

Naveau A, Monteil-Rivera F, Guillon E, Dumonceau J. 2007. Interactions of aqueous selenium (- II) and (IV) with metallic sulfide surfaces. *Environ Sci Technol* 41:5376–5382.

Nguyen VNH, Beydoun D, Amal R. 2005. Photocatalytic reduction of selenite and selenate using TiO_2 photocatalyst. *J Photochem Photobiol A* 171:113–120.

Niss ND, Schabron JF, Brown TH. 1993. Determination of selenium species in coal fly ash extracts. *Environ Sci Technol* 27:827–829.

Nriagu JO. 1989. Global cycling of selenium. In: Ihnat M, editor. Occurrence and distribution of selenium. Boca Raton (FL, USA): CRC Pr. p 327–240.

Nriagu JO. 1991. Heavy metals in the environment, Vol. 1. Edinburgh (UK): CEP Consultants.

Ohlendorf HM. 2003. Ecotoxicology of selenium. In: Hoffman DJ, Rattner BA, Burton GA Jr, editors. Handbook of ecotoxicology. Boca Raton (FL, USA): CRC Pr. p 465–500.

Oram LL, Strawn DG, Marcus MA, Fakra SC, Möller G. 2008. Macro- and microscale investigation of selenium speciation in Blackfoot River, Idaho sediments. *Environ Sci Technol* 42:6830–6836.

Oremland RS, Hollibaugh JT, Maest AS, Presser TS, Miller LG, Culbertson CW. 1989. Selenate reduction to elemental selenium by anaerobic bacteria in sediments and culture: biogeochemical significance of a novel, sulfate-independent respiration. *Appl Environ Microbiol* 55:2333–2343.

Palace VP, Spallholz JE, Holm J, Wautier K, Evans RE, Baron CL. 2004. Metabolism of selenomethionine by rainbow trout (*Oncorhynchus mykiss*) embryos can generate oxidative stress. *Ecotoxicol Environ Saf* 58:17–21.

Parida KM, Gorai B, Das NN, Rao SB. 1997. Studies on ferric oxide hydroxides III. Adsorption of selenite (SeO_2^{-3}) on different forms of iron oxyhydroxides. *J Colloid Interface Sci* 185:355–362.

Pedrero Z, Madrid Y. 2008. Novel approaches for selenium speciation in foodstuffs and biological specimens: a review. *Anal Chim Acta* 634:135–152.

Peters GM, Maher WA, Krikowa F, Roach AC, Jeswani HK, Barford JP, Gomes VG, Reible DD. 1999a. Selenium in sediments, pore waters and benthic infauna of Lake Macquarie, New South Wales, Australia. *Mar Environ Res* 47:491–508.

Peters GM, Maher WA, Jolley D, Carroll BI, Gomes VG, Jenkinson AV, McOrist GD. 1999b. Selenium contamination, redistribution and remobilisation in sediments of Lake Macquarie, NSW. *Org Geochem* 30:1287–1300.

Petrovic M, Kastelan-Macan M. 1996. Uptake of inorganic phosphorus by insoluble metal-humic complexes. *Water Sci Technol* 34:253–258.

Pickering IJ, Brown GE Jr, Tokunaga TK. 1995. Quantitative speciation of selenium in soils using X-ray absorption spectroscopy. *Environ Sci Technol* 29:2456–2459.

Pickering IJ, Prince RC, Salt DE, George GN. 2000. Quantitative, chemically specific imaging of selenium transformation in plants. *Proc Natl Acad Sci USA* 97:10717–10722.

Pickering IJ, Wright C, Bubner B, Ellis D, Persans MW, Yu EY, George GN, Prince RC, Salt DE. 2003. Chemical form and distribution of selenium and sulfur in the selenium hyperaccumulator *Astragalus bisulcatus*. *Plant Physiol* 131:1460–1467.

Pickett T, Sonstegard J, Bonkoski B. 2006. Using biology to treat selenium: Biologically treating scrubber wastewater can be an attractive alternative to physical-chemical treatment. *Power Engineer* 110:140–141.

Pilon-Smits EAH, De Souza MP, Hong G, Amini A, Bravo RC, Payabyab ST, Terry N. 1999. Selenium volatilization and accumulation by twenty aquatic plant species. *J Environ Qual* 28:1011–1018.

Polatajko A, Jakubowski N, Szpunar J. 2006. State of the art report of selenium speciation in biological samples. *J Anal At Spectrom* 21:639–654.

Post DM. 2002. Using stable isotopes to estimate trophic position: models, methods, and assumptions. *Ecology* 83:703–718.

Presser TS, Swain WC, Tidwell RR, Severson RC. 1990. Geologic sources, mobilization, and transport from California Coastal ranges to the western San Joaquin Valley: a reconnaissance study. Menlo Park (CA, USA): US Geological Survey. Water-Resources Investigations Report 90-4070.

Pyrzynska K. 1998. Speciation of selenium compounds. *Analyt Sci* 14:479–483.

Pyrzynska K. 2002. Determination of selenium species in environmental samples. *Mikrochim Acta* 140:55–62.

Ralston NVC, Unrine J, Wallschläger D. 2008. Biogeochemistry and analysis of selenium and its species. Washington (DC, USA): North American Metals Council, Selenium Working Group. 58 p; www.namec.org/Selenium.

Raptis SE, Kaiser G Tölg G. 1983. A survey of selenium in the environment and a critical review of its determination at trace levels. *Anal Bioanal Chem* 316:105–123.

Reinfelder JR, Fisher NS, Luoma SN, Nichols JW, Wang WX. 1998. Trace element trophic transfer in aquatic organisms: a critique of the kinetic model approach. *Sci Tot Environ* 219:117–135.

Rickard D, Morse JW. 2005. Acid volatile sulfide (AVS). *Mar Chem* 97:141–197.

Roach AC, Maher W, Krikowa F. 2008. Assessment of metals in fish from Lake Macquarie, New South Wales, Australia. *Arch Environ Contam Toxicol* 54:292–308.

Roberts B. 1994. The accumulation and distribution of selenium in fish from Lake Macquarie, NSW [BSc (Hons) thesis]. Canberra (Australia): Univ of Canberra. 91 p.

Ross HB. 1984. Atmospheric selenium. Springfield (VA, USA): National Technical Report Library (NTIS). Report No. CM-66.

Rovira M, Giménez J, Martinez M, Martínez-Lladó X, de Pablo J, Marti V, Duro L. 2008. Sorption of selenium (IV) and selenium (VI) onto natural iron oxides: goethite and hematite. *J Hazard Mater* 150:279–284.

Ryser AL, Strawn DG, Marcus MA, Fakra S, Johnson-Maynard JL, Möeller G. 2006. Microscopically focused synchrotron X-ray investigation of selenium speciation in soils developing on reclaimed mine lands. *Environ Sci Technol* 40:462–467.

Saeki K, Matsumoto S, Tatsukawa R. 1995. Selenite adsorption by manganese oxides. *Soil Sci* 160:265–272.

Salazar M, Salazar SM. 2005. Field experiments with caged bivalves to assess chronic exposure and toxicity. In: Ostrander GK, editor. Techniques in aquatic toxicology. Boca Raton (FL, USA): CRC Pr. p 117–135.

Sayre WG. 1980. Selenium: a water pollutant from flue gas desulfurization. *J Air Pollut Control Assoc* 30:1134–1143.

Schink B. 1997. Energetics of syntrophic cooperation in methanogenic degradation. *Microbiol Mol Biol Rev* 61:262–280.

Schwartz SE. 2008. Uncertainty in climate sensitivity: causes, consequences, challenges. *Energy Environ Sci* 1:430–453.

Scott MJ, Morgan JJ. 1996. Reactions at oxide surfaces. 2. Oxidation of Se (IV) by synthetic birnessite. *Environ Sci Technol* 30:1990–1996.

Séby F, Potin-Gautier M, Giffaut E, Borge G, Donard OFX. 2001. A critical review of thermodynamic data for selenium species at 25°C. *Chem Geol* 171:173–194.

Seiler RL, Skorupa JP, Naftz DL, Nolan BT. 2003. Irrigation-induced contamination of water, sediment, and biota in the western United States-synthesis of data from the National

Irrigation Water Quality Program. Menlo Park (CA, USA): US Geological Survey. Report No. 1655. USGS Professional Paper. 123 p.

Selivanova NM, Kapustinskii AF, Zubova GA. 1959. Thermochemical properties of sparingly soluble selenates and entropy of aqueous selenate ion. *Russ Chem Bull* 8:174–180.

Sharmasarkar S, Reddy KJ, Vance GF. 1996. Preliminary quantification of metal selenite solubility in aqueous solutions. *Chem Geol* 132:165–170.

Sharmasarkar S, Vance GF, Cassel-Sharmasarkar F. 1998. Analysis and speciation of selenium ions in mine environments. *Environ Geo* 34:31–38.

Shaw PJ, Jones RI, De Haan H. 2000. The influence of humic substances on the molecular weight distributions of phosphate and iron in epilimnetic lake waters. *Freshwater Biol* 45:383–393.

Siddique T, Zhang Y, Okeke BC, Frankenberger WT Jr. 2006. Characterization of sediment bacteria involved in selenium reduction. *Bioresour Technol* 97:1041–1049.

Simpson SL, Angel BM, Jolley DF. 2004. Metal equilibration in laboratory-contaminated (spiked) sediments used for the development of whole-sediment toxicity tests. *Chemosphere* 54:597–609.

Skorupa JP. 1998. Selenium poisoning of fish and wildlife in nature: lessons from twelve real-world examples. In: Frankenberger WT Jr, Engberg RA, editors. Environmental chemistry of selenium. Boca Raton (FL, USA): CRC Pr. p 315–354.

Sonstegard J, Pickett T, Harwood J, Johnson D. 2008. Full scale operation of biological technology for the removal of selenium from FGD wastewaters. Presented at the 69th International Water Conference; San Antonio (TX, USA).

Sors TG, Ellis DR, Salt DE. 2005. Selenium uptake, translocation, assimilation and metabolic fate in plants. *Photosynth Res* 86:373–389.

Spooner DR, Maher W, Otway N. 2003. Trace metal concentrations in sediments and oysters of Botany Bay, NSW, Australia. *Arch Environ Contam Toxicol* 45:92–101.

Stewart R, Luoma S, Schlekat C, Doblin M, Hieb K. 2004. Food web pathway determines how selenium affects aquatic ecosystems: a San Francisco Bay case study. *Environ Sci Technol* 38:4519–4526.

Stolz J, Basu P, Oremland R. 2002. Microbial transformation of elements: the case of arsenic and selenium. *Int Microbiol* 5:201–207.

Stolz JF, Basu P, Santini JM, Oremland RS. 2006. Arsenic and selenium in microbial metabolism. *Annu Rev Microbiol* 60:107–30.

Stone DM. 1964. Beach stability in Lake Macquarie entrance. Report submitted to Gutteridge, Haskins and Davey, Consulting Engineers. Sydney (NSW, Australia): Univ of New South Wales, Water Research Laboratory. 33 p.

Su Y, Chen H, Gao Y, Li X, Hou X, Lv Y. 2008. A novel HPLC-UV/nano-TiO$_2$-chemiluminescence system for the determination of selenocystine and selenomethionine. *J Chromatogr B* 870:216–221.

Sutton SR, Bajt S, Delaney J, Schulze D, Tokunaga T. 1995. Synchrotron X-ray fluorescence microprobe: quantification and mapping of mixed valence state samples using micro-XANES. *Rev Sci Instrum* 66:1464–1467.

Suzuki KT, Kurasaki K, Okazaki N, Ogra Y. 2005. Selenosugar and trimethylselenonium among urinary Se metabolites: dose- and age-related changes. *Toxicol Appl Pharmacol* 206:1–8.

Terry N, Zayed AM. 1994. Selenium volatilization by plants. In: Frankenberger WT Jr, Benson S, editors. Selenium in the environment. New York (NY, USA): Marcel Dekker. p 343–367.

Terry N, Zayed AM, De Souza MP, Tarun AS. 2000. Selenium in higher plants. *Ann Rev Plant Physiol Plant Mol Biol* 51:401–432.

Thompson-Eagle ET, Frankenberger WT Jr, Karlson U. 1989. Volatilization of selenium by *Alternaria alternata*. *Appl Environ Microbiol* 55:1406–1413.

Tokunaga TK, Brown GE Jr, Pickering IJ, Sutton SR, Bajt S. 1997. Selenium redox reactions and transport between ponded waters and sediments. *Environ Sci Technol* 31:1419–1425.

Tokunaga TK, Sutton SR, Bajt S, Nuessle P, Shea-McCarthy G. 1998. Selenium diffusion and reduction at the water-sediment boundary: micro-XANES spectroscopy of reactive transport. *Environ Sci Technol* 32:1092–1098.

Toy TJ, Foster GR, Renard KG. 2002. Soil erosion: processes, prediction, measurement and control. New York (NY, USA): J Wiley. 64 p.

Twidwell LG, McCloskey J, Miranda P, Gale M. 1999. Technologies and potential technologies for removing selenium from process and mine wastewater. Proceedings of the REWAS '99: Global Symposium on Recycling, Waste Treatment and Clean Technology; 1999 Sep 5–9; San Sebastian, Spain. p 1645–1656.

[USEPA] United States Environmental Protection Agency. 1999. Document compilation of air pollutant emission factors, 5th Edition, Volume 1: Stationary point and area sources. Supplement E (AP-42). Washington, DC, USA.

Vandermeulen JH, Foda A. 1988. Cycling of selenite and selenate in marine phytoplankton. *Mar Biol* 98:115–123.

Wallschläger D, Bloom NS. 2001. Determination of selenite, selenate and selenocyanate in waters by ion chromatography-hydride generation-atomic fluorescence spectrometry (IC-HG-AFS). *J Anal At Spectrom* 16:1322–1328.

Wang T, Wang J, Burken JG, Ban H, Ladwig K. 2007. The leaching characteristics of selenium from coal fly ashes. *J Environ Qual* 36:1784–1792.

Wells JM, Richardson DHS. 1985. Anion accumulation by the moss *Hylocomium splendens*: uptake and competition studies involving arsenate, selenate, selenite, phosphate, sulphate and sulphite. *New Phytol* 101:571–583.

Wen HJ, Carignan J. 2007. Reviews on atmospheric selenium: emissions, speciation and fate. *Atmos Environ* 41:7151–7165.

West JM, Williams GD, Madon SP, Zedler JB. 2003. Integrating spatial and temporal variability into the analysis of fish food web linkages in Tijuana Estuary. *Environ Biol Fishes* 67:297–309.

White PJ, Bowen HC, Marshall B, Broadley MR. 2007a. Extraordinarily high leaf selenium to sulfur ratios define 'Se-accumulator' plants. *Ann Bot (Lond)* 100:111–118.

White PJ, Broadley MR, Bowen HC, Johnson SE. 2007b. Selenium and its relationship with sulfur. In: Hawkesford MJ, Kok LJ, De Kok LJ, editors. Sulfur in plants: an ecological perspective. New York (NY, USA): Springer-Verlag. p 225–25.

Windisch W. 2002. Interaction of chemical species with biological regulation of the metabolism of essential trace elements. *Anal Bioanal Chem* 372:421–425.

Woock SE, Garrett WR, Partin WE, Bryson WT. 1987. Decreased survival and teratogenesis during laboratory selenium exposures to bluegill, *Lepomis macrochirus*. *Bull Environ Contam Toxicol* 39:998–1005.

Wrench JJ. 1978. Selenium metabolism in the marine phytoplankters *Tetraselmis tetrathele* and *Dunaliella minuta*. *Mar Biol* 49:231–236.

Wright MT, Parker DR, Amrhein C. 2003. Critical evaluation of the ability of sequential extraction procedures to quantify discrete forms of selenium in sediments and soils. *Environ Sci Technol* 37:4709–4716.

Xie R, Seip HM, Wibetoe G, Nori S, McLeod CW. 2006. Heavy coal combustion as the dominant source of particulate pollution in Taiyuan, China, corroborated by high concentrations of arsenic and selenium in PM10. *Sci Total Environ* 370:409–415.

Yan R, Gauthier D, Flamant G, Wang Y. 2004. Behavior of selenium in the combustion of coal or coke spiked with Se. *Combust Flame* 138:20–29.

Zhang YQ, Frankenberger WT Jr. 2003. Determination of selenium fractionation and speciation in wetland sediments by parallel extraction. *Int J Environ Anal Chem* 83:315–326.

Zhang YQ, Frankenberger WT Jr. 2006. Removal of selenate in river and drainage waters by *Citrobacter braakii* enhanced with zero-valent iron. *J Agric Food Chem* 54:152–156.

Zhang YQ, Moore JN. 1997. Environmental conditions controlling selenium volatilization from a wetland system. *Environ Sci Technol* 31:511–517.

Zhang YQ, Moore JN, Frankenberger WT Jr. 1999. Speciation of soluble selenium in agricultural drainage waters and aqueous soil-sediment extracts using hydride generation atomic absorption spectrometry. *Environ Sci Technol* 33:1652–1656.

Zhang YQ, Zahir ZA, Frankenberger WT Jr. 2004. Fate of colloidal-particulate elemental selenium in aquatic systems. *J Environ Qual* 33:559–564.

Zheng J, Shibata Y, Furuta N. 2003. Determination of selenoamino acids using two-dimensional ion-pair reversed phase chromatography with on-line detection by inductively coupled plasma mass spectrometry. *Talanta* 59:27–36.

4.7 ATTACHMENT 1: THE CHEMICAL STRUCTURES OF THE MAJOR Se SPECIES IN THE CATEGORIES SHOWN IN FIGURE 4.2.

Se-methylselenomethionine
MeSeMet

dimethyl selenoniopropionate
DMSeP

adenosine 5′-phosphoselenate
APSe

3′-phosphoadenosine 5′-phosphoselenate
PAPS

Se-methylselenocysteine selenoxide
SeCysSeO

selenocystamine
SeCya

selenocystathionine
SeCyst

selenobetaine
SeB

selenocystine
SeCystine

selenohomocysteine
SehoCys

glutamyl-methylselnocysteine
GMSeC

5 Bioaccumulation and Trophic Transfer of Selenium

Robin Stewart, Martin Grosell, David Buchwalter,
Nicholas Fisher, Samuel Luoma, Teresa Mathews,
Patricia Orr, Wen-Xiong Wang

CONTENTS

5.1   INTRODUCTION

In this chapter we identify the state-of-the-science regarding selenium (Se) bioaccumulation and trophic transfer. We discuss Se bioaccumulation and how its unique attributes tie bioaccumulation to toxicity. We identify biodynamic modeling as a promising approach that can provide a unified view of the processes contributing to bioaccumulation and illustrate how this kinetic modeling can be used and improved. We also discuss the most important uncertainties that need to be addressed if we are to better understand and model Se bioaccumulation.

5.1.1   AREAS OF GENERAL SCIENTIFIC AGREEMENT

Concerns about environmental contaminants stem from their potential negative impacts on populations or even whole ecosystems. Potential impacts begin when individual organisms accumulate the contaminant, because organisms respond only to chemicals that are somehow associated with them (i.e., bound to membrane components or transported into cells). Contaminants that remain in abiotic environmental compartments (e.g., sediments, water) will not have direct effects. This fact is as true for biologically essential elements as for nonessential elements: Only the accumulated element can provide nutritional support for the organism. Thus, it is critical to understand the accumulation of these substances in order to evaluate their biological effects (nutritive or toxic). Selenium, which is both essential at low concentrations and toxic at elevated concentrations, is not unlike other essential but potentially toxic elements (e.g., Cu, Zn) where the organism must accumulate it from the environment in order to perform normal physiological functions but must regulate or otherwise refrain from accumulating too much to avoid toxicity.

Understanding bioaccumulation and trophic transfer is central to managing ecological risks from Se. Selenium bioaccumulation is relatively well studied, but there remain many research areas where advances in understanding could aid better management of ecological risks. Managing ecological risks from this element has been controversial in part due to the fact that at least some of the relatively well-established principles contradict the conventional preconceptions

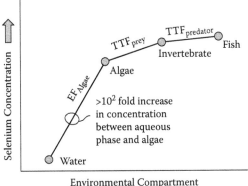

FIGURE 5.1 Selenium enrichment and trophic transfer in aquatic food webs. Enrichment function represents the increase in Se concentration between water and the base of the aquatic food web, which often is algae (EF_{algae}; range 10^2 to 10^6). Trophic transfer function represents the increase in Se concentration between algae and invertebrates (TTF_{prey}; range 0.6 to 23) and invertebrates and fish ($TTF_{predator}$; range 1 to 3).

and traditional approaches that are often used to manage metal contaminants. But the greatest challenge with managing risks is that neither Se bioaccumulation nor toxicity can be predicted from environment to environment based solely upon Se concentrations in water. This inability to predict is likely the basic reason for international incoherence and differences between freshwater and marine waters in water quality guidelines.

Ecological risks from Se are affected by uptake at the base of the food web, dietary exposure, dietary toxicity, and transfer through the food web. The direct toxicity of waterborne Se alone (the basis of most traditional risk assessment and risk management activities) tells us little if anything about ecological risks of exposure to Se. There is strong evidentiary support for the fundamental concept that the concentration of Se taken up into primary producers and microbes at the base of the food web is preserved and/or further concentrated as it is passed on to consumer organisms and their predators (Presser and Luoma in press; Luoma and Presser 2009) (Figure 5.1).

Ultimately, the poor linkage between dissolved Se and either bioaccumulated Se in the food web or Se toxicity is the reason that new risk assessment and risk management strategies are necessary for this element. That poor linkage at least partly reflects variability at each of the intervening steps in the conceptual model linking dissolved Se to its effects (Chapters 3 and 4). Uptake at the base of the food web is species- and environment-specific. Differences among species and environments also occur in the efficiency with which Se is assimilated and retained by consumer organisms at the second trophic level. Differences occur in the types of prey eaten by predators. Finally, risk assessments need to consider the toxic effects of the dose of Se achieved by each species (cf. Chapters 6 and 7).

5.1.2   Ecosystem-Scale and Biodynamic Models

Ecological risks from Se are distinguished by a complex interplay between bio-geochemical, biological, and ecological considerations. Biodynamic models can be useful in explaining the interactions among the biology- and environment-specific functions that ultimately define bioaccumulation and therefore ecological risks from Se, especially if the model is used as part of the process of defining linkages in the ecosystem-scale conceptual model for Se risks (Chapter 3). Biodynamic modeling can help risk managers understand their ecosystem and forecast the outcomes of risk management decisions. But use of those models requires application of biological and ecological principles heretofore underutilized in ecotoxicology, risk assessment, and risk management.

Selenium partitioning to biological material in the first step of the food web is difficult to predict and sometimes to measure. Selenium partitioning from water to tissues is not adequately described by useful constructs (e.g., fugacity) for understanding organic contaminant behavior, nor are geochemical or thermodynamic equilibrium approaches even remotely predictive. Rather, Se partitioning is primarily a biologically mediated process. For this reason, water–organism bio-accumulation factors (BAFs) have limited predictive value because critical intervening steps in the food web between water and higher organisms that vary from environment to environment cannot be considered in this simple ratio. However, a slightly more complex model can consider those factors. For example, the concentration of Se in plants or microbes at the base of the food web is a crucial input to such a model, although isolation of algae and microorganisms from whole seston or sediment samples from the field can be difficult. If it is not possible or practical to isolate biotic and abiotic components of particulate phases, an environment-specific EF may be operationally based on the relationship between Se concentration in water and whole seston or sediment. But better strategies for quantifying uptake at this first step of the food web are an important research need.

The combined influence of environmental Se concentrations and physiological processes on bioaccumulation can be integrated in a biodynamic or biokinetic model (Luoma et al. 1992; Wang et al. 1996a; Luoma and Fisher 1997; Wang and Fisher 1999; Luoma and Rainbow 2005; Wang and Rainbow 2008). These models provide a broad framework for addressing controls of contaminant bioaccumulation and can be used for evaluating contaminant bioavailability and determining the relative importance of different routes of contaminant accumulation (Wang et al. 1996a; Wang and Fisher 1999). They are flexible enough to incorporate environmental variability in contaminant sources, contaminant concentrations, food availability, and organism growth rates in their predictions of the concentrations of Se accumulated by an organism. One widely used version of these models treats contaminant accumulation as a first-order function of contaminant concentrations in particles and water and is expressed as

$$\frac{dC}{dt} = (k_u \times C_w) + (\text{AE} \times \text{IR} \times C_f) - (k_e + k_g) \times C \tag{1}$$

where C is the contaminant concentration in the animals (mg/kg), t is the time of exposure (d), k_u is the uptake rate constant from the dissolved phase (L/g/d), C_w is the contaminant concentration in the dissolved phase (µg/L), AE is the assimilation efficiency from ingested particles (%), IR is the ingestion rate of particles (mg/g/d), C_f is the contaminant concentration in ingested particles (µg/mg), k_e is the efflux rate constant (/d), and g is the growth rate constant (/d). At steady state, this equation simplifies to

$$C_{ss} = \frac{(k_u \times C_w) + (AE \times IR \times C_f)}{(k_e + k_g)} \qquad (2)$$

where C_{ss} is the steady-state concentration of contaminant in the organism (mg/kg). The efflux parameter, k_e, can be further split into solute (k_{ew}) and food (k_{ef}) components if the loss rates from these exposure regimes differ, where

$$C_{ss} = \frac{k_u \times C_w}{k_{ew} + k_g} + \frac{(AE \times IR \times C_f)}{k_{ef} + k_g} \qquad (3)$$

If it is assumed that food is the dominant source of uptake of Se, a particularly useful format from the model is derivation of a trophic transfer function (TTF), where

$$TTF = \frac{(AE \times IR)}{(k_{ef} + k_g)} \qquad (4)$$

The TTF is species-specific, may vary with dietary Se concentration, and is affected by factors that affect AE and IR. It will be affected by factors that affect AE and IR. Indeed, a number of environmental and biological factors can influence each of the parameters used in the equations (AE, k_e, k_u, etc.) and that caution should be used in applying just a single value. In Section 5.3 we consider the elements of the biodynamic model in detail and discuss ways to improve application of this concept (see also Text Box 1). We also evaluate sensitivities to uncertainties in model parameters and conduct some simple forecasts to demonstrate uses of the model.

Lastly, in some situations it may not be practical or even possible to derive a food web model based on kinetic data. For example, kinetic data for Se bioaccumulation in freshwater invertebrates are largely lacking (Section 5.2.6.2). Also, some biota consume such a wide variety of prey species that generating kinetic data for each food web linkage could be time consuming and costly. In such cases, it may be more appropriate to generate a TTF based on a ratio of the field-measured Se concentration in a consumer to that found in known or inferred dietary organisms. Such values could be used in addition to, or as an alternative to, laboratory-derived kinetic data for construction of a site-specific food web model.

TEXT BOX 5.1: NEW TERMINOLOGY

Traditionally the term "K_d" has been used to describe the relationship between contaminant concentration in water and accumulation by particles, including living cells. In addition, the term "uptake rate constant" (K_u) has been used to describe the relationship between contaminant concentration in water and uptake rate by single-celled organisms, invertebrates, and fish. The relationship between prey and predatory contaminant concentration is traditionally referred to as the "trophic transfer factor." In this book we refer to enrichment of Se by particles, including single-celled organisms, such as algae, as "enrichment function" (EF); uptake from the dissolved phase is referred to as "uptake rate function"; and trophic transfer is referred to as "trophic transfer function" (TTF). These new terms recognize that entry into the food web from the dissolved phase, and likely also transfer from prey to predator, are dependent on concentration in a non-linear manner.

For Se and many other elements, uptake by microorganisms and multicellular organisms is governed by specific transport pathways that facilitate the movement of the element in question from the environment and across cell membranes or epithelia into the organism. These transport pathways, which consist of trans-membrane proteins, may differ with respect to specificity but typically display high affinity for the element and a limited maximal capacity for uptake (Section 5.2.2). An important consequence of such high-affinity, limited-capacity uptake pathways is that elemental uptake from low ambient concentrations, is highly efficient but becomes less efficient at higher concentrations due to saturation of the uptake pathway. This non-linear relationship between ambient concentration and uptake rates is better described by a Michaelis-Menten relationship (Section 5.3.1.1) than by a single constant or factor. Recognizing that non-linear relationships exist between uptake and ambient concentrations, we recommend the use of "enrichment function" and "uptake rate function" as terms to describe uptake from water by microorganisms and multicellular organisms, respectively.

Elemental uptake from dietary sources is also conducted via more or less specific uptake pathways (Section 5.2.2), resulting in relatively high uptake efficiency (or assimilation efficiency [AE]) from low dietary elemental concentrations and less efficient uptake at higher concentrations. This non-linear relationship between dietary exposure concentration and AE (Section 5.3.1.2) has implications for trophic transfer. These implications are particularly relevant for Se for which dietary sources in general dominate its accumulation by animals; the concentration dependence of AE leads to predictions of a non-linear relationship between dietary exposure and trophic transfer. For this reason, we suggest "trophic transfer function" rather than "trophic transfer factor" to describe how biomagnification may depend on elemental (in this case, Se) concentrations in prey organisms.

5.2 PROCESSES THAT CONTROL Se CONCENTRATIONS IN FOOD WEBS

5.2.1 PHYSIOLOGICAL REQUIREMENTS FOR SE

A total of 30 seleno-protein families are known, and seleno-proteins are found in all lineages of life illustrating the essentiality of Se (Kryukov et al. 2003; Kryukov and Gladyshev 2004; Castellano et al. 2005; Zhang et al. 2005). Specifically, proteins containing the 21st natural amino acid selenocysteine (Sec) are found in all 3 major forms of life (bacteria, Archaea, and eukaryotes) (Hatfield et al. 1999). Fish possess the most prolific selenoproteomes, with as many as 30 individual selenoproteins, but in general selenoproteomes are small (Vanda Papp et al. 2007). Although the function of many selenoproteins remains to be described, some, such as glutathione peroxidases, thioredoxin reductases, iodothyronine deiodinases, and selenophosphate synthetases, have been ascribed physiological functions. Studies with mice have illustrated the essentiality of at least some thioredoxin reductases and glutathione peroxidases (Vanda Papp et al. 2007). All the above proteins have oxidoreductase functions; a process as fundamental as DNA synthesis, for example, depends on Se in the catalytic site of thioredoxin reductases (Vanda Papp et al. 2007).

A minimum Se intake (or uptake) is required for normal physiological function. The recommended daily Se intake in humans is ~0.6 μg Se/kg/d as organic Se, a dose that is hypothesized to have primarily antioxidant and immune-strengthening effects, while doses as much as 10-fold higher have been reported to have specific cancer-preventive properties (Rayman 2002; Bügel et al. 2008). In comparison, channel catfish have been reported to require 0.1 to 0.5 mg Se/kg diet (ww) as inorganic Se, which, with an assumed feeding ration of 5% body weight per day, translates to 5 to 25 μg/kg/d (Gatlin and Wilson 1984), and rainbow trout requires a minimum of 3.5 μg/kg/d as inorganic Se (Hilton et al. 1980). In agreement with these studies are more recent studies of juvenile grouper fed a diet containing selenomethionine indicating a requirement of 0.7 mg Se/kg or, at 5% body weight daily ration, 35 μg/kg/d for optimal growth (Lin and Shiau 2005). Aquatic birds show deficiencies below dietary concentrations of 0.3 to 1.1 mg Se/kg (Puls 1988; Ohlendorf 2002), suggesting similar Se demands in most vertebrates examined to date. Limited information is available about Se requirements for aquatic invertebrates, which clearly marks a subject in need of study. However, a single study demonstrates that 20 mg/kg Se as inorganic Se appears optimal for shrimp (Tian and Lui 1993), a value that falls well above the range reported for optimal vertebrate physiology. Selenium requirements have been documented for the unicellular freshwater green alga *Chlamydomonas reinhardtii* in which 3.9 μg/L enhanced growth compared to "Se" conditions (Novoselov et al. 2002). *Dunaliella viridis*, a green alga typically found in saline systems, showed increased growth with increasing Se concentrations up to 18 μg/L (Martin Grosell, University of Miami, personal communication). In comparable experiments, diatoms in general appear to accumulate substantially more Se than green algae (Table 5.1), but interestingly seemed to have a lower Se requirement of 0.7 μg Se/L for 50% maximal growth (Price et al. 1987; Harrison et al. 1988). For the freshwater cladoceran *Daphnia magna*, Keating and Dagbusan (1984) suggested that 1 μg Se/L was sufficient to satisfy minimal needs.

TABLE 5.1

Degrees of Enrichment (× 10⁴) of Se in Algal Cells Relative to Ambient Seawater (Dissolved) Following Exposure to Either 0.15 nM (0.012 µg/L) or 4.5 nM (0.36 µg/L) Selenite for 4 Days. Values are Calculated as Moles Se/µm³ Cell Volume Divided by Moles Se/µm³ in Water. nd: Not Determined. Data from Baines and Fisher (2001)

Algal Species	Selenite Concentration (nM)	
	0.15	4.5
Diatoms		
Chaetoceros gracilis	2.8	0.21
Thalassiosira pseudonana	45	2.5
Skeletonema costatum	0.004	0.01
Chlorophytes		
Chlorella autotrophica	0.4	0.003
Dunaliella tertiolecta	1.0	0.04
Nannochloris atomus	0.5	0.002
Cryptophytes		
Chroomonas sp.	nd	0.02
Cryptomonas sp.	41	1.1
Rhodomonas salina	nd	0.007
Dinoflagellate		
Prorocentrum minimum	26	1.3
Prasinophytes		
Pycnococcus provasolii	nd	15
Tetraselmis levis	nd	1.0
Prymnesiophytes		
Emiliania huxleyi	280	8.2
Isochrysis galbana	nd	6.8

5.2.2 Cellular Se Uptake Pathways

The essentiality of Se means that specific cellular uptake pathways have evolved to facilitate high-affinity Se uptake. Unicellular algae possess Se uptake pathways, allowing for accumulation during exposure to low ambient concentrations (see Section 5.2.3), and can absorb inorganic as well as organic Se. Freshwater green algae display uptake of selenite, selenate, and selenomethionine, with the uptake rates for selenomethionine exceeding those for inorganic Se. Uptake of both selenate and selenomethionine shows saturation kinetics illustrating the involvement of specific transmembrane transport proteins (i.e., carriers) (Fournier et al. 2006). In contrast, uptake of selenite in freshwater green algae was found to be a linear function of ambient concentration, showing no evidence for carrier-mediated uptake (Fournier et al. 2006). Such Se uptake patterns are not unique to green algae and have been reported also for cyanobacteria and diatoms, along with strong evidence for non-passive, carrier-mediated uptake of selenate, selenomethionine, and also selenite (Riedel et al. 1991). To our knowledge, the molecular identity of carriers involved in uptake of inorganic Se in algae remains to be revealed, but it is clear there is a competitive interaction between sulfate (or phosphate) and selenate. Such competitive interactions are not unique to algae and have been documented by either sulfate-induced differences in Se toxicity or Se uptake or accumulation for higher plants (Hurd-Karrer 1938), yeast (Fels and Cheldelin 1949), bacteria (Brown and Shrift 1980, 1982), and a range of invertebrates from freshwater and saline environments (Hansen et al. 1993; Maier et al. 1993; Forsythe and Klaine 1994; Ogle and Knight 1996; Brix et al. 2001). At least in humans, it appears that the interaction between cellular sulfate and Se uptake pivots around a NaS_2 transporter capable of cellular uptake of oxyanions of Se and chromium (Miyauchi et al. 2006) and is thus dependent on the electrochemical gradient for Na^+.

While the pathway of seleno-amino acid absorption by algae displays high affinity, its molecular identity is unknown. However, recent studies on Se uptake by human intestinal and renal cells reveal the nature of carrier-mediated Se uptake by vertebrate intestinal epithelia and the nature of renal seleno-amino acid re-absorption (Nickel et al. 2009). The major route for cellular selenomethionine uptake by intestinal cells and kidney cells appears to be a b^0 family of amino acid transporters, which show high substrate affinity. In kidney cells the b^0 transporter, which is an electrogenic Na^+:amino-acid co-transporter, transporting uncharged amino acids, confers high affinity selenomethionine uptake. In intestinal cells, the b^0 amino-acid transporter, $b^{0,+}$ rBAT which is a Na^+-independent, high-affinity transport system for neutral and dibasic acids, is responsible for selenomethionine uptake. The $b^{0,+}$ rBAT, transporter functions as an exchanger and is found both in the kidney and intestine (Nickel et al. 2009). Both intestinal and kidney b^0 seleno-amino acid transporters display affinity constants in the sub-mM range, which makes them likely candidates for Se uptake from diets containing Se concentrations in the low mg/kg range. Notably, only seleno-aminoacids, and not seleno-derivates like selenobetaine and selenocystamine, are transported by the b^0 amino acid transporters.

The generality of b^0 transporters being involved in dietary Se uptake in lower vertebrates and in invertebrates remains to be examined. However, studies of Se assimilation efficiency by *Artemia fransiscana* fed Se-enriched green algae demonstrates that part of the intestinal Se absorption in this invertebrate is by transport systems with an affinity for Se in the low mg/kg range (Martin Grosell, University of Miami, personal communication). The similarity between intestinal affinity for dietary Se uptake in *Artemia*, dietary Se requirements in the low mg/kg range for fish and birds (Section 5.2.1), and the affinity constant for the mammalian seleno-amino acid transport systems in the b^0 family suggests a widespread distribution of seleno-amino acid transport systems. It thus appears that cellular selenoproteomes, as well as membrane-associated Se uptake pathways, are evolutionarily conserved to support homeostasis of the essential element Se.

5.2.3 IMPLICATIONS OF CELLULAR UPTAKE PATHWAYS FOR SE ACCUMULATION IN FOOD WEBS

As a consequence of high-affinity Se uptake pathways, aquatic organisms display an ability to accumulate Se concentrations sufficient for normal physiological function even in the presence of low ambient and dietary Se concentrations. The implications of saturable, high-affinity uptake systems for Se accumulation in aquatic organisms might then include expectations of apparent bioconcentration, bioaccumulation, and TTFs to be highest at the lowest ambient and dietary Se concentration and decline as Se exposure concentrations increase. A relationship of decreasing bioconcentration factors (BCFs) or BAF with increasing aqueous exposures has been documented for a number of metals and Se on the basis of field observations (McGeer et al. 2003; DeForest et al. 2007). However, no such relationship has been documented for TTFs for metals relative to dietary exposures. Indeed, laboratory experiments to test these mechanisms are difficult to conduct because at the highest concentrations, factors other than physiology, such as behavior (e.g., feeding inhibition), become more important in regulating uptake by organisms (Croteau and Luoma 2008). In the field, determining such relationships becomes even more difficult due to shifting biological species found across the spectrum of dietary exposures being tested and the paucity of organisms found at the highest concentrations. Finally, multiple transport systems are known for other constituents wherein a higher-capacity, lower-affinity transport process takes over once the high-affinity, low-capacity system is saturated (i.e., at higher concentrations in the gut or, for algae, in the water). One result is the perception of linear uptake over a wide range of concentrations, albeit at a lower slope than in the low concentration system. Much remains to be learned about Se transport and its implications for bioaccumulation across wide concentration ranges.

5.2.4   FOOD WEB BASE

5.2.4.1   Accumulation of Inorganic Se by Algae

Understanding the extent to which Se builds up in aquatic food webs necessarily starts at the base of each food web because diet comprises the largest, and often nearly the entire, source of Se for most aquatic animals, and because by far the largest bioconcentration step of Se from the aqueous phase into organisms is its bioconcentration by the microorganisms (algae and bacteria) that serve as the food web base. Considerably more studies have been conducted to assess the factors that govern the bioaccumulation of Se into algae than have addressed the bioaccumulation of Se into bacteria or other microorganisms (e.g., protozoa, fungi).

Like many other inorganic and organic contaminants, the microorganisms at the base of the food web bioconcentrate Se up to 10^6-fold from ambient water (Baines and Fisher 2001), but there are several key factors that distinguish the accumulation of Se from that of most other contaminants. First, Se is an essential element for algae (Doucette et al. 1987; Price et al. 1987) and its uptake is a non-passive, carrier-mediated process (Section 5.2.2). That is, cells need to expend energy to take up dissolved Se, and dead cells display negligible uptake of Se (Fisher and Wente 1993). Further, the various dominant species of aqueous Se — selenite, selenate, and organic selenides — can be accumulated at significantly different rates, and can be greatly influenced by water chemistry. This is particularly true for selenate because this form of Se is taken into algal cells through the sulfate uptake pathway (Shrift 1954; Fisher and Wente 1993). Indeed, when excess concentrations of Se are taken into a cell, Se can behave as an S analog in algae and other plants; the proteins and enzymes that have Se substituting for S may not function properly and this may account for

Se's toxicity. Because sulfate concentrations in seawater are 7 orders of magnitude greater than selenate concentrations, the uptake rate of selenate by marine algae is particularly low and often unmeasurable, unlike in fresh waters where sulfate levels are far lower. Selenite, by contrast, is rapidly accumulated by these same cells and is generally the preferred form over selenate taken up by diverse algal cells (Riedel et al. 1991; Hu et al. 1997).

Selenium exists primarily in anionic form and does not appreciably sorb to suspended particles (which carry a negative surface charge), so mixtures of living phytoplankton and non-living material that commonly compose seston (especially in coastal waters) would be expected to display lower degrees of Se enrichment than in pure phytoplankton assemblages, which is consistent with field observations (Cutter 1989; Doblin et al. 2006). Because the uptake of Se is carrier mediated and follows typical Michaelis-Menten kinetics, Se concentrations in algae do not linearly reflect ambient concentrations, particularly as ambient concentrations approach those that saturate carrier systems (approximately 10 nM [0.79 µg/L] for selenite for diatoms) (Baines and Fisher 2001). Consequently, increases in dissolved selenite concentrations in the 0.1 to 10 nM (0.0079 to 0.79 µg/L) range result in a 3.5-fold increase in marine algal Se levels.

Because Se uptake requires energy, equilibrium partitioning between dissolved and particulate phases does not apply for Se. Further, once cells take up Se, it is rapidly converted to organic selenides, so the concept of equilibrium partitioning between organic selenides and ambient inorganic Se is inappropriate. Thus, the term "distribution coefficient" (K_d) as a descriptor of the enrichment of Se in particulate matter relative to ambient water, is misleading. Still, it is recognized that microorganisms such as bacteria and algae can become greatly enriched relative to ambient water for selenite. Degrees of enrichment can exceed 10^6 in axenic cultures, with most algal species exceeding 10^4 (Baines and Fisher 2001). This initial bioconcentration step (from the dissolved phase into living cells) is clearly the greatest of any of the accumulation steps in an aquatic food chain. Therefore, the extent to which algal or bacterial cells are enriched with Se is a major determinant of Se contamination throughout a food web.

Another significant difference between Se bioconcentration in algae and that of most cationic metals is the large inter-specific variations among algal taxa (Wrench and Measures 1982; Harrison et al. 1988; Vandermeulen and Foda 1988; Riedel et al. 1991; Fisher and Wente 1993; Baines and Fisher 2001; Wang and Dei 2001), not unlike the variability in terrestrial plant Se requirements (Brown and Shrift 1982). In fact, degrees of Se enrichment relative to ambient water among different marine algae vary up to 5 orders of magnitude, from about 30 to well over 10^6 under the same environmental conditions (Baines and Fisher 2001). Such differences probably result from inter-specific differences in Se cellular requirements but possibly also from different capabilities of cells to regulate Se uptake. In contrast, inter-specific differences in uptake for metals that require no energy expenditure typically display less than 1 order of magnitude variation in cell volume-normalized concentration factors (Fisher and Reinfelder 1995). Much of the inter-specific variability for cationic metal concentration factors can be attributed to cell size, with highest concentration factors associated with the smallest cells and thus the highest surface-to-volume ratios (Fisher and Reinfelder 1995). This pattern is not seen for Se.

Because of the high inter-specific variability in Se bioconcentration, it follows that the degree to which organisms at the base of the food web can be enriched sources of Se could vary tremendously with algal species composition (Table 5.1). Spatial and temporal variability in algal community structure could therefore have a pronounced effect on the bioavailable Se in algal cells. Typically, among marine forms, the chlorophytes show the lowest degree of enrichment (Wang and Fisher 1996a; Baines and Fisher 2001; Wang and Dei 2001), possibly reflecting lower cellular requirements or, alternatively, greater regulation of Se uptake. While there is some regulation of Se uptake in algal cells, there also appears to be "luxury" uptake, in excess of requirements, of this nutrient in at least some algal species (Harrison et al. 1988; Baines and Fisher 2001), as has been noted for many other nutrients.

In any case, bodies of water that are dominated by chlorophytes could be expected to have lower algal Se available for herbivores than waters dominated by other algae (e.g., diatoms, prasinophytes, dinoflagellates, prymnesiophytes). Even within the same taxonomic group, Se uptake can also vary greatly. For example, the diatom *Skeletonema costatum* accumulates much less Se than other diatom species such as *Thalassiosira pseudonana* (Table 5.1). The species composition of algal communities varies with nutrient concentrations and ratios (Chisholm 1992), vertical stratification and mixing (Margalef 1978), selective grazing pressure (Smetacek et al. 2004), and salinity variations (Cloern and Dufford 2005). Thus, a body of water could have seasonally variable Se concentrations in herbivores that reflect these changes in algal communities. As will be evident later in this chapter, such differences in algal composition could result not just in differences in herbivore Se levels but in differences in Se tissue concentrations in organisms higher in the food web.

5.2.4.2   Accumulation of Se by Bacteria

Marine and freshwater bacteria have also been shown to bioconcentrate selenite from water (Foda et al. 1983; Riedel and Sanders 1996; Baines et al. 2004). For example, in California's San Francisco Bay Delta waters, Se uptake in the 0.2 to 1.0 μm size fraction accounted for 34% to 67% of the Se uptake in the dark, and bacterial Se:C ratios were up to 13 times those of phytoplankton (Baines et al. 2004). Consequently, bacterial cells may serve as especially enriched sources of organic selenides for bacterivores. This is an understudied aspect of the biogeochemical cycling of Se in aquatic food webs that deserves further study.

5.2.4.3   Organic Selenide Uptake and Cycling

Once selenite is taken into a cell, it is readily converted to organo-selenium compounds such as selenomethionine and selenocysteine, as well as to polypeptides (Shrift 1954; Wrench 1978; Wrench and Campbell 1981; Bottino et al. 1984; Fisher and Reinfelder 1991; Besser et al. 1994; Riedel et al. 1996), such as glutathione peroxidase (Price and Harrison 1988). Thus, animals that ingest algae are exposed primarily to organic forms of Se rather than inorganic forms. The assimilation of ingested selenides by herbivores grazing on phytoplankton is considered in Section 5.2.5. For some algal species, such as the common centric diatom *Skeletonema costatum*, the degree of enrichment of Se varies with the physiological state of the cell, where rapidly growing cells are far less enriched than cells entering senescence; many other species, including other diatoms, do not display this pattern (Baines and Fisher 2001). Thus, biological variability in Se demand, perhaps in response to oxidative stress, is likely to account for some of the pronounced differences in algal Se concentrations.

Plants, phytoplankton, and epilithic organisms release their organic selenides into ambient water through excretion, through cell lysis, or when grazed upon by herbivores. Once cells die and decompose, Se is released rapidly into ambient water, at rates comparable to that of organic carbon (Lee and Fisher 1992a, 1993; Fisher and Wente 1993). Similarly, for dead zooplankton, Se is lost from copepod carcasses and fecal pellets at rates similar to that of carbon with half-lives of only about 1 day in copepod carcasses (Lee and Fisher 1992b). In general, bacterial decomposition enhances Se loss from decomposing phyto- and zooplankton. Viral lysis of algal cells has also been shown to enhance Se release rates into seawater, and this released Se is as highly bioavailable as selenite to other algal cells (Gobler et al. 1997). This finding is consistent with observations that organic selenides (i.e., lysates of diatoms) are accumulated by marine phytoplankton at rates and to extents comparable to those for selenite (Baines et al. 2001). As with selenite, the chlorophytes display significantly lower accumulation of organic selenides than other algal forms (Baines et al. 2001). Similarly, Riedel et al. (1991) showed that selenomethionine can be readily accumulated by freshwater phytoplankton. Thus, models that consider the bioaccumulation of Se in aquatic food webs must take into consideration the high bioavailability of organic selenides, especially at the base of the food web. Given that organic selenide concentrations can approach those of inorganic Se forms and can account for 80% of the dissolved Se in open ocean surface waters (Wrench 1983; Cutter 1989; Cutter and Cutter 1995), bioaccumulation of organic selenides by algae and bacteria is arguably important and has largely been under-studied. Although much of the organic selenide pool in ocean surface waters is surely more refractory than the labile forms released by algal cell lysis (Cutter and Bruland 1984; Cutter and Cutter 1998), the cycling of dissolved organic carbon (DOC) and dissolved organic Se (DOSe) in the oceans suggest that DOC has a much longer residence time and is probably more resistant to biological degradation and uptake than organic selenides (Baines et al. 2001). The release of organic selenides and their subsequent bioaccumulation by plankton (i.e., biological recycling) helps explain the nutrient-type vertical profile seen for Se in oceanic water columns (Measures and Burton 1980; Cutter and Bruland 1984) and its relatively long residence times in ocean surface waters (Broecker and Peng 1982). Enhanced recycling of organoselenium released by decomposing biological material likely also contributes to elevated Se in biota inhabiting lentic compared to lotic freshwater environments (Orr et al. 2006).

5.2.5 IMPORTANCE OF DIETARY INTAKE OF SE

If only contaminants that are associated with an organism can elicit toxic effects, then understanding the extent to which Se can be accumulated by different aquatic organisms under different environmental conditions is of toxicological relevance.

In addition, toxicity is dependent on the exposure route (aqueous vs. dietary; Hook and Fisher 2001), so delineating the sources of Se accumulation is important for the toxicological interpretation of contamination to aquatic organisms. Water quality criteria or guidelines for Se recognize that chronic toxicity tests in which organisms were exposed to Se only through water require unrealistically high aqueous concentrations to reach body burdens and elicit chronic responses seen in nature (USEPA 2004). Dietary exposures (and trophic transfer) are important pathways for Se accumulation in aquatic invertebrates and fish (Lemly and Dimmick 1982; Luoma et al. 1992; Besser et al. 1993).

Figure 5.2 shows the relative contribution of dietary trace element uptake in various aquatic invertebrates, which can be inferred based on Equation 5:

$$\% \text{ Dietary uptake} = \frac{(AE \times IR \times C_f)/(k_{ef} + g)}{C_{ss}} \times 100 \tag{5}$$

It is clear from this figure that Se, more than any other trace element considered, is accumulated overwhelmingly from dietary exposure. The relative importance of dietary exposure for other trace elements varies among metals and with such factors as aqueous metal concentration, metal content in food, and food quantity. For Se, the contribution of dietary intake in a variety of aquatic organisms (i.e., marine worms, bivalves, crustaceans) that consume very different diets was consistently high, with more than 90% of Se body burdens derived from dietary exposure. This difference in the relative importance of metal uptake pathways is likely due to relatively high assimilation efficiencies for Se and low uptake rates from solution.

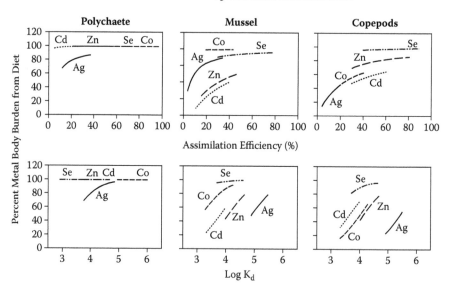

FIGURE 5.2 Percent of uptake from diet of different metal species by marine invertebrates as a function of assimilation efficiency and K_d. Reprinted from Wang and Fisher 1999, with permission from Elsevier.

For organisms higher in the food web (e.g., fish, birds) more kinetic data are needed, but nevertheless, there is evidence from both laboratory and field studies that dietary exposure is a major route for Se accumulation (Besser et al. 1993). In the mangrove snapper, *Lutjanus argentimaculatus*, nearly all Se in the fish was due to dietary uptake primarily because of the extremely low aqueous uptake rate (lowest $k_u = 0.0008$ L/g/d; Xu and Wang 2002a). Presser and Luoma (in press) examined the relationship between predicted (from food alone) and measured Se body burdens in fish collected in the field (both freshwater and marine). They observed a striking 1:1 relationship between model predictions and measured Se body burdens in fish, suggesting that Se body burdens in fish can be accurately predicted solely on the basis of dietary intake. Indeed, field studies have shown that dietary pathways (rather than aqueous Se concentrations) can explain differences in Se body burdens between predator fish (Stewart et al. 2004).

Presser and Luoma (in press) further examined the relationship between modeled and observed Se body burdens in aquatic invertebrates, neglecting aqueous intake of Se. Predicted Se concentrations in invertebrates were lower than observed by about 15% (slope $= 0.86$). This prediction suggests that while dietary intake is still the most important exposure route for Se accumulation in invertebrates, aqueous uptake may not be negligible for certain species. For example, *Dreissena polymorpha* (zebra mussel), *Artemia fransiscana*, and *Daphnia magna* show very high uptake from water ($k_u = 0.05$ to 0.43). In addition to the high uptake rates of dissolved Se, the zebra mussel exhibits low dietary Se assimilation efficiency (<46%) leading to an estimated 24% to 61% contribution from dissolved Se to the Se burden in this bivalve (Roditi and Fisher 1999; Roditi et al. 2000a, 2000b). Tsui and Wang (2007) predicted that about 20% to 40% of Se body burden in the freshwater cladoceran *D. magna* was due to uptake from the aqueous phase; this percentage was noted to be higher than predicted for a variety of marine animals. A higher contribution of aqueous Se was mainly due to moderate dietary assimilation efficiency (20% to 60%), as well as relatively high aqueous uptake of Se in daphnids (Yu and Wang 2002). In contrast, relatively high dietary Se assimilation efficiencies (>80%) in *A. fransiscana* still amount to a relatively modest dietary Se uptake when feeding on green algae, because the green algae contain relatively low Se concentrations compared to diatoms (Section 5.2.4.1) (Martin Grosell, University of Miami, personal communication). Thus, the relatively low dietary Se uptake in *Artemia* combined with a high K_u results in a contribution from dissolved Se of >50% (Martin Grosell, University of Miami, personal communication).

A partial explanation for the relatively high contribution of dissolved Se uptake in the 3 species discussed in the previous paragraph could be that green algae, in general, accumulate less Se than other unicellular algae (Sections 5.2.2 and 5.2.5) and that dietary uptake, even with a high assimilation efficiency, is limited due to low dietary Se concentrations. Under conditions of modest dietary Se availability, uptake from the dissolved phase across respiratory surfaces (uptake rate function) may be elevated to meet requirements for essential Se. However, low Se accumulation in green algae cannot be the only universal explanation for the above 3 examples of relatively high contributions from the dissolved phase. Zebra mussels, for example, show low assimilation efficiency of Se regardless of whether they are feeding on green algae, diatoms, cyanobacteria, or bacteria (Roditi and Fisher 1999). Thus, it appears that, in addition to the limited Se availability in diets consisting of green algae, overall low assimilation efficiency in the zebra mussel may be part of the explanation for the relatively high contribution from dissolved Se. The form of Se may also play a role in the importance of dissolved uptake. Where bioaccumulation studies have distinguished between inorganic and organic Se uptake, it has been found that organic forms of Se are more readily accumulated from aqueous exposure (Besser et al. 1993; Baines et al. 2001), so in water bodies where organic Se concentrations are elevated, aqueous Se uptake may not be negligible.

5.2.6 Invertebrates

Invertebrates form a critical link in ecosystems between primary producers and higher-level consumers and can play an important role in the trophic transfer of many contaminants. In the case of Se, where dietary exposure pathways tend to dominate, the ways in which invertebrates differ in their accumulation are central to the discussion of Se behavior and effects in ecosystems. While we cannot ignore that there are cases where accumulation of Se from water can be important in invertebrates, most of this discussion will focus on dietary exposure pathways. With this in mind, there are 2 major ways in which invertebrates can differ in their Se accumulation from the environment: 1) diet choice (different species ingest food items that are differentially loaded with Se) and 2) physiological processing (differential assimilation and retention of dietary Se in tissues). Given the tremendous biodiversity of aquatic invertebrates, these life history (diet choices) and physiological processes (Se assimilation, retention) can and do vary widely.

Differences in Se concentrations in invertebrate tissues among sympatric species can be profound. Until biodynamic modeling approaches provided a mechanistic understanding of why and how inter-specific differences in Se body burdens occur, we were limited to simply describing them from site to site with limited predictive power. Biodynamic modeling now provides a mechanistic basis for understanding and predicting Se bioaccumulation differences among species.

5.2.6.1 Biodynamic Controls on Se Accumulation

Typically, marine invertebrate assimilation of Se from primary producers is an efficient process (usually 70% to 90%) (Table 5.2). One major driver of Se bioaccumulation differences among species is elimination (K_e). For example, in San Francisco Bay, Se concentrations in benthic clams ranged from 5 to 20 mg/kg dw, while crustacean zooplankton ranged from 1 to 4 mg/kg dw over the same exposure period and location (Stewart et al. 2004). This difference was largely explained by the fact that clams tend to eliminate Se at a rate that is 8 to 10 times slower than crustaceans. Variations within taxonomic groups exist, but to a lesser degree (~2-fold differences). The reasons for these differences in loss are not well understood but appear to be related to the efficiency with which organisms recycle proteins that contain Se (Wright and Manahan 1989; Manahan 1990; Wright 1995).

Table 5.2 summarizes the studies that have been conducted on assimilation of Se from algal diets in diverse invertebrates. Presently, marine and estuarine invertebrates

TABLE 5.2

Reported Assimilation Efficiencies (AEs), Efflux Rates Following Dietary Uptake (k_{ef}), Uptake Rate Constants from the Aqueous Phase (k_u), and Elimination Rates Following Uptake from the Aqueous Phase (k_{ew}) for Animals Feeding on Phytoplankton. All Data Are for Marine Animals Except *D. polymorpha*, Which Is a Freshwater Species. When More Than 1 Algal Species Was Used as a Diet, a Range of Values Is Given. When Only 1 Algal Species Was Used, Mean Values + 1 SD Are Given, Where Available. *Denotes Study with *M. edulis* of Different Sizes (1.5 to 5 cm)

Species		AE (%)	k_{ef} (/d)	k_u (L/g/d)	k_{ew} (/d)	Reference
Mytilus edulis	Mussel	74–86; 15–72; 8–86; 72–81*	0.02–0.05; 0.02; 0.055–0.065; 0.048–0.075*	0.035; 0.032–0.039; 0.033–0.082*	0.019 + 0.005; 0.018–0.022*	Wang et al. 1995; Wang and Fisher 1996a, 1996b, 1997; Reinfelder et al. 1997
Dreissena polymorpha	Mussel	18–46	0.022–0.026	0.05–0.10	0.035 + 0.001	Roditi and Fisher 1999
Macoma balthica	Clam	74–78; 86	0.03–0.03; 0.01			Luoma et al. 1992; Reinfelder et al. 1997
Mercenaria mercenaria	Clam	92 + 2	0.01 + 0.004			Reinfelder et al. 1997
Mercenaria mercenaria	Clam larvae	100 + 2				Reinfelder and Fisher 1994a
Crassostrea virginica	Oyster	70 + 6	0.07 + 0			Reinfelder et al. 1997
Crassostrea virginica	Oyster larvae	105 + 6				Reinfelder and Fisher 1994a
Artemia salina	Brine shrimp nauplii	60 + 6	0.60			Mathews and Fisher 2008
Acartia tonsa and *Temora longicornis*	Copepods	97 + 2; 71–94; 58–59	0.42–1.14; 0.26–0.34; 0.25–0.29	0.024	0.155	Fisher and Reinfelder 1991; Reinfelder and Fisher 1991; Wang et al. 1996b; Wang and Fisher 1998
Sesarma reticulatum	Crab larvae	63				Anastasia et al. 1998
Dyspanopeus sayi	Crab larvae	78				Anastasia et al. 1998
Uca pugnax	Crab larvae	61				Anastasia et al. 1998
Balanus amphitrite	Barnacle	62–79	0.0141			Wang et al. 1999
Elminius	Barnacle	34–66				Rainbow and Wang 2001
Semibalanus balanoides	Barnacle larvae	93				Anastasia et al. 1998

comprise the vast majority of invertebrate taxa for which biodynamic models have been developed.

More work has considered bivalves, in part because these animals (especially the blue mussel *Mytilus edulis*) are used worldwide as bioindicator organisms of coastal contamination (Phillips and Rainbow 1993).

It is noteworthy that assimilation efficiencies of ingested Se tend to be high (often >60%) for most herbivore and algal species combinations, although exceptions are noted. The exceptions are most commonly found for animals consuming chlorophytes, which themselves tend to have lower degrees of Se enrichment (Table 5.1). These assimilation efficiencies tend to be higher than for all metals except methylmercury, for which they are approximately comparable. In contrast to methylmercury, however, the efflux rates of Se from invertebrates tend to be relatively fast, with rate constants of loss of about 2% to 6%/d for bivalves and values exceeding 25%/d for crustacean zooplankton (Table 5.2).

Uptake of selenite from the aqueous phase tends be slow, which helps explain the predominance of the dietary pathway as a Se source for the marine invertebrates (Wang and Fisher 1999). In contrast, uptake rates from the aqueous phase for freshwater zebra mussels are considerably higher, which can help explain the higher fraction of Se taken into zebra mussels from the aqueous phase than is commonly observed for marine mussels (Wang et al. 1996a; Roditi et al. 2000b). The kinetic parameters given in Table 5.2 have been used in biodynamic models to predict steady-state Se concentrations in mussels and copepods in Long Island Sound, San Francisco Bay, diverse freshwater systems in New York State, and in the western Mediterranean; in every case predicted values closely matched field measurements of Se concentrations in these animals (Wang et al. 1996a; Fisher et al. 2000; Roditi et al. 2000a). The close match of model-predicted concentrations and independent field measurements suggests that we can account for the major processes governing Se concentrations in these animals, and that laboratory-derived kinetic parameters are applicable to field conditions.

Biodynamic parameters can be used to estimate TTFs for Se in invertebrates and provide clarity in understanding and predicting Se movement through food webs (Table 5.3).

TABLE 5.3
Trophic Transfer Functions Derived from Laboratory Studies with Marine and Estuarine Invertebrates. Data from Presser and Luoma (in press)

Taxon	Group	TTF (Range)	Nr of Studies
Bivalves	Clams	3.6–23.0	7
	Oysters	1.6–2.5	2
	Mussels	3.8–8.8	2
Crustaceans	Copepods	1.3–3.1	4
	Mysids	1.1– 1.3	1
	Amphipods	0.6	1
	Barnacles	9.9–22.6	2

For example, Se efflux rates for marine bivalves are typically lower than for zooplankton, so bivalve Se concentrations are generally higher than zooplankton in the same body of water (despite AEs that are less than or equal to those in zooplankton). Thus, marine bivalves should be a more enriched source of Se for their predators than for predators of zooplankton. Indeed, TTFs calculated from laboratory studies showed clams having TTFs up to 7-fold higher than those for copepods (Table 5.3).

The importance of differential Se bioaccumulation in invertebrates and their trophic transfer is clearly demonstrated in San Francisco Bay, where Se concentrations of the clam *Corbula amurensis* are 6- to 8-fold higher than in amphipods (Stewart et al. 2004). These differences are propagated up food webs, resulting in differential Se concentrations in apex predators. We can similarly use experimentally derived TTF values to predict that predators consuming barnacles, for example, are much more likely to have higher Se concentrations in their tissues than those consuming amphipods.

5.2.6.2 Freshwater Environments

Relative to our mechanistic understanding of Se bioaccumulation in marine invertebrates, our understanding of freshwater invertebrates is limited. Very few laboratory studies exist that have quantified biodynamic parameters (e.g., freshwater cladocerans, zebra mussels), so most of our inferences must be drawn from field data. Relying on field studies alone is limiting in 3 major ways. First, accurate assessment of food concentrations at the base of food webs is difficult. The separation of algae and/or bacteria from other particulate material or sediments is extremely challenging and rarely done. Thus, estimates of Se in invertebrate diets are generally crude. Second, in many field studies of Se bioaccumulation, invertebrates are often pooled and measured as composite benthic samples, or grossly separated to extremely coarse taxonomic levels (orders or higher). This type of composite sampling obscures species-specific patterns of Se bioaccumulation and trophic transfer; however, this may not matter in the case of non-selective consumers (e.g., some benthivorous fish) for which samples of pooled taxa may suffice to describe dietary concentrations. Finally, the dietary preferences of many invertebrates are not well known.

Presser and Luoma (in press) generated estimates of freshwater invertebrate TTFs from an analysis of existing field data (Table 5.4). Note that these estimates cannot be interpreted identically to laboratory-derived TTFs for some of the reasons described above, and do not provide the mechanistic understanding of how and why species vary in their Se content (AE, k_e, etc). Until a more mechanistic understanding of Se bioaccumulation in freshwater invertebrates is established, these values should be viewed as coarse guides.

One important observation to make from field data is that, relative to other taxa, freshwater clams are not necessarily strongly accumulative as is the case with their marine counterparts. In some systems such as the Mud Reservoir (WV, USA), clams tend to be high in Se relative to crayfish and dragonflies, for example (Presser and Luoma in press). In other systems such as the San Diego Creek Watershed (CA, USA), clams are comparable in Se content with zooplankton and are lower than dragonflies (Presser and Luoma in press). These differences could be a result of taxonomic variation or site-specific factors that are not well understood. Also noteworthy is the high TTF measured

TABLE 5.4
Trophic Transfer Function Estimates from Field Data. Data are from Presser and Luoma (in press) and Andrahennadi et al. (2007)

Taxon	TTF
Presser and Luoma (in press)	
Amphipod	0.9
Zooplankton	1.5
Crayfish	1.6
Daphnia sp.	1.9
Aquatic Insects – bulk	2.1– 3.2
Clam (*C. fluminea*)	1.4–4.0
Zebra mussel	4.5–7.0
Andrahennadi et al. 2007	
Aquatic insects – species	
Rhithrogena sp.	1.6–2.7
Drunella sp.	1.5–1.6
Epeorus sp.	1.2–1.5

in the zebra mussel. These organisms filter water at a substantial rate, and thus, uptake of dissolved Se can be an important route of exposure in this organism (Section 5.2.6). Similar to other invasive marine bivalves, this invasive freshwater species could be particularly problematic in Se-rich ecosystems to species that consume it.

Another key observation is the relatively high TTFs estimated for aquatic insects. This is important both in terms of the trophic transfer of Se from insects to fish and birds, as well as potential risks to the insects themselves. Insects are fundamentally important components of freshwater food webs, particularly in lotic systems, because they process organic materials and are key to nutrient dynamics. Because insects are the primary food source of many socially important fish species (e.g., salmonids) and birds, understanding of Se dynamics in insects in a comparative context is critical. In a rare study of streams in Alberta, Canada, TTFs were estimated for periphyton-grazing mayflies (Andrahennadi et al. 2007). Within a given genus, TTFs varied slightly among sites, which the authors suggest may be a function of periphyton community structure. A recent laboratory study examined selenite bioaccumulation in natural biofilms and subsequent transfer to the grazing mayfly *Centroptilum triangulifer* (Conley et al. 2009). In that study TTFs in adult mayflies (post–egg release) ranged from 1.9 to 2.4. Importantly, these mayflies transferred significant proportions of their body burdens to eggs. It is not yet clear how the inclusion of egg Se would modify these TTF estimates because reliable measures of egg weights could not be obtained in this study. However, Se loads of gravid adults were 36% to 51% higher than Se loads in those same animals post–egg release.

It is important to reiterate that we lack the mechanistic understanding of inter-specific Se bioaccumulation in freshwater invertebrates, and thus, to the extent possible, the lowest achievable taxonomic units should be used in reports of field data. For example, there are >1,500 species of Trichoptera described to date in the Americas north of Mexico, and we should expect to see variation among families and genera. Differences in Se bioaccumulation patterns among species are not idio-syncratic. Rather, they are mediated by evolutionarily derived life history (ecological) and physiological processes, which may ultimately prove to be predictable using comparative phylogenetic approaches (Buchwalter et al. 2008). Because of the tendency for closely related species to resemble one another, we might expect to find major phylogenetic differences among taxa in terms of propensity to bioaccumulate Se in tissues. Such phylogenetically based patterns could be extremely helpful in determining which species to sample in an assessment or monitoring context.

5.2.6.3 Intra-specific Differences: Size Influences Se Enrichment in Bivalves

There is evidence that size can play a role in the uptake of Se by bivalves. Selenium concentrations in the estuarine clam *C. amurensis* exposed to dissolved sources of Se(IV) in the laboratory were found to decrease by 50% as mean shell length of the clams increased by ~30% (Lee et al. 2006). Smaller clams had higher Se concentrations than larger clams. In the marine black mussel *Septifer virgatus*, Se uptake decreased with increasing body size (quantified as tissue dry weight), with a power coefficient of −0.317 (Wang and Dei 1999a). A similar response was observed in replicate composites of field-collected *C. amurensis* exposed to aqueous sources of Se (Stewart et al. 2004). The cause of the size-specific difference may be specific to bivalves and size-specific filtration rates. Wang and Dei (1999b) found that the power coefficient of the Se uptake as a function of tissue dry weight (−0.317) in the black mussels was directly comparable to the power coefficient of the mussel filtration rate as the function of tissue dry weight (−0.32), which strongly suggested that the allometric change of Se uptake in the mussels was controlled by the same process as the filtration activity, such as the gill surface to volume ratios. Consequences of size-specific differences in Se uptake in bivalves are not trivial. Minor differences in clam size can modify apparent Se concentrations by up to 50%, creating problems in interpretation of monitoring data collected spatially and temporally. Further, shifts in size distributions of the bivalve community due to food availability and predation may alter Se exposures to higher trophic levels and their risks of Se toxicity.

5.2.6.4 Subcellular Distribution in Controlling Se Trophic Transfer

There has been substantial interest in understanding the various processes controlling dietary Se AE in a variety of marine herbivores and carnivores. Reinfelder and Fisher (1991) found that the Se AE in marine copepods was nearly comparable to the Se distribution in diatom (prey) cytoplasm, implying that the assimilation was controlled by the Se cytosolic fraction. Marine copepods (*Acartia spincauda*) were able to assimilate the Se-associated diatom detritus (either freshly prepared from the cellular debris of diatoms or the decomposed products) at an efficiency of 44% to 57%, which indicated that the Se associated with the diatom cell walls might also be available to copepods (Xu and Wang 2002b). The assimilation processes of marine predators may be even more complicated than those of herbivores. For Se, Dubois and Hare (2009) quantified the subcellular Se distributions in the oligochaete *Tubifex tubifex* and in the insect *Chironomus riparius* and how they affected Se trophic transfer to a predatory insect (the alderfly *Sialis velata*). In their study, the predator assimilated about 66% of the Se from the prey, which was similar to the Se distribution of 62% in the protein and organelle fractions. In the marine fish grunt, *Terapon jarbua*, the Se AEs varied by prey (copepods, barnacles, clams, mussels, and fish viscera) over a range of 13% to 36% (Zhang and Wang 2006). Such variation was significantly related to the heat-stable protein fraction of Se in prey. Again, subcellular forms of Se in the fish prey similarly affected Se assimilation by the predator. Further experiments using purified subcellular fractions of copepods and mussels as fish diets suggested that Se bound with the insoluble fraction (including metal-rich granules [MRG], cellular debris, and organelles) had a much lower AE (29% to 33%) than Se bound with the protein fractions (41% to 54%) (Zhang and Wang 2006). However, feeding processes also affected the Se assimilation in fish. Selenium AE was significantly dependent on the ingestion rate of fish and gut passage time of Se.

5.2.7 Fish

Elevated Se concentrations in fish found in contaminated areas have raised environmental as well as public health concerns (Lawrence and Chapman 2007) because the consumption of fish may represent a significant source of Se to humans (Thompson et al. 1975; Schubert et al. 1987). Predicting Se accumulation from the aqueous Se concentration in a given system is not straightforward because Se body burdens in fish and other aquatic animals may vary widely among species within the same water body (Stewart et al. 2004). Fish are often considered to be the most sensitive group of organisms to chronic Se exposure (Hamilton et al. 1990; Hermanutz 1992; Hermanutz et al. 1992; Coyle et al. 1993). A quantitative understanding of the variables affecting Se accumulation in fish is therefore needed to properly evaluate the biological and ecosystem-level effects of Se contamination, and to set appropriate environmental quality criteria or guidelines.

As with invertebrates, toxicity to fish can occur when Se is present at levels above the concentrations that are required for metabolic functions. Differences in choice of diet (prey selectivity), seasonal movements or migration, habitat utilization, and tissue allocations that occur both within and among species are sources of variability that are important to consider in interpreting Se levels in fish tissues. Ecological impacts to fish are usually associated with effects on early life stages as a result of maternal transfer of Se to eggs (Lemly 1993; Holm 2002). Therefore, fish are particularly vulnerable during the period when eggs are being formed, although the precise timing and duration of this vulnerability varies based on differences in reproductive characteristics among species (Section 5.2.7.4).

5.2.7.1 Trophic Transfer Patterns

Relative to other trace elements, Se is efficiently assimilated into fish from diet (Reinfelder and Fisher 1994b; Baines et al. 2002; Xu and Wang 2002a), and where

TABLE 5.5
Trophic Transfer Functions (TTFs) for Fish from Laboratory Studies

Species		Type of Prey	TTF	Reference
Juvenile striped bass	*Morone saxatilis*	Crustacean zooplankton	0.94–2.8	Baines et al. 2002
Juvenile sea bream	*Sparus auratus*	Crustacean zooplankton	0.46–0.69	Mathews and Fisher 2008
Juvenile black sea bream	*Acanthopagrus schlegeli*	Crustacean zooplankton	0.5–1.5	Zhang and Wang 2007
Mangrove snapper	*Lutjanus argentimaculatus*	Crustacean zooplankton and bivalve	1.07–2.09	Xu and Wang 2002a
Intertidal mudskipper	*Periophthalmus cantonensis*	Polychaetes	1.13–1.68	Ni et al. 2005
Juvenile sea bass	*Dicentrarchus labrax*	Juvenile fish	0.5–1.3	Mathews and Fisher 2008

loss rates are slow, Se has the potential to biomagnify in aquatic food chains (TTF > 1) (Wang 2002; Zhang and Wang 2007). Both laboratory and field studies used to determine TTFs for fish report remarkably similar results, despite the fact that they are derived through different methods. Trophic transfer functions are sometimes derived from field studies by dividing the Se concentrations in consumers by that measured in known or presumed prey (Presser and Luoma in press). Laboratory studies are usually based on food chains that are short and linear to generate the kinetic data used to calculate TTF values (Table 5.5). Laboratory-derived TTFs for Se have been reported for marine fish that were fed crustacean, bivalve, and fish diets (Xu and Wang 2002a; Zhang and Wang 2007; Mathews and Fisher 2008). The TTF values varied with AE and IR, resulting in a range of TTF values rather than a single best estimate. Although TTF values were lower for predatory fish fed a crustacean diet than for fish fed either bivalve or fish diets, the spread of TTF values was relatively narrow regardless of the prey consumed, ranging from 0.5 to 1.5 for the intermediate ingestion rates considered.

Compared to the amount of data available on Se trophic transfer from invertebrate prey to fish, there are relatively few data in the laboratory or in the field that describe Se trophic transfer to piscivorous fish species. This is likely because of intrinsic difficulties in the measurement of whole-body Se levels in such large organisms. Mathews and Fisher (2008) report TTF values of 0.5 to 1.3 for sea bass fed on juvenile sea bream (for IR = 0.1 g/g/d). These laboratory-derived TTFs for piscivorous fish are comparable to the TTF values reported for fish feeding on invertebrates (Table 5.5).

In contrast to the simple, controlled food chains typical of laboratory-based studies, field-derived TTFs represent time-integrated Se accumulation and loss processes for fish consuming a varied diet, for which the specific composition of diet, food chain length, and food web pathway may be unknown or uncertain. Comparisons between field- and laboratory-derived TTFs are further complicated by the fact that, while kinetic studies of Se accumulation in laboratory studies are based on whole-body analyses, field studies typically report as whole-body Se burdens for invertebrates while Se levels in field-collected fish are often reported with respect to specific tissues (muscle, liver, gonad). Despite this, the mean TTF for 15 freshwater and marine fish species feeding on invertebrate prey was estimated at 1.2, within a remarkably narrow range of 0.51 to 1.8 (Presser and Luoma in press) and showed close agreement with laboratory-derived values (Table 5.5). Muscatello et al. (2008) and Muscatello and Janz (2009) suggested that TTFs of up to 10 or more could occur in benthivorous fish (spottail shiner *Notropis hudsonius*, white sucker *Catostomus commersoni*, and stickleback *Pungitius pungitius*) through selective feeding on specific invertebrate guilds. However, actual fish diets were not determined. Lack of precise information about dietary habits of fish is a common limitation of field studies, for which it is difficult to track specific feeding habits over time. Assuming an average invertebrate concentration in the diet of fish based on data presented by the same authors, results in a TTF of ≤3, which is much more consistent with the data summarized by Presser and Luoma (in press) from other studies. Burbot (*Lota lota*), also classified as benthivorous by Muscatello and Janz (2009), had a TTF of ~7 based on an average of the Se concentrations in available invertebrate prey, but at 2 to 3 years of age the burbot diet may have included other fish (Scott and Crossman 1973). A piscivorous diet for the burbot in that study would yield TTF values up to 8.7 or higher, depending on choice of prey (e.g., 107 for stickleback). Analyses of gut contents in addition to stable isotopes of carbon, nitrogen, and/or sulfur can be useful in elucidating trophic relationships (Vander Zanden and Rasmussen 1999; Stewart et al. 2004; Orr et al. 2006) and, thus, provide for more precise TTF estimates. Nevertheless, the substantially higher TTF values for burbot are notable compared to other species. The higher values may be a consequence of the relatively larger contribution of liver tissue (known to concentrate and metabolize Se) compared to other tissues on a whole-body basis. Indeed, the hepatosomatic index (HSI, percent mass contribution of liver tissue to whole body) for burbot is up to 6% higher than for other freshwater species, including trout and salmon, and may lead to a higher TTF (Tom Johnston, Ontario Ministry of Natural Resources, personal communication).

Despite the confounding influences associated with field studies and differences between field- and laboratory-derived methods for determining trophic transfer of Se to fish, TTF values for fish appear to be relatively consistent and low relative to trophic transfer at lower food chain steps.

5.2.7.2 Other Factors That Influence Se Enrichment

5.2.7.2.1 Lentic vs. Lotic Habitats

Lentic systems, which are characterized by long hydraulic retention times, low oxygen content, and high carbon content, favor reducing conditions. In these environments, Se is often found as selenite, reflecting the recycling in which Se is progressively reduced to more bioavailable organic forms. In lotic systems that have higher flushing rates and lower productivity, Se is found in the more oxidized form of selenate, which does not easily migrate to sediments and thus is not as rapidly recycled. The reduced Se found in lentic sediments is readily accumulated at the base of the food web, passed on to benthic organisms, and transferred through the food chain to fish in

TABLE 5.6

Mean Se Concentrations Observed in Different Media Sampled from the Elk River Watershed, British Columbia, Spring 2006 (Minnow Environmental Inc. et al. 2007)

Habitat	Status	Location	Mean Water 2004–2006 (µg/L)	Sediment (mg/kg dw)	Composite Benthic Invertebrates (mg/kg dw)	Whole-Body Cutthroat Trout (mg/ kg dw)	Trophic Transfer Function (Trout and Invertebrate)
Lotic	Reference	AL4	<1		3.9	4.4	1.1
		EL12	1.4		4.0	6.2	1.5
	Exposed	FO9	16.4		4.4	7.8	1.8
		LI8	22.7		7.8	9.3	1.2
		FO23	19.0		10.0		
		MI5	2.8	5.0	4.0	4.6	1.2
		MI3	1.4		6.2	5.7	0.9
		MI2	7.2		6.7	5.2	0.8
		EL1	5.9		7.1	4.8	0.7
Lentic	Reference	BA6	<1	3.9	3.3	7	2.1
		EL14	<1.5	2.5	4.4	4.5	1.0
	Exposed	FO10	23.2	25.1	17.5	45.9	2.6
		CL11	48.0	6.1	30.9	57.3	1.9
		HA7	25.0	7.9	22.4	21.1	0.9
						Mean	1.4

these systems. Selenium bioaccumulation in fish is significantly higher in lentic than in lotic systems (Orr et al. 2006); BAFs for fish in lentic systems have been reported to be greater than in lotic systems by a factor of 10 or more (Adams et al. 2000).

Trophic transfer functions were calculated for cutthroat trout collected in lotic and lentic habitats of the Elk River watershed (BC, Canada) by dividing whole-body fish Se levels by concentrations measured in benthic invertebrates collected in the same areas (Table 5.6). Trout TTF values ranged from 0.7 to 2.6, with a mean from all sites of 1.4. These values are comparable to the TTFs reported by Presser and Luoma (in press) for the same species (0.93 to 1.25; mean 1.0).

As noted in previous sections, differences in Se concentrations among fish in different locations can be largely ascribed to differences in uptake at the base of the food web and trophic transfer to invertebrates. This linkage between Se concentration in fish and their food web base is illustrated by differences in fish tissue concentrations between areas in Table 5.6, with higher concentrations in fish collected in lentic than in lotic areas, even when water concentrations are relatively similar (e.g., LI8 versus FO10). The higher concentrations in fish from lentic vs. lotic areas might be explained by the fact that organisms associated with lentic area sediments appear to accumulate more Se from water than epilithic organisms in lotic areas (Orr et al. 2006). The higher bioavailability of Se from the water to the base of the food web in lentic environments may be due in part to longer hydraulic retention time in lentic environments allowing for more retention and recycling of organoselenium than in lotic environments (i.e., organoselenium is taken up more efficiently from water than selenate or selenite: Besser et al. 1993; Bowie et al. 1996).

5.2.7.2.2 Size and Age

Unlike other contaminants such as methylmercury, fish size and patterns of growth do not appear to significantly influence Se tissue body burdens, except in juvenile stages. For example, while tissue concentrations of mercury (Hg), cesium (Cs), and thallium (Tl) showed relationships with age, size, and trophic position in Arctic char (*Salvelinus alpinus*), indicating their potential to bioaccumulate and/or biomagnify, Se showed no such relationship (Gantner et al. 2009). Likewise, Se concentrations in adult or juvenile striped bass and white sturgeon collected in San Francisco Bay (CA, USA) showed no significant relationship with total length (Robin Stewart, U.S. Geological Survey, personal communication). In some species and locations, Se concentrations have been shown to vary significantly with length, but the changes are driven by ontenogenetic shifts in fish diets rather than size-specific effects. For example, inverse relationships with fish total length were observed in redear sunfish and Inland silversides in the San Francisco Bay and Delta (CA, USA). The relationships corresponded to shifts in diets of redear sunfish from an open-water–based food source (zooplankton) that was higher in Se to a near-shore–based food source (amphipods) that was lower in Se, while the reverse was true for Inland silversides (Lucas and Stewart 2005). Zhang and Wang (2007) modeled the Se accumulation in the marine juvenile fish *Acanthopagrus schlegeli* and showed that the Se concentration decreased exponentially with an increase in fish length. They also demonstrated that TTF values for these fish decreased (from 1.5 to 0.5) with increasing fish size. The driving force for the observed decreased in TTFs with size appeared to be the decrease in ingestion rates, because the assimilation efficiency of Se increased with increasing fish size (over a size range of 1 to 3 cm) (Zhang and Wang 2007). Currently, the biokinetic parameters available for Se in fish are generally limited to small-sized individuals. This limitation is because these parameters are derived through radiotracer experiments, and the space limits of the gamma detectors used to measure Se radioactivity make it difficult to study larger fish.

5.2.7.2.3 Marine vs. Freshwater

The speciation of trace metals and the permeability of biological membranes change with salinity. For this reason it might be expected that Se toxicity, uptake from water, assimilation from food, and elimination from both exposure routes may be different in marine and freshwater systems. This is the reasoning typically given for the pronounced differences between Se guidelines in freshwater and saline waters. Studies that systematically compare tissue levels in marine and freshwater fish are lacking, but several studies have examined the effects of changes in salinity on Se accumulation and toxicity. For example, Schlenk et al. (2003) showed that increasing salinity (from 0.5 to 13.4 psu) resulted in significantly lower mortality rates in rainbow trout (*Oncorhynchus mykiss*) exposed to dietary seleno-L-methionine. But Ni et al. (2005) showed that salinity affected aqueous, but not dietary, uptake of Se in the intertidal mudskipper (*Periophthalmus cantonensis*). For example, varying the salinity from

10 to 30 psu had no effect on the dietary assimilation of Se or on the efflux rates in these fish. However, concentration factors from aqueous exposure were higher at lower salinities (10 to 20 psu) than at higher salinity (30 psu). Much remains to be learned about this important subject, but at this point it is difficult to support a greater than 10-fold difference in Se guidelines between fresh and salt water based upon Se toxicity differences in fish alone. Indeed, competition from sulfate at uptake sites on membranes in marine systems may be an important factor (Section 5.2.2).

5.2.7.3 Selenium Turnover in Fish Tissues

As noted previously, aqueous Se uptake in fish is so slow as to be negligible ($k_u <$ 0.01 L g^{-1} d^{-1}) (Zhang and Wang 2007). In addition, efflux rates in fish from aqueous exposure are significantly higher than from dietary exposure (Ni et al. 2005), highlighting the importance of dietary exposure in contributing to Se body burdens in fish. Biological half-lives ($t_{1/2}$) for Se in fish can be as short as 7 days, but are more typically on the order of 3 to 4 weeks (Presser and Luoma in press). Zhang and Wang (2007) showed that the biological half-life of Se in the intertidal mudskipper was affected by salinity and exposure route, with higher salinity and dietary exposure leading to the longest $t_{1/2}$ (38.5 d). Trophic transfer studies by Bennett et al. (1986) and Dobbs et al. (1996) showed that, after about 1 week of exposure to Se-enriched rotifers (*Brachionus calyciflorus*), juvenile fathead minnow (*Pimephales promelas*) had Se body burdens approximating that of their diet. Juvenile bluegill consuming Se-laden worms also showed increased tissue Se levels within 1 week, but steady-state equilibration with diet took approximately 100 days (McIntyre et al. 2008). These data suggest that tissue Se levels in fish rapidly begin to reflect dietary levels (e.g., within 1 week), although a steady-state relationship may not be achieved for a period of weeks or months.

5.2.7.4 Periods of Vulnerability

Selenium effects in fish are manifest through maternal transfer to eggs and effects among progeny, so fish are most vulnerable during the period when they are actively developing eggs. This period varies widely in duration and season among fish species (Table 5.7). Many fish species, particularly larger ones, exhibit synchronous

TABLE 5.7
Ovarian Development Periods Relative to Spawning Habitats (Environment Canada 2009)

Reproduction Type	Ovary Development
Synchronous spawners	Starts in late fall for spring spawners. Early to mid-summer for fall spawners.
Multiple spawners, few spawns	Starts 2+ months prior to spawning with maximum development in last 4 weeks.
Multiple spawners, many spawns	Rapid development approximately 4 weeks prior to spawning.
Asynchronous spawners	May occur in as few as 2 weeks prior to initiation of spawning, sometimes longer.
Asynchronous development (year off)	Development only in years when spawning will occur.

spawning based on specific temperature or flow cues (e.g., catastomids, salmo-nids) and initiate ovarian development well in advance (e.g., months) of spawning (Table 5.7). Species exhibiting asynchronous development (e.g., some Arctic char populations) may also show prolonged ovarian development, but only in the years in which spawning occurs. Other species, particularly those that spawn multiple times per year (multiple spawners) or asynchronously, exhibit relatively rapid ovar-ian development just prior to spawning (e.g., within weeks; Table 5.7). Therefore, the period of vulnerability with respect to maternal uptake of Se and transfer to eggs is highly species dependent.

5.2.7.5   Fish Movements

Many types of fish migrate on a regular basis, on time scales ranging from daily to annual, and over distances ranging from a few meters to thousands of kilometers. This migration is often to satisfy dietary or reproductive needs, although the reasons for fish movement are not always known. Such migrations can confound interpreta-tion of dietary contaminant uptake, because measured contaminant body burdens may relate to sources located somewhere other than where the fish are collected. In Se studies, linking tissue Se levels to sources of exposure can be particularly problematic because it is not unusual for migration to occur from preferred forage areas to spawning areas that may be quite distant. Fish may move closer to or far-ther away from Se sources during the period when eggs are being actively devel-oped. Therefore, characterization of Se risks to fish necessitates an understanding of whether sensitive aquatic species are occupying Se-rich habitats at times when eggs are rapidly developing. Species are relatively less vulnerable outside of this period, although juvenile and adult life stages could be affected through direct dietary expo-sure if Se concentrations are sufficiently high.

Residency of fish can be assessed in various ways. Radiotelemetry allows for move-ments of individuals to be tracked over time, but this approach tends to be highly labor intensive and expensive (Brenkman et al. 2007). Comparison of the stable isotope signatures of fish tissues relative to other abiotic and biotic samples collected within an area can also assist in determining site fidelity (Orr et al. 2006). More recently, analysis of life history exposure to Se was assessed by laser ablation–inductively coupled–mass spectrometry of Se in fish otoliths (Palace et al. 2007); concentrations of Se in annual growth zones of the otoliths suggested that fish from a mine-impacted system were recent immigrants from nearby reference streams. Implementation of these or alternative techniques can be highly beneficial in interpretation of data for species for which duration of occupancy in Se-rich areas is uncertain.

5.2.7.6   Tissue Allocations

Relationships among Se concentrations in different fish tissues (e.g., whole body, muscle) can vary widely among fish species within locations and sometimes within species among locations (GEI Consultants et al. 2008). Tissue Se relationships have typically been strong (based on high r^2) for most species within studies, but there have been many exceptions (e.g., almost one-fourth of relationships presented by GEI Consultants et al. (2008) had $r^2 \leq 0.5$). Of 10 fish species for which data were presented, rainbow trout demonstrated the highest concentrations of Se in eggs relative to muscle, while brook trout showed the lowest egg concentrations relative to muscle, indicating that generalizations respecting relative tissue Se allocations cannot be assumed, even between closely related species. Additional data from a study conducted in the Elk River (BC, Canada) gave similar results in that pre-spawning westslope cutthroat trout showed very strong correlations between muscle plug, muscle fillet, ovary, and whole-body Se concentrations, but muscle Se concentrations in pre-spawning mountain whitefish did not strongly correspond to concentrations in either ovary or whole body (Minnow Environmental Inc. et al. 2007). Differences in factors such as the habitats (and therefore diet) utilized during spawning-related migration or the precise stage of egg development among sampled individuals may affect the strength of tissue Se relationships. The rapid uptake of Se (Section 5.2.7.3) from diet and the relatively longer period of time over which Se is redistributed among tissues may influence apparent tissue Se relationships, particularly in settings where there are sharp spatial and temporal gradients.

5.2.7.7 Interactions between Se and Mercury

Significant interaction between Se and Hg was recognized as early as the 1960s (Parizek and Ostadalova 1967; Koeman et al. 1973). It has been known that Se can protect mammals against Hg intoxication (Augier et al. 1993; Glynn et al. 1993; Schlenk et al. 2003), and thus most studies on the interactions between Hg and Se have been conducted in mammalian and fish systems, with few studies on invertebrates. Interaction of Se and Hg may occur in the external environment by complexation or within the metabolic sites after metals are accumulated intracellularly (Amiard-Triquet and Amiard 1998). Possibilities of the protection of Se against Hg toxicity include the redistribution of Hg in the tissues, the competition for binding sites, and the formation of an Hg–Se complex (Cuvin-Aralar and Furness 1991). In addition, the formation of an equimolar Hg–Se complex binding to selenoprotein P may lead to a positive correlation between Hg and Se (Luten et al. 1980; Sasakura and Suzuki 1998).

The interaction of Se and Hg is far from consistent in different studies. For example, there can be either no correlation (Cappon 1981; Lyle 1986; Barghigiani et al. 1991) or a negative correlation (Paulsson and Lundbergh 1991; Chen et al. 2001) between the concentrations of Se and Hg in fish. Sheline and Schmidt-Nielsen (1977) found little effect of Se on the overall body retention of MeHg in the fish *Fundulus heteroclitus*, while Turner and Swick (1983) showed that addition of Se can effectively reduce Hg concentrations in pike (*Esox lucius*) and perch (*Perca flavescens*) with appropriate doses and addition periods. In other field studies, Southworth et al. (2000) found a long-term increase in Hg concentrations in the largemouth bass with a reduction in waterborne Se. A significantly negative correlation between the total Hg concentrations in perch and walleye muscle with the Se concentrations collected from 9 Sudbury (ON, Canada) lakes was also documented (Chen et al. 2001). In the rainbow trout *Oncorhynchus mykiss*, exposure to Se(IV) or Se(VI) in the external medium at 0.075 to 0.75 µM or 5.9 to 59 µg/L did not affect the uptake of MeHg across the perfused gills or its liberation from the gills (Pedersen et al. 1998). However, a Se(IV) or Se(VI) concentration of 7.5 µM or 590 µg/L augmented the MeHg uptake across the gills and internal Se(IV) also increased the efflux of MeHg.

In a more recent study, Mailman (2008) conducted a mesocosm experiment to evaluate the effectiveness of low Se concentrations to lower Hg concentrations in yellow perch. After 8 weeks of exposure, the concentrations of spiked Hg in muscle and liver of fish inversely correlated with Se concentrations in water. Increasing the Se concentrations from about 0.2 to 1.0 μg/L resulted in Hg concentrations in muscle of fish that were 54% lower relative to controls.

Fewer studies have addressed the interaction of Hg and Se in invertebrates (e.g., crabs, starfish, and bivalves; Pelletier 1986; Micallef and Tyler 1987; Sorensen and Bjerregaard 1991; Bjerregaard and Christensen 1993; Larsen and Bjerregaard 1995; Wang et al. 2004) and phytoplankton (Gotsis 1982). Starfish *Asterias rubens* exposed simultaneously to 75 μg/L Se(IV) and 10 μg/ L Hg accumulated more Hg and Se in the tube feet and body wall than did starfish exposed to the two alone, suggesting a synergistic interaction between Se and Hg (Sorensen and Bjerregaard 1991). In the shore crabs *Carcinus maenas*, exposure to Se(IV) either increased the assimilation of MeHg from the food (Bjerregaard and Christensen 1993) or did not consistently alter the AE of MeHg (Larsen and Bjerregaard 1995). In the shrimp *Pandalus borealis*, the biologically incorporated Se in the mussel prey did not apparently affect Hg uptake (Rouleau et al. 1992).

Pelletier (1986) found that Se bioaccumulation in the mussel *Mytilis edulis* increased in the presence of Hg, but Hg bioaccumulation was not affected by various concentrations and chemical species of Se. A synergistic interaction of Se with Hg was documented by Micallef and Tyler (1987). In their study, simultaneous additions of a high concentration of Se were more toxic to the mussel *M. edulis* than the Hg-alone treatment. Patel et al. (1988) found that Se did not offer any protection against the toxic effects of Hg in marine mollusks, and Siegel et al. (1991) demonstrated that there is no consistent difference in the protection against Hg poisoning in a wide variety of organisms (invertebrates, fish, and vascular plants) by different S and Se derivatives. Based on these limited studies, consistent conclusions regarding the interaction of Se and Hg in marine invertebrates are not possible. Reasons for this inconsistency may be the use of different Se species and concentrations in these earlier studies and/or because of the differences in the exposure history of the animals. For phytoplankton, Gotsis (1982) indicated that Se(IV) and Hg(II) interacted in an antagonistic way on the cell growth of alga *Dunaliella minuta* when both were added simultaneously at the beginning of the growth period.

Wang et al. (2004) examined the influences of different concentrations and species of Se (selenite, selenate, seleno-L-methionine) in the ambient environment on the accumulation of Hg(II) and (MeHg) by the diatom *Thalassiosira pseudonana* and the green mussel *Perna viridis*. Aqueous uptake and dietary assimilation of both Hg species were not significantly affected by the different Se(VI) and Se(IV) concentrations (<500 μg/L). In contrast, seleno-L-methionine significantly inhibited the uptake of MeHg and enhanced the uptake of Hg(II) by the diatoms and the mussels at a relatively low concentration (2 μg/L), but did not affect assimilation from the ingested diatoms. One possible reason for the increasing Hg(II) uptake with seleno-L-methionine could be complexation of the Hg(II). The complex may be transported across the membrane at a faster rate because Hg(II) has one of the highest binding affinities with sulfur-containing compounds. The green mussels were exposed to Se(IV) and selenomethionine for different time periods (1 to 5 weeks) to allow the build up of different tissue body burdens of Se, and the accumulation of Hg(II) and MeHg was then quantified. Tissue Se concentrations did not significantly affect the dietary assimilation of Hg, but the influences on the aqueous uptake were variable. These data thus indicated the specificity of the Se–Hg interaction in marine mussels for different Se and Hg species. Heinz and Hoffman (1998) also found that selenomethionine and MeHg interacted in opposite ways in adult and young mallards (*Anas platyrhynchos*). Selenomethionine protected against MeHg toxicity to adult males, but it worsened the effects of MeHg to the young individuals (hatching, survival, and growth).

5.2.8 BIRDS

Birds have been shown to be highly sensitive to Se exposure. Many of the processes that modify exposures in fish apply to birds, but there are important differences that can lead to variable Se exposures in nature. There is limited information on the biodynamics of Se in birds due to the difficulties of conducting laboratory exposures of large organisms. Thus, the vast majority of estimates of uptake and loss have been inferred from field assessments, dietary toxicity tests, or captive breeding studies. Selenium accumulation in birds was not a primary focus of this chapter but is examined in detail relative to toxic effects in Chapter 6. Here we highlight a few of the critical mechanisms known to influence Se accumulation in birds.

5.2.8.1 Feeding Behavior Influences Se Enrichment

In comparing birds from the same geographical region, breeding and developmental stage Se concentrations often vary the most among those species with different diets. In the northern reach of San Francisco Bay (CA, USA), diving ducks feeding on bivalves (surf scoters, scaup) had higher liver Se values than shorebirds feeding on invertebrates in tidal mud flats (avocets, *Recurvirostra americana*) and higher yet than shorebirds feeding on invertebrates in vegetated edge marsh habitats (black-necked stilts, *Himantopus mexicanus*) or Bay piscivorous species (Forester's tern, *Sterna forsteri* and Caspian terns, *Hydroprogne caspia*) (Ackerman and Eagles-Smith 2009). Ackerman and Eagles-Smith (2009) suggest that differences in exposure likely originate from site-specific Se concentrations determined by the habitat and Se concentrations of the prey items known to vary in the region (Stewart et al. 2004).

Bioenergetic models suggest that the relative caloric content of a bird's diet and the bird's metabolic requirements may further play a role in determining Se uptake. DuBowy (1989) used field Se values for a variety of bird diets and estimated ingestion rates based on data in Heinz et al. (1987) to calculate relative exposures from different diets. He predicted that birds consuming vascular plants and algae, which typically have low Se levels relative to invertebrates or fish, are expected to have the highest exposures, due to the higher intake rates of plant material to meet caloric requirements. These patterns appear to be supported to some degree by field data from a closed system that show the highest Se levels in herbivorous marsh birds (e.g., American coots) and lower levels in insectivorous ducks and shorebirds (Ohlendorf et al. 1986).

5.2.8.2 Time Scales of Exposure

Uptake rates and tissue turnover times appear to be substantially higher in bird tissue and eggs than in fish tissues. When birds fed on Se-contaminated diets during the laying season, the exposure was quickly reflected in elevated levels of Se in eggs (Latshaw et al. 2004). Similarly, when the birds were switched to a clean diet, Se concentrations in eggs declined quickly. When mallard hens were fed a diet containing Se at 15 mg/kg dw (as selenomethionine), levels peaked in eggs (to about 43 to 66 mg/kg dw) after about 2 weeks on the treated diet and leveled off at a relatively low level (<16 mg/kg dw) about 10 days after switching to an untreated diet (Heinz 1993).

Because it is the Se in the egg, rather than in the parent bird, that causes developmental abnormalities and death of avian embryos, Se in the egg gives the most sensitive measure for evaluating hazards to birds (Skorupa and Ohlendorf 1991). Given the rapid accumulation and loss patterns of Se in birds (Heinz et al. 1990; Heinz 1993; Heinz and Fitzgerald 1993; Latshaw et al. 2004), Se concentrations in eggs also probably best represent contamination of the local environment. Additional advantages of measuring Se in eggs are that eggs are frequently easier to collect than adult birds, the loss of one egg from a nest probably has little effect on a population, and the egg represents an integration of exposure of the adult female during the few days or weeks before egg laying.

5.2.9 REPTILES AND AMPHIBIANS

There is limited information on the accumulation of Se by reptiles and amphibians. However, available data suggest that amphibians, in particular, accumulate Se efficiently and transfer it to reproductive tissues where reproductive effects have been observed.

In a study of Se accumulation in the food web of a swamp located near a power generation facility in South Carolina (USA), bullfrog larvae (*Rana catesbeiana*) had the highest Se burden (~28 mg/kg dw) of all species examined, followed by clams (*Corbicula fluminea*; 20 mg/kg dw) and other taxa (~5 to 15 mg/kg dw), including snails, aquatic insects, and fish (Unrine et al. 2007). It is unclear why the bullfrog larvae were enriched relative to the other taxa. Bullfrog larvae are thought to be omnivorous feeding on sediments, biofilms (bacteria, diatom, algae, and detritus), and animal tissues, diets that would have been shared to some degree with other species. Presently, there are no biodynamic data that could help determine whether the higher accumulation rates in the bullfrog larvae are due to their foraging ecology or physiology. Other studies on adult anurans from the same site show that not only do adult toads accumulate high levels of Se but they also transfer it efficiently to their eggs (Hopkins et al. 2006). Toads collected from reference and contaminated sites were brought in from the field and placed in experimental uncontaminated mesocosms to document the effects of Se exposures on the hatching success and development of toad larvae (*Gastrophryne carolinensis*). Hopkins et al. (2006) found that

Se concentrations in the tissues of the female toads (up to 42 mg/kg dw) strongly influenced their egg Se concentrations (Log $[Se]_{eggs} = 1.0255 \times Log [Se]_{female} - 0.0448$, $r^2 = 0.94$). Further, the females transferred roughly 53% of their Se body burden to their eggs upon oviposition. This pattern of partitioning was independent of exposure history and was roughly equivalent to the proportion of reproductive tissue to whole-body mass. The maternal transfer of Se by *G. carolinensis* far exceeded that observed for other reptiles, birds, and fish from the same site. Slightly lower percentages (~33%) were observed in female fence lizards (*Sceloporus occidentali*) relative to their ovaries (eggs were not measured) in experimental exposures (Hopkins et al. 2005). As with the bullfrog larvae, it is unclear why the adult toads had such high Se concentrations relative to other species. *Gastrophryne carolinensis* are predaceous, feeding on terrestrial insects, including ants and beetles, which may lead to a more efficient trophic transfer of Se, if these insects accumulate substantial amounts of Se. In another high-Se area (Elk River, BC, Canada), Columbia spotted frog eggs had Se concentrations (12 to 38 mg/kg dw) that were comparable to or lower than concentrations measured in sediments (62 mg/kg dw) or other biota (composited benthic invertebrates 21 mg/kg dw, whole-body longnose sucker 9 to 80 mg/kg dw, red-winged black bird eggs 18 to 20 mg/kg dw) collected from the same area (Minnow Environmental Inc. et al. 2007).

Selenium concentrations appear to be relatively consistent among life stages of anuran species collected from contaminated sites and show patterns that are unlike many other trace elements (Snodgrass et al. 2004; Roe et al. 2005). Selenium concentrations of the southern toad (*Bufo terrestris*) showed slight (but not significant) increases in concentrations moving from larval to metamorph to adult stages, despite shifts in feeding behavior among life stages. Selenium concentrations in southern leopard frog (*Rana sphenocephala*) were similar for larvae and metamorphs and then significantly declined to reference site levels in adults. It is unknown what drives these species-specific differences in Se accumulation among life stages.

Reptiles have also been shown to accumulate high levels of Se from their diet (Hopkins et al. 2004, 2005) and to transfer it to their eggs. In a long-term laboratory exposure study, seleno-D,L-methionine–spiked diets had a significant influence on tissue Se levels of exposed brown house snakes (Hopkins et al. 2004). At the highest Se exposure levels (20 mg/kg dw), snakes had Se concentrations up to 20 mg/kg dw in liver and ovary and up to 30 mg/kg dw in kidney. Selenium content of the eggs was significantly related to dietary exposures and Se tissue levels of the females (Hopkins et al. 2004). Gopher snakes collected from the Se-enriched area of the Kesterson Reservoir (CA, USA) had Se concentrations in their livers ranging from 4.7 to 32 mg/kg dw (mean 11.4), levels significantly higher than those from the nearby reference site in the Volta Wildlife Area (range 1.3 to 3.6 mg/kg dw, mean 2.14) (Ohlendorf et al. 1988). The site-specific differences in Se concentrations of the snakes reflected the site-specific differences in Se concentrations in the bird eggs, a known prey item (Ohlendorf et al. 1988). Selenium concentrations were not significantly correlated with body size, sex, or date of collection. Further, it is unclear whether the large range in Se concentrations was in response to the movement of the snakes between reference and contaminated environments.

Until biodynamic information is obtained for a variety of reptile and amphibian species, it will be difficult to resolve critical mechanisms (i.e., physiology vs. foraging ecology) driving inter- and intra-specific patterns in accumulation. The elevated concentrations observed in this group of organisms highlights the need for further study.

5.3 BIODYNAMIC MODELING OF Se

5.3.1 MODEL IMPROVEMENTS

The biodynamic modeling approach for predicting and understanding trace element accumulation is powerful and accurate when employed with appropriate model parameters (Luoma and Rainbow 2005; Wang and Rainbow 2008). Even simplified model versions predicting trophic Se transfer from IR and AE and/or field-derived data are relatively successful (Presser and Luoma in press). Nevertheless, advances in our understanding of Se uptake from the aqueous phase and from diets as well as dietary relationships (prey selectivity), present an opportunity to further refine biodynamic modeling of Se. In particular, non-linear relationships between dissolved Se and accumulation in microorganisms, non-linear relationships between dissolved Se and uptake in invertebrates, and finally concentration-dependent AEs must be studied further. In particular, such refinements could improve site-specific application of models, or at least should be evaluated with regard to whether they do result in such improvements.

The ubiquitous essentiality of Se is associated with high-affinity transport systems for cellular Se uptake (Section 5.2.2), which effectively ensures sufficient Se uptake even at low ambient or dietary Se concentrations. High-affinity Se carrier systems are also characterized by a limited capacity for uptake which, at least in theory and over a narrow concentration range, results in hyperbolic relationships between uptake and concentration. These qualities of Se uptake pathways may contribute to observed negative correlations between BAFs and exposure concentrations and between BCFs and exposure concentrations (McGeer et al. 2003; DeForest et al. 2007).

5.3.1.1 Uptake Terms K_d and K_u

Uptake of dissolved Se by algae is traditionally described by a K_d relating the ambient and accumulated Se concentrations. This approach assumes a constant enrichment of Se in microorganisms regardless of dissolved Se concentration. At very low Se concentrations, perhaps similar to uncontaminated conditions, this assumption may not accurately reflect Se entry into the base of food webs, one of the most significant parameters for Se accumulation at higher trophic levels. Algae display a hyperbolic (Michaelis-Menten) relationship with higher apparent "K_d" at low concentrations than at higher concentrations (Table 5.1; Riedel et al. 1991; Fournier et al. 2006), illustrating the potential error associated with the concept of constants describing the enrichment of Se by microorganisms.

Similar considerations apply to the uptake of dissolved Se (and most other trace elements) by invertebrates. Although ingestion and assimilation of dietary Se sources dominates Se accumulation in most invertebrates and aquatic vertebrates, some exceptions may exist. In these cases, uptake of dissolved Se by invertebrates has typically been described by a rate constant (K_u), which may not accurately reflect Se accumulation over broad concentration ranges. Indeed, saturation kinetics (hyperbolic, Michaelis-Menten relationships) seems to accurately describe uptake from the water at least by some invertebrates (Martin Grosell, University of Miami, personal communication). For metals such non-linearities typically occur at concentrations that exceed those expected in even contaminated environments (Luoma and Rainbow 2008), but no such systematic evaluation is available for Se.

It is recommended that, when available information allows, an EF rather than a constant K_d is employed to characterize the relationship between ambient Se and that accumulated by microorganisms, and that an uptake rate function rather than a constant K_u is used to describe uptake of dissolved Se at higher trophic levels.

To describe both uptake of dissolved Se by microorganisms and higher trophic levels ($\mu g/g/h$), the following relationship is recommended:

$$\text{Dissolved Se uptake} = \frac{a \times [\text{ambient Se}]}{b + [\text{ambient Se}]} \tag{6}$$

where a and b denote maximal uptake rate ($\mu g/g/h$) and affinity ($\mu g/L$), respectively, of the Se uptake pathway. It is important to apply EFs and uptake rate functions only to a concentration range for which they have been derived. It follows from this recommendation that laboratory experiments to assess the relationship between ambient concentrations and uptake rates should be designed to span the concentrations relevant to the environmental situation of interest.

5.3.1.2 Assimilation from the Diet (AE) and Ingestion Rate (IR)

Limited information is available regarding the relationship between AE and prey Se concentration. However, expectation of interactions between AE and prey Se concentration is dictated by the transport kinetics of intestinal Se transport pathways leading to non-linearity at the lowest (environmentally relevant) concentrations. In theory, at low dietary Se concentrations, high affinity uptake pathways could result in high AE. In contrast, as dietary Se concentrations increase and intestinal carrier systems become saturated, overall AE could decrease. Nevertheless, non-linear relationships between Se concentrations in prey organisms and dietary uptake are not captured in biodynamic modeling efforts that assume AE is a simple constant regardless of exposure concentration. The inverse, non-linear, relationship between dietary Se concentration and AE, predicted from saturable intestinal Se uptake pathways, was first demonstrated for *Daphnia* feeding on 2 different algae species (Guan and Wang 2004). In addition, a recent study of Se assimilation by *A. franciscana* fed a green algal diet revealed very high AEs (~95% at low algal Se concentration), which gradually declined to a minimum (~75%) as dietary Se concentrations increased (Martin Grosell, University of Miami, personal communication).

In addition to the influence of prey Se concentrations, IR and AE are likely to show interactions. Specifically, for constant prey Se concentrations, it can be expected that higher IRs, which may lead to shorter gut passage time, may impose a limitation on Se assimilation from the diet. Limited information is available about the interactions between AE and IRs, which clearly points to an area in need of further attention. The most important research needs with regard to model refinements lie in understanding the concentration ranges over which differences occur relative to concentrations in nature (contaminated and uncontaminated waters) and in considering the magnitude of the differences with regard to how they might affect model predictions.

5.3.2 MODEL OUTCOMES: HYPOTHETICAL SCENARIOS ILLUSTRATE PRINCIPLES

One value of any modeling is to illustrate principles in quantitative terms. Figure 5.3 contrasts Se uptake in 2 food webs: one in which the phytoplankton are purely made up of a species of chlorophyte and the other in which the phytoplankton are purely a species of dinoflagellate. Using data from Baines and Fisher (2001), the concentration of Se in the chlorophyte would be 0.2 mg/kg dw and in dinoflagellate would be 20 mg/kg dw at a selenite concentration of 2 µg/L. In general chlorophytes have among the lowest concentration factors for Se and some dinoflagellates have among the highest concentration factors. In one case, we employ a bivalve with a TTF of 6 (similar to zebra mussels and marine mussels) as the invertebrate and use the average fish TTF of 1.1 (typical of many marine and freshwater fish that have been studied). The model suggests that Se concentrations in mussels could range from 1.2 to 120 mg/kg dw for different algal community situations and in the fish they would range from 1.3 to 132 mg/kg dw. The difference is driven by biological differences in the predominant algal species.

Figure 5.3 also contrasts a mussel-based marine/estuarine food web with a copepod-based food web. The scale is log–log, so the difference between these two is minimized visually. But the calculations show that at 2 mg/kg dw in the dinoflagellate, one would expect 13 mg/kg dw in the estuarine fish feeding on mussels and 4.4 mg/kg dw in the estuarine fish feeding on copepods. These differences are similar to those seen in San Francisco Bay (CA, USA; Stewart et al. 2004).

These simple simulations show that at similar Se concentrations in water, outcomes for fish can differ widely, driven by differences in enrichment at the base

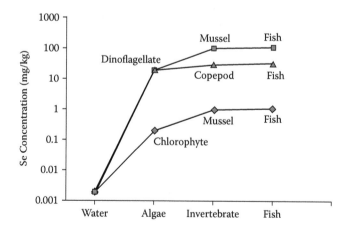

FIGURE 5.3 Selenium accumulation in different species of algae, invertebrates, and fish. Data (TTFs) are for a chlorophyte food web in fresh waters and a dinoflagellate food web in an estuary. Both food webs have a bivalve as the invertebrate and use an average fish TTF of 1.1. The estuarine food web also illustrates the outcome for a copepod with a lower TTF from algae than a mussel.

of the food web and differences in trophic transfer to invertebrates rather than any change in Se concentrations. A choice of a single, universal water quality guideline for Se in these situations would under-protect some aquatic environments and over-protect others in terms of food web exposure to Se.

5.4   SUMMARY

Understanding bioaccumulation and trophic transfer is central to managing ecological risks from Se. The dietary route of exposure generally dominates bioaccumulation processes. This fact has practical implications because the traditional ways of predicting bioaccumulation in animals on the basis of exposure to water concentrations do not work for Se. Further, the predominance of dietary Se exposure pathways mandates that we understand fundamental aspects of Se bioaccumulation in key components of the ecosystems we are trying to protect, from primary producers to top predators. Biodynamic modeling provides a unifying basis for understanding and quantifying dietary uptake and the linkages among food web components.

The single largest step in the bioaccumulation of Se occurs at the base of food webs (Figure 5.1). Primary producers generally concentrate Se from 10^2- to 10^6-fold above ambient dissolved concentrations. We have termed this initial concentrating process the "enrichment function" (EF) because thermodynamic or equilibrium-based constants are not appropriate for describing Se bioaccumulation at the base of food webs. Concentration-dependent EFs are specific to each plant or microbe (particulate material). Uptake of Se by phytoplankton is unlike uptake of trace metals (or organic contaminants). The fact that dead cells do not accumulate or appreciably sorb Se implies that Se bioaccumulation is a non-passive, carrier-mediated process.

Potential to bioaccumulate Se in consumer and predatory animals can be described by a trophic transfer function (TTF; Figure 5.1). TTFs can be derived from established laboratory experimental protocols (biodynamics) or, perhaps with more uncertainty, by using field data to calculate a ratio of the Se concentrations in an animal to Se concentrations in its assumed food. Further, it should be recognized that the TTF can vary with the concentration of Se in the diet due to transport processes in the gastrointestinal tract.

Selenium bioaccumulation by primary producers, invertebrates, and predators varies widely among species. This variation, for animals, is a function of food choice and physiological processes, which can be fundamentally different among taxonomic groups (Figure 5.3). Selenium accumulated by consumer organisms is passed on efficiently to their predators. This finding implies that higher-trophic organisms could be at greater risk in Se-contaminated environments. However, relative to the initial large Se incorporation step at the base of the food web, subsequent transfers to higher trophic levels tend to be smaller. Depending on relative sensitivity to effects, protection of top predators may not guarantee protection of all biota situated lower in the food web.

In light of all of these factors, a single, universal water quality criterion cannot be derived for Se that will protect all aquatic environments with any degree of certainty. Aqueous concentrations of Se that are considered protective in one system may not be protective or attainable in another.

The following knowledge gaps were identified:

1) TTFs in freshwater environments have a relatively high degree of uncertainty because biodynamic parameters for invertebrates and vertebrates are lacking. Therefore, the application of established experimental protocols for dominant freshwater groups (insects and fish) would be highly beneficial. Additionally, there is relatively little information available for fish-to-fish TTFs in both freshwater and marine environments.

2) The variability of TTFs as a function of taxonomy is unclear. Some trends have been identified in marine species, but no such understanding occurs for freshwater taxa. Additional data representing a broad taxonomic range from different ecosystems are required.

3) We need to better understand enrichment at the base of food webs. Specific areas of weakness include our understanding of kinetic processes, particularly saturation kinetics at environmentally relevant concentrations in a wide variety of basal species. Additionally, data for Se uptake into and trophic transfer from bacteria are practically absent for both freshwater and marine systems. Finally, protocols for isolating biotic from abiotic components of suspended particles and bottom sediments would improve model inputs representing Se concentrations at the base of the food web.

4) The bioavailability of selenate to freshwater primary producers deserves more study. In marine systems, the relative abundance of sulfate makes selenate uptake into primary producers relatively unimportant. In freshwaters, this may not be the case.

5) Inter-organ transfers and thus distributions of Se in fish are obviously key mediators of toxicity, but inter-species differences in inter-organ distributions, their variability, and their relevance to reproductive toxicity, remain poorly understood.

5.5   REFERENCES

Ackerman JT, Eagles-Smith CA. 2009. Selenium bioaccumulation and body condition in shorebirds and terns breeding in San Francisco Bay. *Environ Toxicol Chem* 28: 2134–2141.

Adams WJ, Toll JE, Brix KV, Tear LM, DeForest DK. 2000. Site-specific approach for setting water quality criteria for selenium: differences between lotic and lentic systems. Proceedings of the 24th Annual British Columbia Mine Reclamation Symposium; 2000 Jun 21–22. Williams Lake (BC, Canada): Ministry of Energy and Mines. p 231–141.

Amiard-Triquet A, Amiard J-C. 1998. Influence of ecological factors on accumulation of metal mixtures. In: Langston WJ, Bebianno M-J, editors. Metal metabolism in aquatic environments. London (UK): Chapman and Hall. p 351–386.

Anastasia JR, Morgan SG, Fisher NS. 1998. Tagging crustacean larvae: assimilation and retention of trace elements. *Limnol Oceanogr* 43:362–368.

Andrahennadi R, Wayland M, Pickering IJ. 2007. Speciation of selenium in stream insects using X-ray absorption spectroscopy. *Environ Sci Technol* 41:7683–7687.

Augier H, Benkoel L, Chamlian A, Park WK, Ronneau C. 1993. Mercury, zinc and selenium bioaccumulation in tissues and organs of Mediterranean striped dolphins *Stenella coeruleoalba meyen*. Toxicological result of their interaction. *Cell Mol Biol* 39:621–634.

Baines SB, Fisher NS. 2001. Interspecific differences in the bioconcentration of selenite by phtyoplankton and their ecological implications. *Mar Ecol Progr Ser* 213:1–12.

Baines SB, Doblin MA, Fisher NS, Cutter GA. 2001. Uptake of dissolved organic selenides by marine phytoplankton. *Limnol Oceanogr* 46:1936–1944.

Baines SB, Fisher NS, Stewart R. 2002. Assimilation and retention of Se and other trace elements from crustacean food by juvenile striped bass (*Morone saxatilis*). *Limnol Oceanogr* 47:646–655.

Baines SB, Fisher NS, Doblin MA, Cutter GA, Cutter LS, Cole B. 2004. Light dependence of selenium uptake by phytoplankton and implications for predicting selenium incorporation into food webs. *Limnol Oceanogr* 49:566–578.

Barghigiani G, Pellegrini D, D'Ulivo A, De Ranieri S. 1991. Mercury assessment and its relation to selenium levels in edible species of the northern Tyrrhenian sea. *Mar Pollut Bull* 22:406–409.

Bennett WN, Brooks AS, Boraas ME. 1986. Selenium uptake and transfer in an aquatic food chain and its effects on fathead minnow larvae. *Arch Environ Contam Toxicol* 15:513–517.

Besser JM, Canfield TJ, La Point TW. 1993. Bioaccumulation of organic and inorganic selenium in a laboratory food chain. *Environ Toxicol Chem* 12:57–72.

Besser JM, Huckins JN, Clark RC. 1994. Separation of selenium species released from Se-exposed algae. *Chemosphere* 29:771–780.

Bjerregaard P, Christensen L. 1993. Accumulation of organic and inorganic mercury from food in the tissues of *Carcinus maenas*: effect of waterborne selenium. *Mar Ecol Progr Ser* 99:271–281.

Bottino NR, Banks CH, Irgolic KJ, Micks P, Wheeler AE, Zingaro RA. 1984. Selenium containing amino acids and proteins in marine algae. *Phytochemistry* 23:2445–2452.

Bowie GL, Sanders JG, Riedel GF, Gilmour CC, Breitburg DL, Cutter GA, Porcella DB. 1996. Assessing selenium cycling and accumulation in aquatic ecosystems. *Water Air Soil Pollut* 90:93–104.

Brenkman SJ, Corbett SC, Volk EC. 2007. Use of otolith chemistry and radiotelemetry to determine age-specific migratory patterns of anadromous bull trout in the Hoh River, Washington. *Trans Am Fish Soc* 136:1–11.

Brix KV, Volosin JS, Adams WJ, Reash RJ, Carlton RG, McIntyre DO. 2001. Effects of sulfate on the acute toxicity of selenate to freshwater organisms. *Environ Toxicol Chem* 20:1037–1045.

Broecker WS, Peng TH. 1982. Tracers in the sea. New York (NY, USA): Eldigio Pr.

Brown TA, Shrift A. 1980. Assimilation of selenate and selenite by *Salmonella typhimurium*. *Can J Fish Aquat Sci* 26 671–675.

Brown TA, Shrift A. 1982. Selenium: toxicity and tolerance in higher plants. *Biol Rev Cambr Philos Soc* 57:59–84.

Buchwalter DB, Cain DJ, Martin CA, Xie L, Luoma SN, Garland T Jr. 2008. Aquatic insect ecophysiological traits reveal phylogenetically based differences in dissolved cadmium susceptibility. *Proc Natl Acad Sci USA* 105:8321–8326.

Bügel S, Larsen EH, Sloth JJ, Flytlie K, Overvad K, Steenberg LC, Moesgaard S. 2008. Absorption, excretion, and retention of selenium from a high selenium yeast in men with a high intake of selenium. *Food Nutr Res* 52. DOI: I0.3402/fnr.v52iO.I642.

Cappon CJ. 1981. Mercury and selenium content and chemical form in vegetable crops grown on sludge-amended soil. *Arch Environ Contam Toxicol* 10:673–689.

Castellano S, Lobanov AV, Chappel C, Novoselov SV, Albrecht M, Hua D, Lescura A, Lengauer T, Krol A, Gladyshev VN, Guiro R. 2005. Diversity of functional plasticity of eukaryotic selenoproteins: identification and characterization of the *SelJ* family. *Proc Natl Acad Sci USA* 102:16188–16193.

Chen YW, Belzile N, Gunn JM. 2001. Antagonistic effect of selenium on mercury assimilation by fish populations near Sudbury metal smelters? *Limnol Oceanogr* 46:1814–1818.

Chisholm SW. 1992. Phytoplankton size. In: Falkowski G, Woodhead AD, editors. Primary productivity and biogeochemical cycles in the sea. New York (NY, USA): Plenum. p 213–237.

Cloern JE, Dufford R. 2005. Phytoplankton community ecology: principles applied in San Francisco Bay. *Mar Ecol Progr Ser* 285:11–28.

Conley JM, Funk DH, Buchwalter DB. 2009. Selenium bioaccumulation and maternal transfer in the mayfly *Centroptilum triangulifer* in a life-cycle, periphyton-biofilm trophic assay. *Environ Sci Technol* 43: 7952–7957.

Coyle JJ, Buckler DR, Ingersoll CG, Fairchild JF, May TW. 1993. Effect of dietary selenium on the reproductive success of bluegills (*Lepomis macrochirus*). *Environ Toxicol Chem* 12:551–565.

Croteau MN, Luoma SN. 2008. A biodynamic understanding of dietborne metal uptake by a freshwater invertebrate. *Environ Sci Technol* 42:1801–1806.

Cutter GA. 1989. The estuarine behavior of selenium in San Francisco Bay. *Estuar Coast Shelf Sci* 28:13–34.

Cutter GA, Bruland KW. 1984. The marine biogeochemistry of selenium: a re-evaluation. *Limnol Oceanogr* 29:1179–1192.

Cutter GA, Cutter LS. 1995. Behavior of dissolved antimony, arsenic, and selenium in the Atlantic Ocean. *Mar Chem* 49:295–306.

Cutter GA, Cutter LS. 1998. Metalloids in the high latitude north Atlantic Ocean: sources and internal cycling. *Mar Chem* 61:25–36.

Cuvin-Aralar MLA, Furness RW. 1991. Mercury and selenium interaction: a review. *Ecotoxicol Environ Saf* 21:348–364.

DeForest DK, Brix KV, Adams WJ. 2007. Assessing metal bioaccumulation in aquatic environments: the inverse relationship between bioaccumulation factors, trophic transfer factors and exposure concentrations. *Aquat Toxicol* 84:236–246.

Dobbs MG, Cherry DS, Cairns J Jr. 1996. Toxicity and bioaccumulation of selenium to a three-trophic level food chain. *Environ Toxicol Chem* 15:340–347.

Doblin MA, Baines SB, Cutter LS, Cutter GA. 2006. Sources and biogeochemical cycling of particulate selenium in the San Francisco Bay estuary. *Estuar Coast Shelf Sci* 67:681–694.

Doucette GJ, Price NM, Harrison PJ. 1987. Effects of selenium deficiency on the morphology and ultrastructure of the coastal marine diatom *Thalassiosira pseudonana* (Bacillariophyceae). *J Phycol* 23:9–17.

Dubois M, Hare L. 2009. Selenium assimilation and loss by an insect predator and its relationship to Se subcellular partitioning in two prey types. *Environ Pollut* 157:772–777.

Dubowy PJ. 1989. Effects of diet on selenium bioaccumulation in marsh birds. *J Wildl Manage* 53:776–781.

Environment Canada. 2009. Technical guidance document for environmental effects monitoring for pulp and paper mills. Ottawa, ON, Canada.

Fels IG, Cheldelin VH. 1949. Selenate inhibition studies. III. The role of sulfate in selenate toxicity in yeast. *Arch Biochem* 22:402–405.

Fisher NS, Reinfelder JR. 1991. Assimilation of selenium in the marine copepod *Acartia tonsa* studied with a radiotracer ratio method. *Mar Ecol Progr Ser* 70:157–164.

Fisher NS, Wente M. 1993. The release of trace elements by dying marine phytoplankton. *Deep-Sea Res* 40:671–694.

Fisher NS, Reinfelder JR. 1995. The trophic transfer of metals in marine systems. In: Tessier A, Turner DR, editors. Metal speciation and bioavailability in aquatic systems. Chichester (UK): J Wiley. p 363–406.

Fisher NS, Stupakoff I, Sanudo-Wilhelmy S, Wang WX, Teyssie JL, Fowler SW, Crusius J. 2000. Trace metals in marine copepods: a field test of a bioaccumulation model coupled to laboratory uptake kinetics data. *Mar Ecol Progr Ser* 194:211–218.

Foda A, Vandermeulen JH, Wrench JJ. 1983. Uptake and conversion of selenium by a marine bacterium. *Can J Fish Aquat Sci* 40:215–220.

Forsythe BL, Klaine SJ. 1994. The interaction of sulfate and selenate (Se+6) effects on brine shrimp, *Artemia* spp. *Chemosphere* 29:789–800.

Fournier E, Adam C, Massabuau JC, Garnier-Laplace J. 2006. Selenium bioaccumulation in *Chlamydomonas reinhardtii* and subsequent transfer to *Corbicula fluminea*: role of selenium speciation and bivalve ventilation. *Environ Toxicol Chem* 25:2692–2699.

Gantner N, Power M, Babaluk JA, Reist JD, Köck G, Lockhart LW, Solomon KR, Muir DCG. 2009. Temporal trends of mercury, cesium, potassium, selenium, and thallium in Arctic char (*Salvelinus alpinus*) from Lake Hazen, Nunavut, Canada: effects of trophic position, size, and age. *Environ Toxicol Chem* 28:254–263.

Gatlin DM, Wilson RP. 1984. Dietary selenium requirement of fingerling channel catfish. *J Nutr* 114:627–633.

GEI Consultants, Golder Associates, Parametrix & University of Saskatchewan Toxicology Center. 2008. Selenium tissue thresholds: tissue selection criteria, threshold development endpoints, and potential to predict population or community effects in the field. Report prepared for the North American Metals Council – Selenium Working Group, Washington, DC, Canada.

Glynn AW, Ilback NG, Brabencova D, Carlsson L, Enqvist EC, Netzel E, Oskarsson A. 1993. Influence of sodium selenite on 203Hg absorption, distribution, and elimination in male mice exposed to methyl203Hg. *Biol Trace Elem Res* 39:91–107.

Gobler CJ, Hutchins DA, Fisher NS, Cosper EM, Sanudo-Wilhelmy SA. 1997. Release and bioavailability of C, N, P, Se, and Fe following viral lysis of a marine chrysophyte. *Limnol Oceanogr* 42:1492–1504.

Gotsis O. 1982. Combined effects of selenium/mercury and selenium/copper on the cell population of the alga *Dunaliella minuta*. *Mar Biol* 71:217–222.

Guan R, Wang WX. 2004. Dietary assimilation and elimination of Cd, Se, and Zn by *Daphnia magna* at different metal concentrations. *Environ Toxicol Chem* 23:2689–2698.

Hamilton SJ, Buhl KJ, Faerber NL, Wiedmeyer RH, Bullard FA. 1990. Toxicity of organic selenium in the diet to chinook salmon. *Environ Toxicol Chem* 9:347–358.

Hansen LD, Maier KJ, Knight AW. 1993. The effect of sulfate on the bioconcentration of selenate by *Chironomus decorus* and *Daphnia magna*. *Arch Environ Contam Toxicol* 25:72–78.

Harrison PJ, Yu PW, Thompson PA, Price NM, Phillips DJ. 1988. Survey of selenium requirements in marine phytoplankton. *Mar Ecol Progr Ser* 47:89–96.

Hatfield DL, Gladyshev VN, Park SI. 1999. Biosynthesis of selenocystein and its incorporation into protein as the 21st amino acid. *Comp Nat Prod Chem* 4:353–380.

Heinz GH. 1993. Selenium accumulation and loss in mallard eggs. *Environ Toxicol Chem* 12:775–778.

Heinz GH, Fitzgerald MA. 1993. Reproduction of mallards following overwinter exposure to selenium. *Environ Pollut* 81:117–122.

Heinz GH, Hoffman DJ. 1998. Methylmercury chloride and selenomethionine interactions on health and reproduction in mallards. *Environ Toxicol Chem* 17:139–145.

Heinz GH, Hoffman DJ, Krynitsky AJ, Weller DMG. 1987. Reproduction in mallards fed selenium. *Environ Toxicol Chem* 6:423–433.

Heinz GH, Pendleton GW, Krynitsky J, Gold LG. 1990. Selenium accumulation and elimination in mallards. *Arch Environ Contam Toxicol* 19:374–379.

Hermanutz RO. 1992. Malformation of the fathead minnow (*Pimephales promelas*) in an ecosystem with elevated selenium concentrations. *Bull Environ Contam Toxicol* 49:290–294.

Hermanutz RO, Allen KN, Roush TH, Hedtke SF. 1992. Effects of elevated selenium concentrations on bluegills (*Lepomis macrochirus*) in outdoor experimental streams. *Environ Toxicol Chem* 11:217–224.

Hilton JW, Hodson PV, Slinger SJ. 1980. The requirement and toxicity of selenium in rainbow trout (*Salmo gairdneri*). *J Nutr* 110:2527–2535.

Holm J. 2002. Sublethal effects of selenium on rainbow trout (*Orcornhynchus mykiss*) and brook trout (*Salvelinus fortinalis*) [MSc thesis]. Winnipeg (MB, Canada): Univ of Manitoba.

Hook SE, Fisher NS. 2001. Sublethal effects of silver in zooplankton: importance of exposure pathways and implications for toxicity testing. *Environ Toxicol Chem* 20:568–574.

Hopkins WA, Staub BP, Baionno JA, Jackson BP, Roe JH, Ford NB. 2004. Trophic and maternal transfer of selenium in brown house snakes (*Lamprophis fuliginosus*). *Ecotoxicol Environ Saf* 58:285–293.

Hopkins WA, Staub BP, Baionno JA, Jackson BP, Talent LG. 2005. Transfer of selenium from prey to predators in a simulated terrestrial food chain. *Environ Pollut* 134:447–456.

Hopkins WA, DuRant SE, Staub BP, Rowe CL, Jackson BP. 2006. Reproduction, embryonic development, and maternal transfer of contaminants in the amphibian *Gastrophryne carolinensis*. *Environ Health Perspect* 114:661–666.

Hu MH, Yang YP, Martin JM, Yin K, Harrison PJ. 1997. Preferential uptake of Se(IV) over Se(VI) and the production of dissolved organic Se by marine phytoplankton. *Mar Environ Res* 44:225–231.

Hurd-Karrer AM. 1938. Relation of sulfate to selenium absorption by plants. *Am J Bot* 25:666–675.

Keating KI, Dagbusan BC. 1984. Effect of selenium deficiency on cuticle integrity in the Cladocera (Crustacea). *Proc Natl Acad Sci USA* 81:3433–3437.

Koeman JH, Peeters WHM, Koudstaal Hol CHM. 1973. Mercury selenium correlations in marine mammals. *Nature* 245:385–386.

Kryukov GV, Gladyshev VN. 2004. The prokaryotic seleno-proteome. *EMBO J* 5:538–543.

Kryukov GV, Castellano S, Novoselov SV, Lobanov AV, Zehtab O, Guigo R, Gladyshev VN. 2003. Characterization of mammalian selenoproteomes. *Science* 300:1439–1443.

Larsen LF, Bjerregaard F. 1995. The effect of selenium on the handling of mercury in the shore crab *Carcinus maenas*. *Mar Pollut Bull* 31:78–83.

Latshaw JD, Morishita TY, Sarver CF, Thilsted J. 2004. Selenium toxicity in breeding ring-necked pheasants (*Phasianus colchicus*). *Avian Dis* 48:935–939.

Lawrence GS, Chapman PM. 2007. Human health risks of selenium-contaminated fish: a case study for risk assessment of essential elements. *Human Ecol Risk Assess* 13:1192–1213.

Lee BG, Fisher NS. 1992a. Degradation and elemental release rates from phytoplankton debris and their geochemical implications. *Limnol Oceanogr* 37:1345–1360.

Lee BG, Fisher NS. 1992b. Decomposition and release of elements from zooplankton debris. *Mar Ecol Progr Ser* 88:117–128.

Lee BG, Fisher NS. 1993. Release rates of trace elements and protein from decomposing planktonic debris. 1. Phytoplankton debris. *J Mar Res* 51:391–421.

Lee BG, Lee JS, Luoma SN. 2006. Comparison of selenium bioaccumulation in the clams *Corbicula fluminea* and *Potamocorbula amurensis*: a bioenergetic modeling approach. *Environ Toxicol Chem* 25:1933–1940.

Lemly AD. 1993. Teratogenic effects of selenium in natural populations of freshwater fish. *Ecotox Environ Saf* 26:181–204.

Lemly AD, Dimmick JF. 1982. Phytoplankton communities in the littoral zone of lakes: observations on structure and dynamics in oligotrophic and eutrophic systems. *Oecologia* 54:359–369.

Lin YH, Shiau SY. 2005. Dietary selenium requirements of juvenile grouper, *Epinephelus malabarious*. *Aquaculture* 250:356–363.

Lucas LV, Stewart AR. 2005. Transport, transformation, and effects of selenium and carbon in the delta of the Sacramento-San Joaquin Rivers: implications for ecosystem restoration. Final Report. Project No. ERP-01-C07. Sacramento (CA, USA): California Bay Delta Authority.

Luoma SN, Fisher N. 1997. Uncertainties in assessing contaminant exposure from sediments: bioavailability. In: Ingersoll CG, Dillon T, Biddinger GR, editors. Ecological risk assessment of contaminated sediments. Pensacola (FL, USA): SETAC Pr. p 211–237.

Luoma SN, Rainbow PS. 2005. Why is metal bioaccumulation so variable? *Environ Sci Technol* 39:1921–1931.

Luoma SN, Rainbow PS. 2008. Metal contamination in aquatic environments. Cambridge (UK): Cambridge Univ Pr.

Luoma SN, Presser TS. 2009. Emerging opportunities in management of selenium contamination. *Environ Sci Technol* 43: 8483–8487.

Luoma SN, Johns C, Fisher NS, Steinberg NA, Oremland RS, Reinfelder JR. 1992. Determination of selenium bioavailability to a benthic bivalve from particulate and solute pathways. *Environ Sci Technol* 26:485–491.

Luten JB, Ruiter A, Ritskes TM, Rauchbaar AB, Riekwel-Booy G. 1980. Mercury and selenium in marine and freshwater fish. *J Food Sci* 45:416–419.

Lyle JM. 1986. Mercury and selenium concentrations in sharks from northern Australian waters. *Austr J Mar Freshw Res* 37:309–321.

Maier KJ, Foe CE, Knight AW. 1993. Comparative toxicity of selenate, selenite, seleno-DL-methionine and seleno-DL-cyctine to *Daphnia magna*. *Environ Toxicol Chem* 12:72–78.

Mailman M. 2008. Assessment of mercury and selenium interactions in freshwater [PhD dissertation]. Winnipeg (MB, Canada): Univ of Manitoba.

Manahan DT. 1990. Adaptations by invertebrate larvae for nutrient acquisition from seawater. *Amer Zool* 30:147–160.

Margalef R. 1978. Life-forms of phytoplankton as survival alternatives in an unstable environment. *Oceanol Acta* 1:493–509.

Mathews T, Fisher NS. 2008. Trophic transfer of seven trace metals in a four-step marine food chain. *Mar Ecol Progr Ser* 367:23–33.

McGeer JC, Brix KV, Skeaff JM, DeForest DK, Brigham SI, Adams WJ, Green A. 2003. Inverse relationship between bioconcentration factor and exposure concentration for metals: implications for hazard assessment of metals in the aquatic environment. *Environ Toxicol Chem* 22:1017–1037.

McIntyre DO, Pacheco MA, Garton MW, Wallschläger D, Delos CG. 2008. Effect of selenium on juvenile bluegill sunfish at reducted temperature. Washington (DC, USA): Health and Ecological Criteria Division, Office of Water, US Environmental Protection Agency. EPA-822-R-08-020.

Measures CI, Burton JD. 1980. The vertical distribution and oxidation states of dissolved selenium in the northeast Atlantic Ocean and their relationship to biological processes. *Earth Planet Sci Lett* 46:385–396.

Micallef S, Tyler PA. 1987. Preliminary observations of the interactions of mercury and selenium in *Mytilus edulis*. *Mar Pollut Bull* 18:180–185.

Minnow Environmental Inc, Interior Reforestation Co Ltd, Paine, Ledge and Associates. 2007. Selenium monitoring in the Elk River Watershed, B.C. Sparwood (BC, Canada): Elk Valley Selenium Task Force.

Miyauchi S, Srinivas SR, Fei YJ, Gopal E, Umapathy NS, Wang H, Conway SJ, Ganapathy V, Prasad PD. 2006. Functional characteristics of NaS2, a placenta-specific Na+-coupled transporter for sulfate and oxyanions of the micronutrients selenium and chromium. *Placenta* 27:550–559.

Muscatello JR, Janz DM. 2009. Selenium accumulation in aquatic biota downstream of a uranium mining and milling operation. *Sci Tot Environ* 407:1318–1325.

Muscatello JR, Belknap AM, Janz DM. 2008. Accumulation of selenium in aquatic systems downstream of a uranium mining operation in northern Saskatchewan, Canada. *Environ Pollut* 156:387–393.

Ni IH, Chan SM, Wang WX. 2005. Influences of salinity on the biokinetics of Cd, Se, and Zn in the intertidal mudskipper *Periophthalmus cantonensis*. *Chemosphere* 61:1607–1617.

Nickel A, Kottra G, Schmidt G, Danier J, Hofmann T, Daniel H. 2009. Characterization of transport of selenoamino acids by epithelial amino acid transporters. *Chem Biol Interact* 177:234–241.

Novoselov SV, Roa M, Onoshko NV, Zhi H, Kryukov GV, Xiang Y, Hatfield DL, Gladyshev VN. 2002. Selenoproteins and selenocysteine insertion systems in the model plant cell system, *Chlamydomonas reinhardtii*. *EMBO J* 21:3681–3693.

Ogle RS, Knight AW. 1996. Selenium bioaccumulation in aquatic ecosystems. 1. Effects of sulfate on the uptake and toxicity of selenate in *Daphnia magna*. *Arch Environ Contam Toxicol* 30:274–279.

Ohlendorf HM. 2002. The birds of Kesterson Reservoir: a historical perspective. *Aquat Toxicol* 57:1–10.

Ohlendorf HM, Hoffman DJ, Saiki MK, Aldrich TW. 1986. Embryonic mortality and abnormalities of aquatic birds; apparent impacts by selenium from irrigation drainwater. *Sci Total Environ* 52:49–63.

Ohlendorf HM, Hothem RL, Aldrich TW. 1988. Bioaccumulation of selenium by snakes and frogs in the San Joaquin Valley, California. *Copeia* 1988:704–710.

Orr PL, Guiguer KR, Russel CK. 2006. Food chain transfer of selenium in lentic and lotic habitats of a western Canadian watershed. *Ecotoxicol Environ Saf* 63:175–188.

Palace VP, Halden NM, Yang P, Evans RE, Sterling G. 2007. Determining residence patterns of rainbow trout using laser ablation inductively coupled plasma mass spectrometry (LA-ICP-MS) analysis of selenium in otoliths. *Environ Sci Technol* 41:3679–3683.

Parizek J, Ostadalova I. 1967. The protective effects of small amounts of selenite in sublimate intoxication. *Experientia* 23:142–143.

Patel B, Chandy JP, Patel S. 1988. Do selenium and glutathione inhibit the toxic effects of mercury in marine lamellibranchs? *Sci Total Environ* 76:147–165.

Paulsson K, Lundbergh K. 1991. Treatment of mercury contaminated fish by selenium addition. *Water Air Soil Pollut* 56:833–841.

Pedersen TV, Block M, Part P. 1998. Effect of selenium on the uptake of methyl mercury across perfused gills of rainbow trout *Oncorhynchus mykiss*. *Aquat Toxicol* 40:361–373.

Pelletier E. 1986. Changes in bioaccumulation of selenium in *Mytilus edulis* in the presence of organic and inorganic mercury. *Can J Fish Aquat Sci* 43:203.

Phillips DJH, Rainbow PS. 1993. Biomonitoring of trace aquatic contaminants. Essex (UK): Elsevier.

Presser T, Luoma SN. In press. A methodology for ecosystem-scale modeling of selenium. *Integr Environ Assess Manage.*

Price NM, Harrison PJ. 1988. Specific selenium-containing macromolecules in the marine diatom *Thalassiosira pseudonana*. *Plant Physiol* 86:192–199.

Price NM, Thompson PA, Harrison PJ. 1987. Selenium: an essential element for growth of the coastal marine diatom *Thalassiosira pseudonana* (Bacillariophyceae). *J Phycol* 23:1–9.

Puls R. 1988. Mineral levels in animal health. Clearbrook (BC, Canada): Sherpa International.

Rainbow PS, Wang WX. 2001. Comparative assimilation of Cd, Cr, Se, and Zn by the barnacle *Elminius modestus* from phytoplankton and zooplankton diets. *Mar Ecol Progr Ser* 218:239–248.

Rayman MP. 2002. The argument for increasing selenium intake. *Proc Nutr Soc* 61:203–215.

Reinfelder JR, Fisher NS. 1991. The assimilation of elements ingested by marine copepods. *Science* 215:794–796.

Reinfelder JR, Fisher NS. 1994a. The assimilation of elements ingested by marine planktonic bivalve larvae. *Limnol Oceanogr* 39:12–20.

Reinfelder JR, Fisher NS. 1994b. Retention of elements absorbed by juvenile fish (*Menidia menidia, M. beryllina*) from zooplankton prey. *Limnol Oceanogr* 39:1783–1789.

Reinfelder JR, Wang WX, Luoma SN, Fisher NS. 1997. Assimilation efficiencies and turnover rates of trace elements in marine bivalves: a comparison of oysters, clams and mussels. *Mar Biol* 129:443–452.

Riedel GF, Sanders JG. 1996. The influence of pH and media composition on the uptake of inorganic selenium by *Chlamydomonas reinhardtii*. *Environ Toxicol Chem* 15:1577–1583.

Riedel GF, Ferrier DP, Sanders JG. 1991. Uptake of selenium by fresh-water phytoplankton. *Water Air Soil Pollut* 57:23–30.

Riedel GF, Sanders JG, Gilmour CC. 1996. Uptake, transformation, and impact of selenium in freshwater phytoplankton and bacterioplankton communities. *Aquat Microb Ecol* 11:43–51.

Roditi HA, Fisher NS. 1999. Rates and routes of trace element uptake in zebra mussels. *Limnol Oceanogr* 44:1730–1749.

Roditi HA, Fisher NS, Sanudo-Wilhelmy SA. 2000a. Field testing a metal bioaccumulation model for zebra mussels. *Environ Sci Technol* 34:2817–2825.

Roditi HA, Fisher NS, Sanudo-Willhelmy SA. 2000b. Uptake of dissolved organic carbon and trace elements by zebra mussels. *Environ Sci Technol* 34:2817–2825.

Roe JH, Hopkins WA, Jackson BP. 2005. Species- and stage-specific differences in trace element tissue concentrations in amphibians: implications for the disposal of coal-combustion wastes. *Environ Pollut* 136:353–363.

Rouleau C, Pelletier E, Pellerin-Massicotte J. 1992. Uptake of organic mercury and selenium from food by nordic shrimp *Pandalus borealis*. *Chem Spec Bioavail* 4:75–81.

Sasakura C, Suzuki KT. 1998. Biological interaction between transition metals (Ag, Cd and Hg), selenide/sulfide and selenoprotein P. *J Inorg Biochem* 71:159–162.

Schlenk D, Zubcov N, Zubcov E. 2003. Effects of salinity on the uptake, biotransformation, and toxicity of dietary seleno-L-methionine to rainbow trout. *Toxicol Sci* 75:309–313.

Schubert A, Holden JM, Wolf WR. 1987. Selenium content of a core group of foods based on a critical evaluation of published analytical data. *J Am Diet Assoc* 87:285–296+299.

Scott WB, Crossman EJ. 1973. Freshwater fishes of Canada. Ottawa (ON, Canada): Fisheries Research Board of Canada. Bulletin 184.

Sheline J, Schmidt-Nielsen B. 1977. Methylmercury-selenium: interaction in the killifish, *Fundulus heteroclitus*. In: Vernberg FJ, Calabrese A, Thurberg FP, Vernberg WB, editors. Physiological responses of marine biota to pollutants. New York (NY, USA): Academic Pr. p 188–193.

Shrift A. 1954. Sulfur-selenium antagonism. 1. Antimetabolite action of selenate on the growth of *Chlorella vulgaris*. *Am J Bot* 41:223–230.

Siegel BZ, Siegel SM, Correa T, Dagan C, Galvez G, LeeLoy L, Padua A, Yaeger E. 1991. The protection of invertebrates, fish, and vascular plants against inorganic mercury poisoning by sulfur and selenium derivatives. *Arch Environ Contam Toxicol* 20:241–246.

Skorupa JP, Ohlendorf HM. 1991. Contaminants in drainage water and avian risk threshold. In: Dinar A, Zilberman D, editors. The economics and management of water and drainage in agriculture. Boston (MA, USA): Kluwer. p 345–368.

Smetacek V, Assmy P, Henjes J. 2004. The role of grazing in structuring Southern Ocean pelagic ecosystems and biogeochemical cycles. *Antarctic Sci* 16:541–558.

Snodgrass JW, Hopkins WA, Broughton J, Gwinn D, Baionno JA, Burger J. 2004. Species-specific responses of developing anurans to coal combustion wastes. *Aquat Toxicol* 66:171–182.

Sorensen M, Bjerregaard P. 1991. Interactive accumulation of mercury and selenium in the sea star *Asterias rubens*. *Mar Biol* 108:269–276.

Southworth GR, Peterson MJ, Ryon MG. 2000. Long-term increased bioaccumulation of mercury in largemouth bass follows reduction of waterborne selenium. *Chemosphere* 41:1101–1105.

Stewart AR, Luoma SN, Schlekat CE, Doblin MA, Hieb KA. 2004. Food web pathway determines how selenium affects aquatic ecosystems: a San Francisco Bay case study. *Environ Sci Technol* 38:4519–4526.

Thompson JN, Erdody P, Smith DC. 1975. Selenium content of food consumed by Canadians. *J Nutr* 105:274–277.

Tian Y, Liu F. 1993. Selenium requirements of shrimp *Panaeus chinensis. Chin J Oceanol Limnol* 11:249–253.

Tsui MTK, Wang WX. 2007. Biokinetics and tolerance development of toxic metals in *Daphnia magna. Environ Toxicol Chem* 26:1023–1032.

Turner MA, Swick AL. 1983. The English-Wabigoon River system: IV. Interaction between mercury and selenium accumulated from waterborne and dietary sources by northern pike (*Esox lucius*). *Can J Fish Aquat Sci* 40:2241–2250.

Unrine JM, Hopkins WA, Romanek CS, Jackson BP. 2007. Bioaccumulation of trace elements in omnivorous amphibian larvae: implications for amphibian health and contaminant transport. *Environ Pollut* 149:182–192.

[USEPA] US Environmental Protection Agency. 2004. Draft Aquatic Life Water Quality Criteria for Selenium – 2004. Washington (DC, USA): Office of Water, Office of Science and Technology, USEPA.

Vanda Papp L, Lu J, Holmgren A, Khanna KK. 2007. From selenium to selenoproteins: synthesis, identity, and their role in human health. *Antioxid Redox Signal* 7:775–806.

Vandermeulen JH, Foda A. 1988. Cycling of selenite and selenate in marine phytoplankton. *Mar Biol* 98:115–123.

Vander Zanden MJ, Rasmussen JB. 1999. Primary consumer $\partial^{13}C$ and $\partial^{15}N$ and the trophic position of aquatic consumers. *Ecology* 80:1395–1404.

Wang WX. 2002. Interactions of trace metals and different marine food chains. *Mar Ecol Progr Ser* 243:295–309.

Wang WX, Dei RCH. 1999a. Factors affecting trace element uptake in the black mussel *Septifer virgatus. Mar Ecol Progr Ser* 186:161–172.

Wang WX, Dei RCH. 1999b. Kinetic measurements of metal accumulation in two marine macroalgae. *Mar Biol* 135:11–23.

Wang WX, Fisher NX. 1996a. Assimilation of trace elements and carbon by the mussel *Mytilus edulis*: effects of food composition. *Limnol Oceanogr* 41:197–207.

Wang WX, Fisher NS. 1996b. Assimilation of trace elements by the mussel *Mytilus edulis*: effects of diatom chemical composition. *Mar Biol* 125:715–724.

Wang WX, Fisher NS. 1997. Modeling the influence of body size on trace element accumulation in the mussel *Mytilus edulis. Mar Ecol Progr Ser* 161:103–115.

Wang WX, Fisher NS. 1998. Accumulation of trace elements in a marine copepod. *Limnol Oceanogr* 43:273–283.

Wang WX, Fisher NS. 1999. Delineating metal accumulation pathways for marine invertebrates. *Sci Total Environ* 237–238:459–472.

Wang WX, Dei RCH. 2001. Influences of phosphate and silicate on Cr(VI) and Se(IV) accumulation in marine phytoplankton. *Aquat Toxicol* 52:39–47.

Wang WX, Rainbow PS. 2008. Comparative approaches to understand metal bioaccumulation in aquatic animals. *Comp Biochem Physiol C Toxicol Pharmacol* 148:315–323.

Wang WX, Fisher NS, Luoma SN. 1995. Assimilation of trace elements ingested by the mussel *Mytilus edulis*: effects of algal food abundance. *Mar Ecol Progr Ser* 129:165–176.

Wang WX, Fisher NS, Luoma SN. 1996a. Kinetic determinations of trace element bioaccumulation in the mussel *Mytilus edulis. Mar Ecol Progr Ser* 140:91–113.

Wang WX, Reinfelder JR, Lee BG, Fisher NS. 1996b. Assimilation and regeneration of trace elements by marine copepods. *Limnol Oceanogr* 41:70–81.

Wang WX, Qiu JW, Qian PY. 1999. The trophic transfer of Cd, Cr, and Se in the barnacle *Balanus amphitrite* from planktonic food. *Mar Ecol Progr Ser* 187:191–201.

Wang WX, Wong RSK, Wang J, Yen YF. 2004. Influences of different selenium species on the uptake and assimilation of Hg(II) and methylmercury by diatoms and green mussels. *Aquat Toxicol* 68:39–50.

Wrench JJ. 1978. Selenium metabolism in marine phytoplankters *Tetraselmis tetrathele* and *Dunaliella minuta*. *Mar Biol* 49:231–236.

Wrench JJ. 1983. Organic selenium in seawater: levels, origins and chemical forms. *Mar Chem* 12:237–237.

Wrench JJ, Campbell NC. 1981. Protein-bound selenium in some marine organisms. *Chemosphere* 10:1155–1161.

Wrench JJ, Measures CI. 1982. Temporal variations in dissolved selenium in a coastal ecosystem. *Nature* 299:431–433.

Wright DA. 1995. Trace metal and major ion interactions in aquatic animals. *Mar Pollut Bull* 31:8–18.

Wright SH, Manahan DT. 1989. Integumental nutrient uptake by aquatic organisms. *Ann Rev Physiol* 51:585–600.

Xu Y, Wang WX. 2002a. Exposure and potential food chain transfer factor of Cd, Se and Zn in marine fish *Lutjanus argentimaculatus*. *Mar Ecol Progr Ser* 238:173–186.

Xu Y, Wang WX. 2002b. The assimilation of detritus-bound metals by the marine copepod *Acartia spinicauda*. *Limnol Oceanogr* 47:604–610.

Yu RQ, Wang WX. 2002. Kinetic uptake of bioavailable cadmium, selenium, and zinc by *Daphnia magna*. *Environ Toxicol Chem* 21:2348–2355.

Zhang L, Wang WX. 2006. Significance of subcellular metal distribution in prey in influencing the trophic transfer of metals in a marine fish. *Limnol Oceanogr* 51:2008–2017.

Zhang L, Wang WX. 2007. Size-dependence of the potential for metal biomagnification in early life stages of marine fish. *Environ Toxicol Chem* 26:787–794.

Zhang Y, Fomenko DE, Gladyshev VN. 2005. The microbial selenoproteome of the Sargasso Sea. *Genome Biol* 6:R37-R37.

6 Selenium Toxicity to Aquatic Organisms

David M. Janz, David K. DeForest, Marjorie L. Brooks, Peter M. Chapman, Guy Gilron, Dale Hoff, William A. Hopkins, Dennis O. McIntyre, Christopher A. Mebane, Vincent P. Palace, Joseph P. Skorupa, Mark Wayland

CONTENTS

6.1   INTRODUCTION

This chapter addresses the characteristics and nature of organic Selenium (Se) toxicity to aquatic organisms, based on the most current state of scientific knowledge.
As such, the information contained in this chapter relates to the toxicity assessment
phase of aquatic ecological risk assessments. While the inorganic forms of Se (e.g.,
selenate, selenite) can be toxic at concentrations in the 10^2 μg/L range via waterborne exposures, dietary exposure to mg/kg concentrations of organic Se poses a
greater hazard to certain classes of aquatic biota, such as fish and birds (Skorupa and
Ohlendorf 1991; USEPA 1998).

In terms of organic Se toxicity to aquatic biota, this chapter specifically addresses

* mechanisms of organic Se toxicity;
* most relevant/indicative toxicity endpoints;
* comparative sensitivity of organic Se to various aquatic species (and factors
  influencing this relative sensitivity);
* factors that modify organic Se toxicity;
* linkages between organic Se toxicity at the suborganismal and organismal
  level to population-level impacts;

- considerations and recommendations for the design and conduct of site-specific effects studies;
- uncertainties associated with Se toxicity; and
- future research needs.

6.2   MECHANISMS OF ACTION

6.2.1   SELENIUM ESSENTIALITY

Selenium was first recognized as an essential element in 1957 (Mayland 1994) and is a key component of a variety of functional selenoproteins in all living organisms, except for higher plants and yeasts (Hesketh 2008). Selenium-containing proteins fall into 3 categories: 1) proteins into which Se is incorporated nonspecifically (mainly as selenomethionine), 2) specific Se-binding proteins, and 3) enzymes that incorporate selenocysteine (the 21st amino acid) into their active site (Patching and Gardiner 1999; Behne and Kyriakopolis 2001; Reilly 2006; Hesketh 2008). Currently characterized selenoproteins catalyze oxidation–reduction reactions (glutathione peroxidases and thioredoxin reductases), activate, or inactivate thyroid hormone (iodothyronine deiodinases), mediate the synthesis of selenocysteine (selenophosphate synthetase), or are involved in Se transport (selenoprotein P) (Behne et al. 2000; Reilly 2006). Selenium is also required for thioredoxin reductase activity, which is involved in DNA synthesis, oxidative stress defense, and protein repair (Arner and Holmgren 2000). In addition, there are at least 20 other selenoproteins identified in vertebrates whose functions remain unclear (Hesketh 2008).

Despite being an essential trace element at dietary concentrations of 0.1 to 0.5 mg Se/kg dry weight (dw) (Mayland 1994), a significant aspect of the toxicological hazard associated with Se is the narrow margin between essentiality and toxicity. In fish, Se toxicity has been reported to occur at dietary concentrations only 7 to 30 times greater than those considered essential for proper nutrition (i.e., > 3 mg Se/kg dw) (Hilton et al. 1980; Hodson and Hilton 1983). In poultry, dietary Se concentrations of less than 0.3 mg/kg dw are considered below the range adequate for good adult health and reproduction, 3 to 5 mg/kg dw are considered high, and above 5 mg/kg dw are considered toxic. In eggs, the tipping point between essentiality and toxicity shifts upward such that Se concentrations lower than 1 mg/kg dw in eggs may indicate inadequate Se in the maternal diet (Puls 1988; Table 6.1). Additional information on physiological requirements is provided in Chapter 5 (Section 5.2.1).

In addition, although Se has important roles in antioxidant defenses at normal dietary levels, at elevated exposure levels it can become involved in the generation of reactive oxygen species, resulting in oxidative stress with increasing exposure. As discussed below, oxidative stress is a key mechanism of toxicity in vertebrate animals.

6.2.2   SELENIUM TOXICITY

Although uncertainties remain, there is a large and growing body of knowledge regarding the toxicity of Se to aquatic biota. Oviparous (egg-laying) vertebrates appear to be the most sensitive taxa, and this section focuses on mechanisms of Se toxicity in these animals. The sequence of mechanistic events involved in Se toxicity

TABLE 6.1

Data Illustrating the Range of Assessment Values for Effects of Egg Se Concentrations in Birds

Status[a]	Concentration (mg Se/kg, dw)	Effects	Comments	References
Adequate	0.66–5.0 (0.20–1.5 ww)	Nutritional needs are met for poultry	Lower dietary concentrations are marginal or deficient, and diets must be fortified	Puls 1988
High	5.0–16 (1.5–5.0 ww)	Levels are excessive and upper end of range may be toxic to poultry	Poultry are relatively sensitive to effects of Se	Puls 1988
Toxic	>8.2 (>2.5 ww)	Reduced egg hatchability and teratogenic effects in embryos/chicks	Poultry are relatively sensitive to effects of Se	Puls 1988
Background	Mean < 3.0 (typically 1.5–2.5); individual eggs <5	None	Concentrations may be higher in some marine birds (Section 6.5.4)	Ohlendorf and Harrison 1986; Skorupa and Ohlendorf 1991; USDOI 1998; Eisler 2000
Reproductive impairment	7.7 (about 2.3 ww)	EC10 for reduced egg hatchability	Based on results of one laboratory study with mallards, assuming hormetic effects	Beckon et al. 2008
Reproductive impairment	9.0	EC8.2 for impaired egg hatchability	Based on results of one laboratory study with mallards, using linear regression analysis	Lam et al. 2005
Reproductive impairment	12 (95% CI = 6.4–16)	EC10 for reduced egg hatchability	Based on results of six laboratory studies with mallards, using logistic regression analysis	Ohlendorf 2003
Reproductive impairment	12 (95% CI = 9.7–14)	EC10 for reduced egg hatchability	Based on results of six laboratory studies with mallards, using hockey stick analysis	Adams (pers. comm.; see Ohlendorf 2007)
Reproductive impairment	14	EC11.8 for reduced clutch viability	Based on results of extensive field studies of black-necked stilts	Lam et al. 2005
Teratogenicity	13–24	Threshold for teratogenic effects on population level	Sensitivity varies widely by species	Skorupa and Ohlendorf 1991

TABLE 6.1 (CONTINUED)
Data Illustrating the Range of Assessment Values for Effects of Egg Se Concentrations in Birds

Status[a]	Concentration (mg Se/kg, dw)	Effects	Comments	References
Teratogenicity	23	EC10 for teratogenic effects in mallard	Mallard is considered a "sensitive" species	Skorupa 1998b; USDOI 1998
Teratogenicity	37	EC10 for teratogenic effects in stilt	Stilt is considered an "average" species	Skorupa 1998b; USDOI 1998
Teratogenicity	74	EC10 for teratogenic effects in American avocet	Avocet is considered a "tolerant" species	Skorupa 1998b; USDOI 1998

[a] 65–80% moisture, varying with species and incubation stage; 70% moisture (i.e., factor of 3.3) used for approximate conversion

Note: Values in the first 3 rows (Puls 1988) are based on domestic poultry rather than wild species.

Source: From information contained in Ohlendorf and Heinz (in press).

to oviparous vertebrates, from molecular to biochemical to subcellular/cellular to individual to population levels of biological organization, is described below.

6.2.2.1 Cellular Mechanisms of Se Toxicity

It has long been thought that the primary initiating event behind the ability of elevated Se concentrations to cause embryo toxicity and teratogenicity comes from its propensity to substitute for sulfur, while protein synthesis is occurring during organogenesis within the embryo. Indeed, while there is a strong body of scientific literature documenting organo-Se residues presumably bound to protein within the eggs and embryos of oviparous vertebrates, there is growing evidence that oxidative stress is also likely to play a role in Se-related teratogenesis.

6.2.2.1.1 Selenium Substitution for Sulfur in Amino Acids: Importance as Mechanism of Toxicity Uncertain

Until recently, researchers had focused on substitution for sulfur as the mechanism for Se toxicity in oviparous vertebrates. Specifically, it was thought that Se obtained from the diet transferred maternally to the developing embryo and assimilated, in place of sulfur, into structural and functional proteins during embryonic development. Since the normal tertiary structure of protein molecules depends upon the formation of S-S linkages, substitution of Se for S in protein synthesis could result in improperly folded or dysfunctional proteins such as enzymes (Diplock and Hoekstra 1976; Reddy and Massaro 1983; Sunde 1984; Maier and Knight 1994). Resultant deformities were believed to result from this nonspecific substitution (Lemly 1997a). However, this proposed mechanism of toxic action has been questioned as discussed below. Definitive studies investigating the mechanistic importance of oxidative stress remain to be conducted in egg-laying vertebrates.

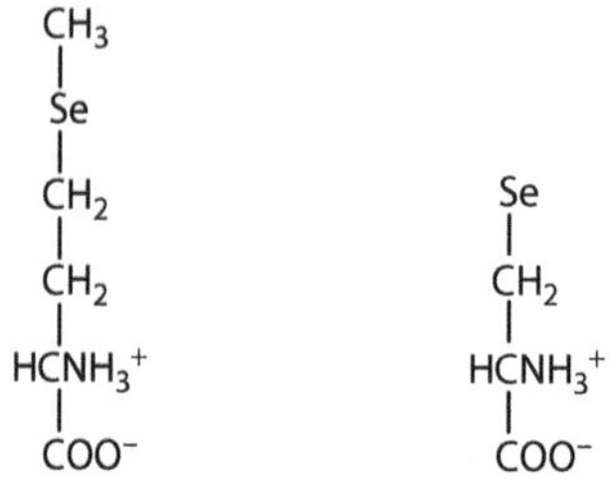

FIGURE 6.1 Schematic structures of the amino acids capable of containing Se: selenomethionine and selenocysteine.

Selenium can be incorporated into two possible amino acids, selenomethionine and selenocysteine. The formation of each relative to their sulfur analogues, methionine and cysteine, appears to use sulfur pathway enzymes and to depend on the relative concentration of Se in the cell (Allan et al. 1999). However, subsequent incorporation of these amino acids into proteins is concentration-dependent only for selenomethionine (Schrauzer 2000). The Se moiety in selenomethionine is insulated by the terminal methyl group in the amino acid structure (Figure 6.1) and so, not surprisingly, substitution of methionine with selenomethionine does not appear to alter either the structure or function of proteins (Yuan et al. 1998; Mechaly et al. 2000; Egerer-Sieber et al. 2006). Conversely, selenocysteine incorporation into proteins is highly regulated at the ribosomal level, by the UGA codon that specifies selenocysteinyl-tRNA (Stadtman 1996). Thus, proteins requiring Se for their structure or function specifically incorporate selenocysteine in the polypeptide via the mRNA sequence. Evidently, cysteine and selenocysteine can randomly be substituted only in some bacteria and plants (Allan et al. 1999). Thus, it appears that neither selenocysteine, which is controlled by the mRNA sequence, nor selenomethionine in which the Se is shielded by the terminal methyl group affect protein structure or function.

6.2.2.1.2 Oxidative Stress Mechanism

More recently, oxidative stress has been proposed as the initiating event of embryo mortality and teratogenic effects from several chemicals (Wells et al. 1997, 2009; Kovacic and Somanathan 2006), including avian species exposed to Se (Hoffman 2002; Spallholz and Hoffman 2002). Interaction with the tripeptide glutathione is apparently critical to propagating oxidative stress in Se-exposed organisms through a variety of mechanisms (Spallholz et al. 2004). Below, we first discuss the normal antioxidative process involving glutathione. Glutathione acting with the enzyme glutathione peroxidase is an intracellular antioxidant with tremendous reducing power that maintains antioxidant enzyme systems. As stated above, the enzyme that catalyzes this critical reaction is glutathione peroxidase (GPx-Se), which contains Se:

$$GPx\text{-}Se + 2GSH + R\text{-}O\text{-}OH \rightarrow GSSG + H2O + ROH + GPx\text{-}Se \qquad (1)$$

Where GSH is the reduced form of glutathione, R-O-OH is a peroxide substrate, ROH is an alcohol product, and GSSG is the oxidized form of glutathione. Intermediate steps, which are not shown, regenerate the glutathione peroxidase (Equation 1). Normally, the ratio of GSH: GSSH is 500: 1 (Stryer 1995).

In a review of the avian Se toxicity literature, Hoffman (2002) documented that exposure to Se caused lower ratios of reduced GSH to oxidized GSSG and increased indices of oxidative cell damage. In feeding studies, mallard ducks (*Anas platyrhynchos*) exposed to elevated levels of selenomethionine as both ducklings (Hoffman et al. 1989, 1991a, 1992a,b, 1996) and adults (Fairbrother and Fowles 1990) demonstrated elevated plasma and hepatic GPx-Se activity as well as increased tissue Se concentrations. The mallard studies also demonstrated that there exists a dose-dependent increase in the hepatic ratio of GSSG to GSH. Such increased GSSG relative to GSH in the presence of elevated Se, apparently increased hydroperoxides responsible for the observed increase in hepatic lipid peroxidation, measured as thiobarbituric-acid reactive substances (TBARS). Consistent results have been obtained in rat (LeBoeuf et al. 1985) and fish models (Holm 2002; Miller et al. 2007; Atencio et al. 2009).

In other instances, glutathione can react with some forms of Se to produce selenopersulfides and thiyl radicals (Spallholz and Hoffman 2002). Selenopersulfides spontaneously produce superoxide anion in the presence of oxygen, or they may react with additional glutathione, producing hydrogen selenide, eventually giving rise to elemental Se and again producing superoxide anion (Lin and Spallholz 1993). Thiyl radicals may react with glutathione to form glutathione disulfide radicals (Arteel and Sies 2001).

The chemical speciation of Se is complex (Chapter 4), and not all forms of Se are capable of associating with glutathione and generating oxidative stress (Spallholz and Hoffman 2002). In fact, the predominant form of Se in the eggs of oviparous vertebrates, selenomethionine, is not highly reactive with glutathione (Spallholz et al. 2001; Spallholz and Hoffman 2002). However, in vivo metabolism of selenomethionine and/or selenocysteine to more reactive Se forms, including methylselenol, could potentiate oxidative stress (Sunde 1997; Miki et al. 2001; Wang et al. 2002; Fan et al. 2002; Palace et al. 2004). In eggs, concentration-dependent incorporation of elevated selenomethionine from exposed adults and subsequent enzymatic cleavage into reactive metabolites in the developing embryo is hypothesized to initiate generation of reactive oxygen species and development of oxidative stress. Furthermore, it has been hypothesized that oxidative stress may be involved in pericardial and yolk sac edema in rainbow trout (*Oncorhynchus mykiss*) embryos exposed to elevated Se (Palace et al. 2004) in a similar manner and etiology as some organic contaminants (Bauder et al. 2005).

Some authors convincingly argue the case for oxidative stress being the mode of action in teratogenesis (Wells et al. 2009). They successfully used rat and mouse models to describe examples where effects from organic xenobiotics well known for teratogenicity (e.g., thalidomide) are ameliorated when embryos are simultaneously exposed to antioxidants (e.g., Vitamin A). However, to date, we are not aware of comparable work conducted with Se and oviparous vertebrates. Currently, there is primarily correlative evidence of oxidative stress and incidence of terata after embryo exposure to Se, and much less in terms of cause–effect relationships.

Aside from oxidative stress related to glutathione homeostasis, a few studies have examined the effect of Se on other antioxidant enzymes and nonenzymatic vitamins in fish. Li et al. (2008) reported a decline in the hepatic activity of superoxide dismutase, which neutralizes the reactive oxygen species superoxide anion, in Japanese medaka (*Oryzias latipes*) exposed to waterborne selenite or nanoparticle Se. Holm (2002) reported that concentrations of the antioxidant vitamins E (tocopherol) and A (retinol) were slightly lower among rainbow trout fed diets enriched with selenomethionine (10 or 20 mg/kg dw) for 302 days. Vitamin E may have been lower because of greater production of Se-dependent glutathione peroxidase, which metabolizes lipid-peroxyl radicals, accounting for the ability of Se to reduce metabolic requirements for vitamin E (Ursini et al. 1985). Histopathological lesions in the livers of splittail (*Pogonichthys macrolepidotus*) fed a selenized yeast diet containing 57.6 mg Se/kg dw exhibited cytoplasmic protein droplets and fatty vacuolar degenerations that the authors speculated could be due to lipid peroxidation (Teh et al. 2004). Miller et al. (2007) found that exposures of juvenile rainbow trout to subacute (to 160 μg Se/L) concentrations of waterborne selenite for 30 days did not alter antioxidant enzyme activities or lipid peroxidation levels. Determining the importance of oxidative stress resulting from Se exposure to different species and life stages is a pressing research need.

6.2.2.1.3 Mechanism of Suppressed Immune Function

Although Se is a well-known antioxidant with positive effects on the immune system (Koller et al. 1986), it has the potential to adversely affect the immune system at elevated concentrations in mammals or birds. Mammals have a slightly reduced immune response at elevated dietary exposures of selenomethionine, sodium selenate, or sodium selenite (Raisbeck et al. 1998). Bird immunity appears to be less sensitive to Se exposure. For example, selenomethionine in drinking water decreased some aspects of mallard immune response, while sodium selenite had no effect (Fairbrother and Fowles 1990). American avocet (*Recurvirostra americana*) chicks hatched from eggs collected from ponds with elevated Se and arsenic (As) concentrations showed reduced responses in some aspects of their immune systems, but elevated activity in others (Fairbrother et al. 1994). Mallard chicks hatched from eggs of ducks feeding in streams contaminated with Se demonstrated increased mortality following infection with duck hepatitis virus (Fairbrother et al. 2004).

The immunocompetence of adult common eiders (*Somateria mollissima*) was impaired when they were fed a diet containing 60 mg Se/kg dw (Franson et al. 2007). Interestingly, thymus glands were absent and cell-mediated immunity was reduced in this group of birds. However, humoral immunity was enhanced in eiders fed a lower concentration of Se (20 mg/kg dw) (Franson et al. 2007). In field-collected birds, cell-mediated immunity was positively correlated, and the ratio of heterophils to lymphocytes was negatively correlated with hepatic Se over a range of concentrations from 9 to 76 mg/kg dw (Wayland et al. 2002). Collectively, the results of the eider studies suggest that Se may enhance immunocompetence at low levels of supplementation or over a range of normal dietary levels in the wild, but that it can impair immunocompetence at elevated dietary levels.

6.2.2.2 Toxicodynamics

6.2.2.2.1 Incorporation of Se into Vitellogenin

Maternal deposition of Se into eggs and its subsequent assimilation by the developing embryo is the key vector for determining the reproductive effects of Se in oviparous vertebrates. However, there are important differences among species in reproductive strategy, reproductive physiology, pattern of oogenesis, biochemical and physical properties of their eggs, and behavior that may affect the deposition of Se into eggs. Vertebrate eggs vary considerably in their anatomical and biochemical composition (Blackburn 1998, 2000; Romano et al. 2004). As discussed below, there appear to be multiple physiological pathways for maternal transfer of Se.

6.2.2.2.2 Fish

Fish exhibit a remarkable range of reproductive strategies, from semelparous species that spawn only once in their lifetime to iteroparous species that spawn multiple times during their lifetime. Even among iteroparous species, strategies may range from taking many years to reach sexual maturity and spawning only every 2 to 3 years, to spawning every year or even multiple times each year (Mommsen and Walsh 1988; Rinchard and Kestemont 2005). While a comprehensive evaluation of the effect of these various reproductive strategies on susceptibility to Se-induced reproductive toxicity has not been conducted, some inferences can be made based on the timing and duration of oogenesis.

In fish, the primary yolk precursor is vitellogenin (VTG), a phospholipoglycoprotein synthesized in the liver under the regulation of the hypothalamic–pituitary–gonadal–liver endocrine axis (Arukwe and Goksøyr 2003). Vitellogenin is exported from the liver, transported in the blood, and incorporated into the developing ovarian follicle by receptor-mediated endocytosis (Kime 1998). In the follicle, VTG is enzymatically cleaved into the primary yolk proteins lipovitellin and phosvitin (Arukwe and Goksøyr 2003). These sulfur-containing proteins can also contain Se, and not surprisingly, Se binding to VTG has been demonstrated in fish (Kroll and Doroshov 1991).

The duration and relative amount of VTG deposited into developing oocytes, and hence the potential for Se incorporation, depends on the reproductive strategy of the fish species in question. For many salmonid fish species, vitellogenesis can occur over several months prior to spawning with a relatively large amount of energy-rich yolk being invested (Estay et al. 2003). As a result, for salmonids, the dietary intake of Se immediately prior to spawning may not have a major impact on egg Se concentrations. Instead, Se from tissue storage sites, including the liver and muscle, will likely contribute proportionally more Se to the oocytes in these fish. In fish species that spawn multiple times in one season, the period of oogenesis can be highly variable with oocyte maturation occurring well before, immediately prior to, or even during the spawning season (Rinchard and Kestemont 2005). In these cases, the immediate diet may be more important for supplying nutrients and trace elements, including Se, for maternal transfer to the developing oocytes. The efficiency of transfer from maternal tissues to eggs is also highly variable among fish species. In their review, deBruyn et al. (2008) report that regression slopes for Se concentrations in muscle versus eggs vary widely among eight fish species, with rainbow trout exhibiting the highest egg muscle ratios, and brook trout (*Salvelinus fontinalis*), the lowest ratio. However, the underlying biochemical reasons for the variable efficiency of Se transfer from tissues to eggs remains uncertain.

6.2.2.2.3   Amphibians and Reptiles

Amphibians and squamate reptiles (lizards and snakes) also incorporate proteins derived from VTG in a similar manner to fish (Unrine et al. 2006). Presumably this means that these groups also mobilize Se from storage tissues to eggs in a similar manner to that described above for fish. Little is known about the mechanisms of maternal transfer of Se in amphibians and reptiles, but their vast diversity in reproductive biology and life history provides opportunities for important comparative research on Se exposure and effects.

Oviparous amphibians produce anamniotic eggs that are most similar in structure to fish eggs (Duellman and Trueb 1986). Like fish, amphibian species vary dramatically in their annual fecundity, ranging from several offspring/year to >80,000/year (Duellman and Trueb 1986) and, as a result, they vary considerably in the proportion of energy, nutrients, and contaminants that they allocate to progeny. Amphibian embryos often undergo development in water where they hatch and transition to a larval stage (i.e., amphibians with complex lifecycles), but others forego the aquatic stage and undergo direct development in the terrestrial environment (e.g., Plethodontid salamanders). Such differences in maternal provisioning and developmental patterns obviously have important implications for understanding maternal transfer of Se and any resultant effects. In the only two studies examining maternal transfer of Se in amphibians, females transferred approximately 28% to 53% of their preoviposition body burden to their eggs (Hopkins et al. 2006; Bergeron, Bodinof, Unrine, and Hopkins, unpublished data).

Similar to amphibians, reptiles (turtles, crocodilians, lizards, snakes, and tuatara) span an incredibly broad range of reproductive strategies, from oviparity to true viviparity (Tinkle and Gibbons 1977; Shine 1985; Thompson et al. 2000). Oviparous species produce an amniotic egg, most similar to that of birds. Although several studies demonstrate maternal transfer of Se in various oviparous reptiles (Nagle et al. 2001; Hopkins et al. 2004, 2005a,b; Roe et al. 2004), selenomethionine is also transferred from low (invertebrate) to high trophic levels (western fence lizard, *Sceloporus occidentalis*) under controlled conditions. During trophic transfer, considerable Se partitions within the lizards' developing follicles and eggs (Hopkins et al. 2005a; Unrine et al. 2007a). Selenium is transported to the egg by vitellogenin, but also via two previously undescribed egg proteins (Unrine et al. 2006).

Among vertebrates, squamate reptiles provide fruitful opportunities for understanding the mechanisms of maternal transfer in placental and nonplacental vertebrates because closely related species (congenerics) span the full spectrum of oviparity to viviparity (Shine 1985; Thompson et al. 2000). Just as squamates have been adopted as models for studying the evolution of viviparity (e.g., Tinkle and Gibbons 1977), similar comparative approaches could prove invaluable for understanding mechanisms of maternal Se transfer.

6.2.2.2.4   Birds

In birds, Se-containing proteins and Se effects differ distinctly from those of fish, amphibians, and squamate reptiles. In contrast to fish, most of the Se in an avian egg is found in the albumin, and therefore the developing chick takes up most of the egg Se before hatching and before yolk sac resorption. For example, in the domestic chicken (*Gallus gallus*) Se is incorporated in ovalbumen, conalbumin, globulin, ovomucoid, and flavoprotein (Jacobs et al. 1993; Davis and Fear 1996). Consequently, also unlike fish, one of the most sensitive toxic endpoints for bird reproduction is reduced egg viability because of hatching failure among full-term, fertile eggs. Because the yolk sac may not be completely metabolized until several days posthatch, the Se dose received from yolk sac resorption can decrease growth rates of newly hatched chicks (Fairbrother et al. 1994) and cause direct chick mortality (Williams et al. 1989; Marn 2003). More research on posthatch reproductive effects (or lack thereof) associated with yolk sac resorption in avian chicks is highly warranted.

The ratio of Se in albumin versus yolk in avian eggs collected in the wild corroborates the ratio observed in controlled feeding studies that supplemented the female's diet with selenomethionine but is not consistent with controlled studies that supplemented the diet with inorganic forms of Se (Latshaw and Osman 1975; Latshaw and Biggert 1981; Moksnes and Norheim 1982; Heinz et al. 1987, 1990; Santolo et al. 1999; Detwiler 2002). Unlike fish, most of the Se in avian eggs is mobilized exogenously from the diet rather than endogenously from maternal tissue (Heinz 1996; DeVink et al. 2008a). Consequently, under controlled feeding conditions (i.e., uniform dietary exposure), there is no laying-order effect (i.e., differences among eggs within the same clutch) as would be expected if an endogenous pool of Se were being increasingly depleted with the production of each successive egg (Heinz 1996). Thus, Se in bird eggs is representative of a relatively short-term (few days) snapshot of a female's dietary exposure during ovulation; moreover, females also return to laying Se-normal ("clean") eggs within days to weeks of switching to a Se-normal diet, depending on the starting point (Heinz 1996). One implication of avian egg Se being derived primarily from the diet over a discrete time window is that eggs within a single clutch in nature can contain Se concentrations that vary to the extent that the female's dietary exposure varies during the ovulation of one egg to the next. For example, among 31 completed 4-egg black-necked stilt (*Himantopus mexicanus*) clutches, total within-clutch variation was typically <1 mg Se/kg dw for clutches averaging <7 mg Se/kg dw (Joe Skorupa, USGS, personal communication). However, variability increased as the mean Se content of each clutch increased, probably reflecting a heterogeneous spatial distribution of Se within an aquatic system with a broad range of Se concentrations (i.e., "hot spots" become more pronounced and the chance of a feeding hen moving in and out of hot spots during ovulation increases). In the most severe case, a clutch that averaged 62 mg Se/kg dw exhibited a 100 mg/kg dw spread between the low egg (9.9 mg Se/kg dw) and the high egg (110 mg Se/kg dw). However, another clutch that averaged 62 mg Se/kg dw exhibited a spread of only 11 mg/kg dw between the low (55 mg Se/kg dw) and high (66 mg Se/kg dw) egg (Skorupa unpublished data). Within-clutch variability can also be exacerbated by landscape scale movement between re-nesting attempts if the first nesting attempt is terminated early because of egg predation, nest flooding, or other sources of early nest failure. Cases of extreme within-clutch variability (such as the 100 mg/kg dw example above) are probably the result of hen movement through a hot spot between successive nesting attempts.

Although captive studies have noted depuration of Se from the liver of laying mallards that were induced to produce artificially excessive numbers of eggs (30 or more as opposed to a normal clutch size of 6 to 8 eggs; Ohlendorf and Heinz, in press), Paveglio et al. (1992) reported that liver Se in field-collected breeding female and male mallards did not follow the pattern observed in captive studies. Given the exogenous source of most avian egg Se, egg production should not normally be a major pathway for excretion of the endogenous pools of tissue Se in breeding females.

6.3   RELEVANT TOXICITY ENDPOINTS

6.3.1   Diagnostic Indicators of Se Toxicity

As indicated earlier, the most important toxicological effects of Se in fish arise following maternal transfer of Se to eggs during vitellogenesis, resulting in Se exposure when hatched larvae undergo yolk absorption. During this life stage (fry), permanent developmental anomalies (e.g., spinal curvatures, missing or deformed fins, and craniofacial deformities) and other effects (e.g., edema) in fish can be related to elevated Se in eggs (Hodson and Hilton 1983; Lemly 1993a; Maier and Knight 1994; Hamilton 2003) (see Figure 6.2 for examples of Se-induced terata in larval fish). Although certain other natural and anthropogenic factors can result in deformities of the spine, fins, and craniofacial structures (Section 6.3.2.1), these terata have been considered diagnostic for Se toxicity (Maier and Knight 1994; Lemly 1997a). As discussed below, various forms of edema (e.g., edema of the pericardium or yolk sac) can arise from exposure to other xenobiotics such as polycyclic and halogenated aromatic hydrocarbons; however, this response is also prevalent in larval fish exposed to elevated concentrations of Se in yolk. In birds, embryonic deformities accompanied by substantively elevated egg Se are perhaps the most unequivocal basis for diagnosing reproductive selenosis (Ohlendorf 1989, 2003; Heinz 1996). However, impaired egg hatchability occurs at distinctly lower egg Se levels than embryo teratogenesis and is therefore a more sensitive effects endpoint (Skorupa 1999).

6.3.2   Types and Severities of Deformities and Edema

6.3.2.1   Fish

In contrast to birds, Se generally does not affect fertility and hatching rates in fish (Gillespie and Baumann 1986; Coyle et al. 1993; Holm et al. 2005; Muscatello et al. 2006). Hermanutz et al. (1992) observed a statistically significant reduction in the hatching rate of Se-exposed bluegill sunfish (*Lepomis macrochirus*) relative to the control, but this effect was atypical in Se toxicity studies. Rather, teratogenesis, edema, and/or larval mortality following hatch are the most sensitive endpoints in fish.

The frequency of teratogenic deformities in early developmental stages of fish is the most useful indicator of Se toxicity. Teratogenesis is a direct expression of Se toxicosis and represents the sum total of parental exposure, regardless of temporal, spatial, or chemical variations in Se exposures. Terata represent a measure of existing, rather than potential, hazard and can be subtle but important causes of recruitment failure in fish populations. Significant loss of the early life stages of a fish

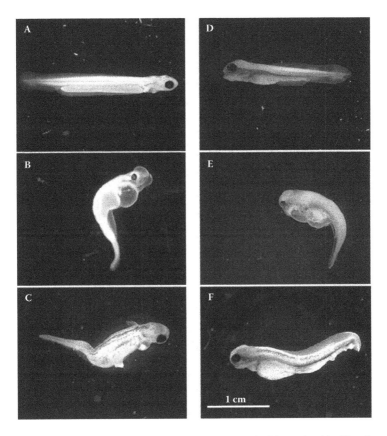

FIGURE 6.2 Examples of Se-induced deformities in larval white sucker (A–C) and northern pike (D–F). Plates A and D are larvae originating from reference sites showing normal morphology. Plate B shows pericardial and craniofacial edema and spinal curvature (scoliosis). Plate C shows spinal curvature (scoliosis) and shortened pelvic fin. Plate E shows craniofacial deformity, microphthalmia, pericardial and yolk sac edema, and spinal curvature (kyphosis). Plate F shows spinal curvature (kyphosis and lordosis) and craniofacial deformity. (Adapted with permission from Muscatello JR. 2009. Selenium accumulation and effects in aquatic organisms downstream of uranium mining and milling operations in northern Saskatchewan. PhD Dissertation. Saskatoon: University of Saskatchewan (Canada).)

population can occur at the same time that adult fish appear healthy. An index for evaluating the impact of Se-induced terata on population mortality (Lemly 1997a) predicts that when <6% terata appear in larvae or fry, less than 5% mortality and negligible impact would result to the population. However, when terata are quantified at rates of between 6% to 25%, mortality would be 5% to 20%, with a slight to moderate impact, and >25% terata would correspond to >20% population loss and a major impact on populations. These relationships are based on data from two families of fishes (Centrarchidae and Cyprinidae), and several studies have questioned the applicability of this index to cold-water fish species, including salmonids and esocids (Kennedy et al. 2000; Holm et al. 2005; Muscatello et al. 2006).

There are at least three ways of determining and categorizing deformities. Simple frequency analysis is scored as either presence or absence for a given category of deformity (e.g., presence of spinal deformity). Graduated severity index (GSI) methods assign a numerical value based on the severity of a given deformity (i.e., 1 = mild, 2 = moderate, and 3 = severe), most effectively with a predetermined criterion established for assignment to each score. Finally, morphometric analyses attempt to make actual quantifiable measures of a given category of deformity, such as the angle of spinal column diversion, or the volume of edematous fluid accumulated or degree of jaw shortening. Holm et al. (2003) evaluated each of these methods by repeated measurement of preserved rainbow trout and brook trout fry. Graduated severity index and frequency analysis provided similar information, but the GSI analysis detected increases in the severity of deformities that simple frequency analysis could not. Morphometric analysis did not provide better information than the previous two methods; however, it required far more analysis effort and specialized analytical instrumentation. Kennedy et al. (2000), also used a GSI approach and a subsequent recommendation has supported the use of this type of analysis (McDonald and Chapman 2009). Quality assurance and quality control (QA/QC) considerations for assessing larval deformities in fish are discussed in Text Box 6.1.

Studies of deformity rates are often hampered by data gaps in the basal rates of deformity in each of the categories (Villeneuve et al. 2005). Clearly, factors other than Se toxicity can lead to the development of larval deformities. For example, skeletal and craniofacial deformities can be influenced by genetic factors (Alfonso et al. 2000), parasite infections (Villeneuve et al. 2005), vitamin or amino acid deficiencies (Dabrowski et al. 1996; Villeneuve et al. 2005), organic contaminants (Mehrle et al. 1982; Tillitt and Papoulias 2002), and elevated water temperatures (Sfakianakis et al. 2006; Georgakopoulou et al. 2007). Spinal deformities can arise from failure of the swim bladder to inflate (Daoulas et al. 1991; Chatain 1994) or when fish develop in high water velocities (Sfakianakis et al. 2006). Spinal deformities associated with vitamin deficiencies, parasitic infections, or contaminants appear at numerous locations along the vertebral column, as is also the case for Se-induced deformities in fish. Conversely, spinal deformities associated with elevated water velocity or swim bladder inflation failure tend to manifest at a consistent spinal location in affected populations (Divanach et al. 1997). Finally, sampling artifacts, including the effects of electrical shock during collection (EVS and PLA 1998) or "packing effects" (where larvae are shaped by other organisms or objects in the fixative) during preservation (Kingsford et al. 1996), influence the final enumeration of skeletal abnormalities.

Baseline deformity rates of 2% to 5% occur in salmonids spawned in the laboratory (Gill and Fisk 1966; Werner et al. 2005); slightly higher rates occur in fish spawned in the wild (Kennedy et al. 2000; de Rosemond et al. 2005; Holm et al. 2005). Villeneuve et al. (2005) reported baseline deformity rates of several Cyprinids and a Catostomid species, as follows: 7% for pikeminnow (*Ptychocheilus oregonensis*); 6% in redside shiner (*Richardsonius balteatus*); 17% in large-scale sucker (*Catostomus macrocheilus*); 8% in peamouth (*Mylocheilus caurinus*); and 13% in chiselmouth (*Acrocheilus alutaceus*) from reference locations in the Willamette River, Oregon. Skeletal, craniofacial, and finfold deformities were generally <10% in northern pike (*Esox lucius*) collected from cold water reference locations in northern

TEXT BOX 6-1: QA/QC FOR ASSESSMENT
OF LARVAL FISH DEFORMITIES

Although a standard operating procedure for conducting deformity (frequency) analysis of field-derived fish larvae has recently been published (Muscatello 2009; Janz and Muscatello 2008), a standardized and validated methodology including quality assurance/quality control (QA/QC) has not yet been formally adopted to assess the frequency and severity of larval fish (or other life stage) deformities. To minimize subjectivity, decrease uncertainty, and provide robust interpretation, we support the following recommendations for Se-induced larval deformity assessments from several studies (Muscatello 2009; Janz and Muscatello 2008; McDonald and Chapman 2009), including

- use of a two-way ANOVA experimental design for embryo incubations (for details, see Muscatello et al. 2006; Muscatello 2009);
- euthanization using overdose of appropriate anesthetic (e.g., 3-aminobenzoic acid [MS-222]) prior to fixation in preservative;
- blind and nonsequential labeling of treatment groups;
- development and application of an a priori framework for deformity analysis;
- internal QC checks to quantify the influence of sample preservatives, observer drift or multiple observers; and
- an external QC check of a minimum of 10% of all larval fish.

Although a standardized, validated methodology is preferred, future reproductive studies with larval fish should, at a minimum, include raw deformity data and assessment, details on all QA/QC elements, and an explicit uncertainty analysis.

Issues with preservatives: Larvae for deformity analysis are typically preserved in formalin together with various buffering agents (Kennedy et al. 2000; Muscatello et al. 2006; de Rosemond et al. 2005) or the Davidson's solution (Holm et al. 2005; Rudolph et al. 2008). Larvae may also be transferred between solutions (e.g., Saiki et al. 2004; transfer from Davidson's solution to isopropyl alcohol). Larval fish morphology (length and weight) is altered by long-term preservation but has not been comprehensively quantified (Paradis et al. 2007; Cunningham et al. 2000; Fey 1999; Fisher et al. 1998). A rigorous evaluation of the effects of preservatives (including both type and duration) on Se-related deformities in larval fish is a clear research need. Larvae should be initially assessed for deformities before preservation, at least in a subsample of larvae. Also, larval fish must be euthanized with an anesthetic overdose before preservation in fixative, since the absence of this step can cause artefactual skeletal curvatures (Janz and Muscatello 2008; Muscatello 2009).

Saskatchewan (Muscatello et al. 2006; Muscatello 2009; Muscatello and Janz 2009). Finally, up to 5% of fish produced in the aquaculture industry have some form of spinal deformity (Andrades et al. 1996). Therefore, and in agreement with Lemly (1997a), deformity rates of less than 5% are not likely ecotoxicologically relevant.

Not all types of deformities have the same ecological relevance. While abnormalities may not be lethal, their persistence in the general population is only likely where there is little threat from predators (Lemly 1997a). It is generally agreed that vertebral deformities are potentially the most critical because they impact the ability of fish to swim to avoid predators or obtain food. Laboratory-reared sea bass (*Dicentrarchus labrax*) that exhibited kyphosis were also lethargic and unresponsive to visual and auditory stimuli. More important, they grew more slowly and did not contribute to population health because their mortality rates were high (Koumondourous et al. 2002). Aside from decreased survival, carp (*Cyprinus carpio*) with spinal deformities exhibited lower growth, further supporting the notion of reduced fitness in affected fish (Al-Harbi 2001). Fernandez et al. (2008) reviewed the literature and found that the operculum complex, premaxilla, maxilla, and dentary bones were the cranial structures most commonly affected when Se-induced deformities were detected. Opercular deformities are often characterized as minor relative to other types of deformities, but Al-Harbi (2001) noted that in a cultured population of carp, while opercular deformities did not impair swimming performance, affected fish exhibited lower growth. Any assessment of Se-induced effects should consider that opercular deformities, in particular, have been linked with disruption of vitamin A and C metabolism and exposure to other contaminants (Lindesjoo and Thulin 1992; Lindesjoo et al. 1994; Fernandez et al. 2008).

One of the most prevalent, and contentious, effects from Se exposure is the appearance of edema in early life stages of fish. Edema is not a true terata because it can be transient and reversible and does not occur solely at the embryo-larval stage (Lemly 1993a). The appearance of edema has also been linked with exposure to organic chemicals (Barron et al. 2004; Billiard et al. 1999). Selenium-related edema in the developing embryo may be mediated by oxidative stress (Palace et al. 2004). Edematous effects from Se toxicity are difficult to dismiss because of strong associations between edema and elevated Se concentrations in fish eggs in a number of studies (Gillespie and Baumann 1986; Pyron and Beitinger 1989; Holm et al. 2005; Muscatello et al. 2006) and because edema is often one of the most sensitive (Muscatello and Janz 2009) and prevalent endpoints (Gillespie and Baumann 1986; Woock et al. 1987; Holm et al. 2005). Pyron and Beitinger (1989) reported that nearly all edematous fathead minnow larvae that were produced by adults exposed to Se did not survive longer than 7 days posthatch. However, Hermanutz (1992) reported that larval fish with edema survived to the juvenile stage in outdoor artificial streams. Additional evaluations regarding the ability of early life stages with edema to survive in the field are required to establish the potential ecological relevance of edema as a marker of Se exposure.

6.3.2.2 Birds

In birds, as in fish, the incidence of embryonic deformities is a commonly measured endpoint. The process for determining whether Se exposure may be implicated in embryonic deformities can be straightforward; often embryos are examined for the presence or absence of deformities and Se concentrations are then measured in those resulting embryos (Seiler et al. 2003). In other instances, a subset of eggs from nests located at a Se-contaminated site is collected and analyzed for Se, and the incidence of deformities in embryos is documented for that site (Ohlendorf et al. 1988; Ohlendorf and Hothem 1995).

Among mallard embryos, a spatulate upper bill is a Se-specific deformity (O'Toole and Raisbeck 1997, 1998). Seiler et al. (2003) illustrated the consistency of this type of deformity across different species of duck embryos and across field study sites (Figure 6.3). Although the exact nature of embryo deformities and the order in which they express themselves may vary among species (e.g., eyes first in shorebirds, bills first in ducks; Skorupa, personal observation), they consistently involve the reduction or

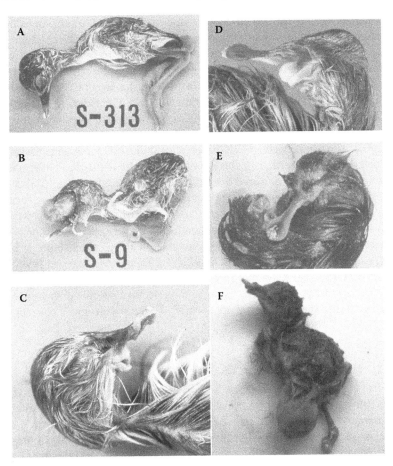

FIGURE 6.3 Examples of Se-induced deformities in bird embryos. (A) Normally developed black-necked stilt; (B) black-necked stilt with missing eyes, malformed bill, limb deformities and exencephaly; (C) gadwall and (D) northern pintail with arrested development of lower bill, spoonbill narrowing of upper bill, and missing eyes; (E) redhead with spoonbill narrowing of upper bill; and (F) American avocet with club foot and malformed bill (adapted from Seiler et al. 2003).

absence of eyes, and/or the reduction or malformation of the upper, lower, or both bills (beak), and/or the reduction or malformation of the limbs, especially the lower limbs (Hoffman and Heinz 1988; Ohlendorf et al. 1988). At very high exposure concentrations, other uncommon deformities (such as the brain protruding from the eye socket) become expressed. Seiler et al. (2003) documented Se-induced embryo deformities in shorebirds. Deformed embryos almost never hatch and likely die quickly in those rare cases when they do hatch. During 2 decades of monitoring more than 5,000 shorebird nests in the Tulare Basin (California), and despite the documentation of hundreds of deformed embryos, only 2 deformed hatchlings were documented, and both had no eyes, were incapable of feeding, and when placed in water could swim only in circles (Skorupa, personal observation). Thus, deformities must be assessed by collecting eggs that are far enough along in incubation to yield embryos and not by surveying hatched broods. Surveying only hatched broods represents a classic case of survivor bias.

Nonteratogenic embryo mortality occurs at substantively lower egg Se concentrations than are required to induce embryo deformity. In one controlled feeding experiment with mallards, a 7 mg Se/kg dietary exposure caused >30% embryo mortality due to impaired egg hatchability but no teratogenic, embryo deformities (Stanley et al. 1996). Based on experimental data, Ohlendorf (2003) estimated the EC10 for *egg hatchability* to be 12 mg Se/kg in mallard eggs in contrast to an EC10 of 23 mg Se/kg for *embryo deformities* in field-collected duck eggs (Seiler et al. 2003). The gap between the EC10 for hatchability and for teratogenesis is even larger based on Beckon et al.'s (2008) estimated EC10 for mallard egg hatchability of 7.7 mg Se/kg. However, compared to teratogenesis, impaired egg hatchability is less Se-specific and more easily induced by many kinds of stressors. Therefore, causation for impaired egg hatchability in the field can be more difficult to establish with high confidence. Artificial incubation of field-collected eggs is one method of reducing the uncertainty of causation potentially associated with egg hatchability as an endpoint (Peakall and Fox 1987; Smith et al. 1988; Hoffman 1990; Skorupa 1999; Henny et al. 2001). A clear exposure–response relationship can also increase diagnostic confidence for field hatchability data, because the effects of confounding stressors and natural stochastic variation should not normally co-vary with egg Se concentration.

6.3.3 MORTALITIES

6.3.3.1 Fish

Adult fish can accumulate sublethal Se concentrations that can cause mortality of their offspring via maternal Se transfer to eggs. This is a severe manifestation of the maternal Se transfer that can result in larval deformities and edema, as discussed above. At sufficiently high Se concentrations in eggs, larvae are unable to survive. Moreover, larvae that initially survive with severe deformities and/or edema will die if these effects impact their ability to adequately feed or escape predators. Hatchability of fish is generally not affected by Se (Gillespie and Baumann 1986; Coyle et al. 1993; Holm et al. 2005; Muscatello et al. 2006), but Rudolph et al. (2008) observed 100% egg mortality at egg Se concentrations ≥86.3 mg/kg dw. In general, larval mortality is not a diagnostic indicator of Se toxicity because the endpoint is not Se specific and most biological surveys would not detect larval mortality (with the exception, perhaps, of a large die-off following a spill). Of course, the absence of early life stages in itself may indicate Se toxicity. At sufficiently high Se concentrations and under the right conditions, adult fish mortality can also be observed. Hermanutz et al. (1992) exposed adult bluegill sunfish to Se in experimental streams dosed with selenite to concentrations of 10 and 30 µg/L. The streams contained well-developed assemblages of fish food organisms, so an environmentally relevant dietary exposure pathway was simulated. After a 356-d exposure, adult survival was significantly reduced from 99% in the control fish to 84% in the 10 µg Se/L treatment and to 0% in the 30 µg Se/L treatment. In this same study, however, effects on offspring (i.e., larval edema and deformities) were a more sensitive endpoint than adult mortality. Therefore, chronic dietary Se exposures can result in adult mortality under certain conditions, and embryo mortality may occur at extremely high egg Se concentrations, but larval mortality resulting from maternal Se transfer is the most sensitive life stage for the mortality end point.

6.3.3.2 Birds

Reproductive impairment is considered the most sensitive indicator of Se toxicity to birds (Ohlendorf 2003; Seiler et al. 2003). The hatchability of eggs incubated to full term is a frequently measured endpoint in birds. Relating egg hatchability to egg Se concentrations is a complicated process, in the field. By the time a clutch of eggs hatches, only the failed eggs remain for chemical analysis, a biased subset of all eggs. An alternative approach is to randomly select a single egg from a clutch for chemical analysis and to use the measured concentration of Se in the egg as a representation of the Se concentrations in all eggs in the clutch. The Se concentration in the egg is then related to the hatchability of the uncollected sibling eggs whose fate must be monitored by several visits to the nest. A similar approach is used in experimental studies in which Se-dosed diets are fed to captive birds. In the case of field studies, suitable reference areas must be included in the study design, whereas in captive feeding trials, a Se-adequate diet must be fed to control birds. In the case of field studies as opposed to feeding trials, the possibility of extreme within-clutch egg Se variability (cf. Section 6.2.2.2.4) must be considered. However, for large sample sizes, such random error should affect the precision but not the accuracy of exposure–response statistical relationships.

For mallards, it has been clinically demonstrated that much higher dietary exposure to Se is required to induce substantive adult mortality than is required to induce substantive embryo mortality (Heinz 1996; O'Toole and Raisbeck 1997, 1998). Accordingly, only one example of substantive Se-induced adult mortality under field conditions has been documented as opposed to numerous cases of embryo mortality (Skorupa 1998a). Specifically, significant Se-induced adult mortality among American coots (*Fulica americana*) occurred at the Kesterson Reservoir (California) (Ohlendorf et al. 1988), possibly due to their greater reliance on aquatic herbivory, as compared to the co-occurring species of water birds (DuBowy 1989).

From a bioenergetic perspective, DuBowy (1989) illustrated that, even if there was no potency difference, the herbivorous diet would be more dangerous. Vegetation has such a low caloric content that coots must consume far more vegetation than do birds feeding on calorie-rich invertebrates to meet their metabolic caloric requirements. As a consequence, coots likely ingested more Se per day than other bird species, even if the concentration of Se in the coot diet was lower than the concentration in co-occurring bird species' diets. This example illustrates that ultimately it is the mass loading of Se that an animal ingests, and not the concentration that determines the dose. However, in most cases, dietary concentrations are an accurate surrogate for mass loading. One noteworthy indicator of severe adult poisoning among Kesterson coots was alopecia (loss of feathers; Ohlendorf et al. 1988), which has also been induced clinically among Se-dosed adult mallards (Albers et al. 1996; O'Toole and Raisbeck 1998).

6.4   COMPARATIVE SENSITIVITY OF AQUATIC ORGANISMS TO Se

6.4.1   BACTERIA

Bacteria are extremely tolerant of metals and metalloids, and bacteria generally have tremendous tolerance for Se. This tolerance may stem from the ability of some bacteria to sequester selenite in insoluble nodules (Sarret et al. 2005). In addition, bacteria are able to eliminate Se through dissimilatory reduction, while anaerobic bacteria can excrete elemental Se as nanospheres (Oremland et al. 2004). Possibly because of chelation with organic acids, siderophores, and phenols produced by bacteria in the rhizosphere, bacteria increase the efficiency of Hg and Se uptake by wetland plants (Pilon-Smits 2005). Bacteria in aquatic systems accumulate approximately twice as much Se as do phytoplankton, and neither appears to be impaired by its Se uptake (Baines et al. 2004).

6.4.2   ALGAE AND PLANTS

Greater concentrations of Se enter marine food webs because some algae have tremendous tolerance for Se, acquiring relatively high tissue burdens without any apparent effect (Baines and Fisher 2001). For example, algae can bioconcentrate Se to a greater extent than any other trophic level. Accumulation can range by 4 to 5 orders of magnitude among algal species when exposed to 40 to 355 ng/L selenite. Chlorophytes typically accumulated the least Se, whereas prymnesiophytes, prasinophytes, and dinoflagellates had the greatest enrichments. The Se by volume per cell of diatoms and cryptophytes can vary by >2 orders of magnitude (Baines and Fisher 2001). However, within species, the Se cell concentrations are not dose dependent and typically vary by only 2- to 3-fold despite exposure to selenite concentrations with as great as 30-fold differences. As would be expected for an essential nutrient, the greatest accumulation occurs at lower concentrations, meaning that algae take up relatively more Se at low concentrations to maintain a consistent body burden (Baines and Fisher 2001; Baines et al. 2004). Clearly, biological mechanisms have evolved that provide algal species with enhanced Se tolerance through elimination or sequestration. They can efficiently volatize selenomethionine as dimethylselenide (Neumann et al. 2003), sequester it as nonprotein seleno-amino acids such as methylselenocysteine (Brown and Shrift 1982), or accumulate it in an insoluble form as Se^0, which has a relatively low toxicity (Wilber 1980). Formation of Se^0 can occur via a reduction reaction from selenocysteine by the selenolyase enzyme, or from selenite (Garifullina et al. 2003). The high tolerance of algal species for Se is of concern because phytoplankton can concentrate Se to an extent that can cause toxicity at higher trophic levels, even at a selenite concentration as low as 0.04 µg/L (Baines and Fisher 2001; Baines et al. 2004).

6.4.3 PROTOZOANS

Relative to the information available on the effects of Se on bacteria, algae, and plants (above), there is a paucity of information on biological effects on protozoans (i.e., nonphotosynthetic unicellular organisms, including free-living amoebae, zooflagellates, and ciliates), despite the fact that these organisms are effective models for evaluating aquatic toxicity (Lynn and Gilron 1993). Two early laboratory studies evaluated the behavioral (swimming speed), growth, and survival effects of Se on the ciliate, *Tetrahymena pyriformis* (Bovee 1978; Bovee and O'Brien 1982). These studies indicated slightly stimulated growth of *T. pyriformis* at Se water concentrations ranging between 5 and 15 µg/L. Moreover, Se inhibited swimming speed of the ciliate at 5 µg/L and stopped it completely at 30 µg/L. Based on comparative experiments with the ciliate, Bovee (1978) concluded that selenite was more toxic than selenous acid, and that overall growth and survival effects were evident in this species at Se water concentrations of >20 µg/L.

In a subsequent laboratory study using a microbial food web model to investigate the potential accumulation of Se, Sanders and Gilmour (1994) reported that population growth rates of the ciliate, *Paramecium putrinum*, were not inhibited when exposed to Se concentrations (as dissolved selenite or selenate) lower than 1,000 µg/L. This study further concluded, based on 5-day feeding experiments with the ciliate and bacteria, that Se was primarily taken up through the diet, and that biomagnification of Se did not occur at the microbial level. In laboratory microcosm experiments, Pratt and Bowers (1990) reported that protozoan species richness could be reduced by 20% when exposed to concentrations of Se >80 µg/L.

There is a high degree of uncertainty with respect to our understanding of potential toxicity of Se to protozoans, based on the relatively small database. Future research should focus on the establishment of acute and chronic water Se concentration thresholds for protozoans, with potential standardized endpoints relating to behavior, growth, and survival.

6.4.4 MACROINVERTEBRATES

Macroinvertebrates have typically only been considered dietary sources of Se to higher trophic levels, in part based on Lemly's (1993b) statement that prey organisms can remain unaffected even when they accumulate relatively high Se body burdens. Cases of major adverse effects to fish and water birds (e.g., Belews Lake, Hyco Reservoir, Kesterson Reservoir) have not coincided with evidence of major adverse effects on macroinvertebrate communities. However, sensitive species within such communities cannot be ruled out. It was concluded by deBruyn and Chapman (2007) that Se may cause toxic effects in some freshwater invertebrate species at concentrations considered "safe" for their predators. Recent studies with the mayfly *Centroptilum triangulifer* report that dietary exposure to 15 to 30 mg Se/kg dw resulted in a 38% reduction in fecundity with significant maternal transfer of Se to eggs (Conley and Buchwalter, North Carolina State University, unpublished). There is also evidence of Se toxicity to planktonic invertebrates in marine ecosystems at environmentally realistic concentrations (Anastasia et al. 1998; Fisher and Hook 2002). Thus, although Se appears not to have community-level impacts to macroinvertebrates, it may adversely affect sensitive species within those communities.

6.4.5 Fish

Since deformed fish and the loss of fish species observed at Belews Lake were linked to Se contamination in the late 1970s, there has been considerable analysis of Se effects on fish. Following the first full year between 1975 and 1976 that the generating units of the power plant began operation at Belews Lake (and the ash pond effluents reached maximal Se loading), largemouth bass (*Micropterus salmoides*), redbreast sunfish (*Lepomis auritus*), and pumpkinseed (*Lepomis gibbosus*) populations crashed (Van Horn 1978; Appendix A). Other centrarchids, including bluegill sunfish, showed dramatic declines in their population by 1977 (Barwick and Harrell 1997). Bluegill sunfish was the focus of early effects testing with Se at the Belews Lake site and another water body receiving effluent from a coal-fired power plant, Hyco Lake. Sexually mature bluegill females and males collected from the Se-enriched Hyco Lake and reference lakes were cross-fertilized and their offspring evaluated for effects (Bryson et al. 1984, 1985a,b; Gillespie and Baumann 1986). Offspring associated with maternal exposure to Se had reduced larval survival and malformations such as edema. This important finding established that Se was maternally transferred to the eggs and that effects in embryos and larvae were considerably more sensitive than effects in adults.

As described above, field investigations discovered a number of fish species that are sensitive to Se. However, field studies also indicate that certain species are relatively insensitive to Se. Following the extirpation of 16 species from Belews Lake due to Se contamination, four species remained: fathead minnow, common carp, eastern mosquitofish (*Gambusia holbrooki*), and black bullhead (*Ameiurus melas*; Lemly 1985). Similarly, following the declines of *Lepomis* spp., largemouth bass, crappie (*Pomoxis annularis*), yellow perch (*Perca flavescens*), and sucker (*Catostomus*) species in Hyco Lake, green sunfish (*Lepomis cyanellus*), satinfin shiner (*Notropis analostanus*), gizzard shad (*Dorosoma cepedianum*), eastern mosquitofish, and redbelly tilapia (*Tilapia zillii*) dominated the fish community (Crutchfield 2000). The western mosquitofish (*Gambusia affinis*) is also relatively insensitive based on reproductive studies by Saiki et al. (2004). Different fish species can thus have differential sensitivity to Se.

When evaluating Se toxicity, it is important to distinguish reproductive from nonreproductive threshold effects for fish because of distinctions in exposure routes.

In reproductive studies, female fish are exposed to Se and the Se is maternally transferred to their eggs during vitellogenesis (see Section 6.2.2.2). In laboratory (Doroshov et al. 1992; Coyle et al. 1993; Hardy et al. 2009) and mesocosm (Hermanutz et al. 1996) studies, parent fish were exposed to Se via the diet and water or diet only. After spawning, the embryos and larvae were monitored for effects. More often, the reproductive effects of Se have been assessed by the field collection of gametes from sites with elevated Se levels and from reference sites (Kennedy et al. 2000; Holm et al. 2005; Muscatello et al. 2006; GEI 2008a; Rudolph et al. 2008; Muscatello and Janz 2009; NewFields 2009). Eggs are most often fertilized in the field and then transported to the laboratory, where embryos and larvae are monitored for effects. In contrast, the larval and juvenile fish in nonreproductive testing had no preexposure to Se from their mother but were fed a series of Se concentrations. Growth and survival were the typical endpoints in such juvenile exposure studies.

The comparative sensitivity of the reproductive endpoints, the EC10 values based on Se concentrations in the egg or ovary, are reasonably similar (Table 6.2). The range of EC10 (or equivalent threshold) values for the fish species shown are within a factor of 1.4 (17 to 24 mg Se/kg dw in eggs; Figure 6.4). Due to testing limitations, lack of response, or excessive variability in response to Se, effect concentrations could not be determined for some studies. One study not listed in Table 6.3 (GEI 2008a) compared larval malformations from fathead minnows collected from streams with elevated Se concentrations and spawned in the laboratory to those of laboratory-reared fathead minnows. Although a wide range of Se concentrations were measured in the female fathead minnows (2 to 47 mg/kg dw whole body) an effect level could not be estimated due to the considerable variation in the endpoint showing the greatest response to Se (graduated severity index of the malformations). The 8 species in Table 6.2 with EC10 values represent both cold water and warm water fish. The overall similarity in reproductive effect levels for these eight species suggests little difference in Se sensitivity between warm water and cold-water fish species. However, further studies are required to confirm this observation.

Fewer EC10 values were determined for nonreproductive endpoints, but their 2-fold range in effect concentration is not particularly large (Table 6.3). Bluegill sunfish and chinook salmon (*Oncorhynchus tshawytscha*) were the more sensitive species with a concentration for winter stress in bluegill at 5.85 mg Se/kg dw whole body and an EC10 for decreased growth in juvenile chinook salmon at 7.34 mg Se/kg dw. As may be the case with the reproductive endpoints, the effect concentration for larval growth in fathead minnows was the greatest, at 51.4 mg Se/kg dw whole body. The relatively higher reproductive and non-reproductive effect values for fathead minnows are consistent with field observations at Belews Lake where fathead minnows were 1 of 4 remaining species after Se contamination extirpated 16 relatively sensitive species (see above).

Hamilton et al. (2005a, 2005b, 2005c) and Beyers and Sodergren (2001a,b) also evaluated Se toxicity to razorback suckers (*Xyrauchen texanus*), but due to uncertainties in whether effects observed in the Hamilton et al. (2005a,b,c) studies could be solely attributed to Se (cf USEPA 2004) and discrepancies in the Se effects levels from the Hamilton et al. (2005a,b,c) and Beyers and Sodergren (2001a,b) studies, Se toxicity data for razorback sucker are not included here.

TABLE 6.2

Data Illustrating the Range of Assessment Values for Reproductive Effects of Se in Fish

Species	Reference	Toxicological Endpoint	Effect conc'n, mg Se/kg dw[a]
Oncorhynchus mykiss Rainbow trout	Holm et al. 2005	EC10 for skeletal deformities	21.1 E
Oncorhynchus clarki Cutthroat trout	Kennedy et al. 2000	NOEC for embryo/larval deformities and mortality	>21.2 E
Oncorhynchus clarki Cutthroat trout	Hardy et al. 2009	NOEC for embryo/larval deformities	>16.04 E
Oncorhynchus clarki Cutthroat trout	Rudolph et al. 2008	EC10 for alevin survival	17–24.1 E
Salvelinus fontinalis Brook trout	Holm et al. 2005	NOEC for craniofacial deformities	>20.5 E
Salmo trutta Brown trout	NewFields 2009	EC10 for larval survival	17.7 E
Esox lucius Northern pike	Muscatello et al. 2006	EC10 larval deformities	20.4 E
Pimephales promelas Fathead minnow	Schultz and Hermanutz 1990	LOEC for larval edema and lordosis	<24 O
Lepomis macrochirus Bluegill sunfish	Bryson et al. 1984	LOEC for larval mortality	<49.65 O
Lepomis macrochirus Bluegill sunfish	Bryson et al. 1985a	Swim-up larvae	<30; >9.1 O
Lepomis macrochirus Bluegill sunfish	Gillespie and Baumann 1986	Larval survival	<46.30 O
Lepomis macrochirus Bluegill sunfish	Doroshov et al. 1992	EC10 larval edema	21.2 E
Lepomis macrochirus Bluegill sunfish	Coyle et al. 1993	EC10 for larval survival	24.10 E
Lepomis macrochirus Bluegill sunfish	Hermanutz et al. 1996	NOAEC and LOAEC for larval survival, edema, lordosis, and hemorrhaging	14.0; 42.1 O
		NOAEC for larval survival, edema, lordosis, and hemorrhaging	≥16.3 O
Micropterus salmoides Largemouth bass	Carolina Power & Light 1997	Threshold for larval mortality and deformities	19 O

Note: Concentrations (dw = dry weight) of Se in fish tissues (E = egg, O = ovary) relative to endpoints.
[a] Effect concentrations based on measured or estimated Se concentration in egg or ovary tissues.

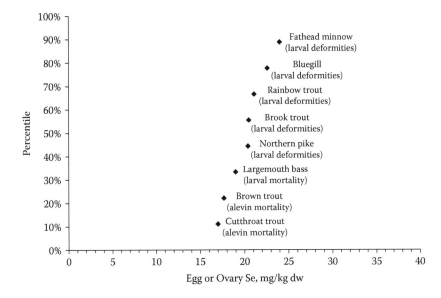

FIGURE 6.4 Distribution of egg- or ovary-based EC10 values (or comparable values if EC10 values could not be calculated). The value shown for brook trout is a NOEC because an EC10 could not be determined (from Holm et al. 2005). The value shown for bluegill is the geometric mean of EC10 values for larval edema (from Doroshov et al. 1992) and larval deformities (from Coyle et al. 1993). The value shown for fathead minnows is an unbounded LOEC associated with 25% larval edema and lordosis (from Schultz and Hermanutz 1990). See Table 6.2 for all toxic effect concentrations.

As discussed above, Se toxicity studies with fish can be broadly classified as either 1) maternal transfer studies in which effects are evaluated in the offspring of Se-exposed fish or 2) dietary Se exposure studies in which effects are evaluated in juveniles. Because these exposure routes are so fundamentally different, from a risk assessment perspective it is important to understand which exposure route is most environmentally relevant and sensitive. Relative sensitivity cannot be inferred by comparing tissue-based toxicity thresholds due to potential differences in bioaccumulation between adults in the maternal transfer studies and juveniles in the direct toxicity studies (DeForest 2008). To truly compare relative sensitivity, the exposure (i.e., dietary Se) concentrations must be compared. Laboratory toxicity data are necessary to make this comparison because in field-based exposure studies the Se concentration in the diet is unknown.

The only species for which both maternal transfer (Woock et al. 1987; Doroshov et al. 1992; Coyle et al. 1993) and juvenile toxicity studies (Cleveland et al. 1993; Lemly 1993c; McIntyre et al. 2008) have been conducted is bluegill sunfish (razorback sucker studies discussed above (Beyers and Sodergren 2001a, 2001b; Hamilton et al. 2005a, 2005b, 2005c) are excluded). As shown in Figure 6.5, the concentration–response relationships were similar among the three maternal transfer studies,

TABLE 6.3

Data Illustrating the Range of Assessment Values for Nonreproductive Effects of Se in Fish

Species	Reference	Toxicological Endpoint	Chronic Value, mg Se/kg dw[a]
Acipenser transmontanus White sturgeon	Tashjian et al. 2006	EC10 juvenile growth	15.08 WB
Oncorhynchus tshawytscha Chinook salmon	Hamilton et al. 1990	EC10 juvenile growth (mosquitofish diet)	11.14 WB
		EC10 juvenile growth (SeMet diet)	7.354 WB
Oncorhynchus mykiss Rainbow trout	Hilton and Hodson 1983	Juvenile growth NOAEC	21.0 L
	Hicks et al. 1984	LOAEC	71.7 L
Oncorhynchus mykiss Rainbow trout	Hilton et al. 1980	Juvenile survival and growth NOAEC	40 L
		LOAEC	100 L
Pogonichthys macrolepidotus Sacramento splittail	Teh et al. 2004	Juvenile deformities NOAEC	10.1 M
		LOAEC	15.1 M
Pimephales promelas Fathead minnow	Bennett et al. 1986	Chronic value for larval growth	51.40 WB
Lepomis macrochirus Bluegill sunfish	Lemly 1993c	LOAEC juvenile mortality at 4 °C	<7.91 WB
		Threshold prior to "winter stress"	5.85 WB
		NOAEC juvenile mortality at 20 °C	>6.0 WB
Lepomis macrochirus Bluegill sunfish	McIntyre et al. 2008	EC10 juvenile survival at 4 °C *Lumbriculus* diet	9.27 WB
		EC10 juvenile survival at 9 °C *Lumbriculus* diet	14.00 WB
		NOAEC juvenile survival at 4 °C	>9.992 WB
		Selenomethionine in Tetramin diet	
Lepomis macrochirus Bluegill sunfish	Cleveland et al. 1993	NOAEC for juvenile survival	>13.4 WB
Morone saxitilis Striped bass	Coughlan and Velte 1989	LOAEC for survival of yearling bass	<16.2 M

Note: Concentrations (dw = dry weight) of Se in fish tissues (WB = whole body; M = muscle; L = liver) relative to endpoints.

[a] All chronic values based on measured Se concentration in whole body, muscle tissues, or liver.

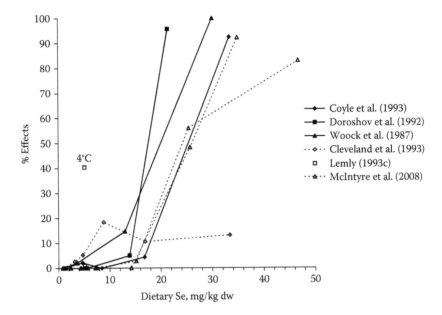

FIGURE 6.5 Effects of Se on bluegill sunfish larvae (from maternal transfer; solid lines) and juveniles (from direct dietary exposures; dashed lines) as a function of dietary Se.

with little effect (larval mortality, edema) up to a dietary threshold of approximately 12 mg Se/kg dw, followed by a rapid increase to a 90% to 100% effect level at concentrations above approximately 21 mg Se/kg dw. The juvenile results were more variable. The toxicity data from Cleveland et al. (1993) did not show the same rapid increase in Se toxicity, and McIntyre et al. (2008) reported a pattern similar to the maternal transfer studies (i.e., it appears that significant juvenile mortality does not begin to occur until dietary Se concentrations reach approximately 14 to 15 mg/kg dw; Figure 6.5). The single data point from Lemly (1993c) at 4 °C suggests high mortality at a dietary concentration of 5 mg Se/kg dw, but a concentration-response could not be evaluated because only a single Se treatment was used.

Overall, the limited data preclude strong conclusions on the relative sensitivity of maternal transfer versus dietary uptake by juveniles. The data for bluegill sunfish suggest that maternal transfer is more toxic or similarly toxic to direct juvenile ingestion. However, more studies are needed to develop predictive capability.

6.4.5.1 Selenium Concentration Relationships between Fish Tissues

Se concentrations in different tissues can be highly variable among species, as shown in Figure 6.6 for egg and muscle Se (deBruyn et al. 2008). Therefore, tissue–tissue relationships should not be used generically to derive tissue-based Se toxicity thresholds. Regression equations for specific species can estimate Se concentrations in one tissue from measurements in another. However, even within a species tissue–tissue extrapolations should ideally be site-specific because individuals show considerable intraspecific variation in the ratio between egg/ovary and whole-body Se.

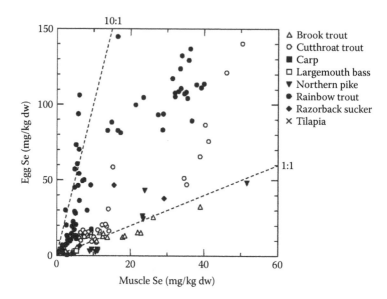

FIGURE 6.6 Egg Se concentrations relative to muscle Se concentrations for 8 fish species (from deBruyn et al. 2008).

For example, the ratio of egg to whole-body Se for black bullhead ranged from 3.1 to 27.9 (average 9.1; $n = 24$) (Osmundson et al. 2007). On the other hand, some species such as bluegill and green sunfish show minimal variation in this relationship (Figure 6.7). After a species-specific tissue–tissue relationship has been developed, any of the candidate tissues should be a reliable surrogate for early life stage Se exposure. If no species-specific tissue–tissue relationship is available, it is not possible to use adult tissue Se to reliably estimate potential early life stage exposure.

6.4.6 AMPHIBIANS

Although amphibians appear to be an extremely ecologically and toxicologically vulnerable class of vertebrates, and pollution has been implicated in some amphibian population declines (Stuart et al. 2004; Hopkins 2007; Wake and Vredenburg 2008), the effects of Se on amphibians are largely unknown. All amphibian Se effects studies currently available were based on complex mixtures, of which Se was only one component. In these situations causal relationships with Se are difficult to establish, but some observed effects are similar to those found in other vertebrates exposed to Se.

When exposed to coal fly ash (containing a complex mixture of trace elements enriched with Se) amphibian larvae (tadpoles) appear to efficiently accumulate Se in their tissues (Unrine et al. 2007a), probably because of their close association with the benthos and their tendency to ingest particulates while grazing biofilms. Associated with elevated concentrations of Se in their tissues, amphibian larvae exhibited increased incidence of axial malformations (Hopkins et al. 2000), similar to those described for fish exposed to Se in Belews Lake (North Carolina;

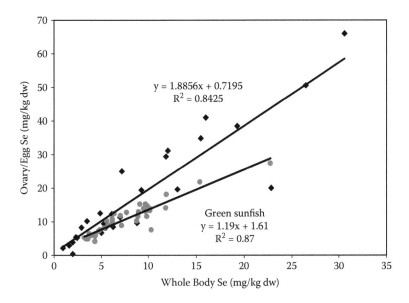

FIGURE 6.7   Ovary/egg Se concentrations relative to whole-body Se concentrations for blue-gill sunfish (data compiled from Osmundson et al. 2007; Doroshov et al. 1992; Coyle et al. 1993; Hermanutz et al. 1996) and green sunfish (Osmundson et al. 2007; Doroshov et al. 1992).

Appendix A). Deformations of the keratinized mouthparts of amphibian larvae also occur (Rowe et al. 1996, 1998), and synchrotron X-ray fluorescence demonstrated colocalization of high Se concentrations in these deformed areas (Punshon et al. 2005). Spinal and oral abnormalities affected swimming and feeding performance (Hopkins et al. 2000; Rowe et al. 1998). Additional studies on amphibian larvae with elevated whole-body Se concentrations documented reductions in growth (Rowe et al. 1998; Snodgrass et al. 2004, 2005), altered predator avoidance capabilities (Raimondo et al. 1998), reduced larval survival (Rowe et al. 2001; Snodgrass et al. 2004, 2005; Roe et al. 2006), and altered time and size at metamorphosis (Snodgrass et al. 2004, 2005; Roe et al. 2006). Studies are needed to determine to what extent Se is responsible for these aberrations.

Unlike most trace elements, Se is retained in amphibian tissues as they undergo metamorphosis (Snodgrass et al. 2003, 2004, 2005). This has important implications for amphibian health during this critical life history transition and during sensitive early terrestrial life stages. Thereafter, terrestrial life stages of amphibians can continue to accumulate Se from invertebrate prey, which are often closely associated with the aquatic environment (Roe et al. 2005; Hopkins et al. 2006). Only two studies have considered the effects of Se on adult amphibians. In the first study, Hopkins et al. (1997, 1998, 1999a) demonstrated that adult male toads (*Bufo terrestris*) with elevated Se and other elements in their tissues exhibited abnormal hormonal profiles during the breeding season. More recently, Hopkins et al. (2006) found that adult narrow-mouthed toads (*Gastrophryne carolinensis*) maternally transfer elevated concentrations of Se (up to 100 mg/kg dw) to their eggs. Compared to reference conditions, eggs from the contaminated site displayed significant reductions in hatching. In addition, there was a higher prevalence of malformed hatchlings from the contaminated site, most of which also exhibited abnormal swimming behavior. Importantly, craniofacial abnormalities, which may be diagnostic of Se teratogenicity, were most common in hatchlings at the contaminated site. This work strongly suggests that additional studies of Se maternal transfer and embryotoxicity are needed and that future studies should be designed to describe Se concentration–response relationships to facilitate comparisons to fish and birds.

6.4.7 REPTILES

Much like amphibians, little is known about the effects of Se on reptiles. Field and laboratory studies on water snakes (*Nerodia fasciata*) demonstrate that elevated concentrations of Se and other contaminants (e.g., As and Cd) are accumulated from ingestion of amphibian and fish prey in a coal ash–contaminated site (Hopkins et al. 1999b, 2001, 2002, 2005a). At lower levels of Se accumulation, exposure to seleniferous prey had no effect on growth, survival, overwinter survival, or metabolism (Hopkins et al. 2001, 2002). However, about one-third of snakes at these same exposure levels exhibited histopathological abnormalities, most notably liver necrosis (Ganser et al. 2003). At higher levels of tissue accumulation (mean liver Se ~140 mg Se/kg dw) in the field, snakes exhibited abnormally high respiratory rates, suggesting significant energetic costs associated with exposure (Hopkins et al. 1999b). At the same field site, maternal transfer of Se was examined in both turtles (red-eared sliders, *Trachemys scripta*) and American alligators (*Alligator mississippiensis*). Despite enormous differences in feeding ecology between these two species (i.e., alligators are top predators that even eat adult turtles), both species maternally transferred approximately 7.5 mg Se/kg dw to their eggs (Nagle et al. 2001; Roe et al. 2004). For comparison, common grackles (*Quiscalus quiscala*), eastern mosquitofish, and narrow-mouthed toads from the same site maternally transferred approximately (means) 6 mg Se/kg, 16 mg Se/kg, and 44 mg Se/kg dw, respectively (Bryan et al. 2003; Staub et al. 2004; Hopkins et al. 2006). This observation further supports the concept that Se does not biomagnify; the majority of food web enrichment in Se occurs at the lowest trophic levels, with comparable exposure possible for secondary (e.g., *Trachemys scripta*) and tertiary (e.g., *Alligator mississippiensis*) consumers (Chapter 5). Neither of these studies on reptiles was designed to rigorously quantify relationships between Se exposure, maternal transfer, and reproductive effects. However, anecdotal observations suggested that hatching success was consistently low for 3 years in the adult alligator monitored at the contaminated site (Roe et al. 2004).

In the laboratory, snakes and lizards have been exposed to Se in isolation from other contaminants, but primarily to study trophic transfer and bioaccumulation. In a simplified laboratory food chain, lizards receiving ~15 mg Se/kg dw for 98 days accumulated whole-body concentrations of ~10 mg Se/kg and exhibited no changes in food consumption, growth, or survival (Hopkins et al. 2005b; Unrine et al. 2006, 2007a). Reproductive effects were not examined, but females partitioned about 33% of their total Se burden into their yolked follicles. Likewise, snakes raised on an experimental diet containing 10 and 20 mg Se/kg dw (as seleno-DL-methionine) for

10 months showed no adverse effects on food consumption, growth, and body condition (Hopkins et al. 2004). Although only limited conclusions can be drawn regarding reproductive effects from this study, available evidence suggested that fewer females exposed to elevated dietary Se were reproductively active. Among individuals that did reproduce, maternal transfer of ~22 mg Se/kg dw to eggs did not adversely affect hatching success or malformation frequency. Clearly, additional studies adopting similar approaches, but with a primary focus on reproductive effects, are critical to advancing our understanding of the sensitivity of reptiles compared to birds and fish (Hopkins 2000, 2006).

6.4.8   BIRDS

The results of field and captive-feeding studies indicate widely variable responses among species of birds to in ovo Se exposure. Field studies in areas receiving irrigation drain water in the western United States have examined the incidence of embryonic deformities in ducks, black-necked stilts, and American avocets and related such information to embryonic Se concentrations (Seiler et al. 2003). Terata considered in that study included major structural deformities that were overtly obvious upon superficial examination. They were limited to major deformities of the eyes, bill, or limbs, whereas nonstructural abnormalities like hydrocephaly and generalized edema were not considered. There were no significant differences between mallards and other duck species, which included gadwalls (*Anas strepera*), pintails (*Anas acuta*), and redheads (*Aythya americana*) in the relationship between embryonic Se concentrations and the frequency of deformities, justifying the pooling of these species in further analyses. The comparison among ducks, black-necked stilts, and American avocets revealed that ducks were more sensitive to in ovo Se exposure than black-necked stilts, which were, in turn, more sensitive than American avocets (Figure 6.8). Fifty percent probabilities of teratogenic effects were calculated to occur at concentrations of 30, 58, and 105 mg Se/kg egg for ducks, stilts, and avocets, respectively. Corresponding Se concentrations related to a 10% probability (i.e., EC10) of teratogenesis were 23, 37, and 74 mg Se/kg, respectively. The results of the Seiler et al. (2003) study suggested that black-necked stilts are about twice as sensitive as avocets and that ducks are about 3.5 times as sensitive as avocets to the teratogenic effects of Se. Teratogenic effects of Se have been reported for killdeer (*Charadrius vociferus*) (Ohlendorf 1989; Skorupa 1998b). Based on data compiled from additional studies (Skorupa, unpubublished data) and controlled for equivalent exposure at a concentration range high enough to quantify a response for the insensitive avocet, killdeer show a degree of sensitivity intermediate between that of avocets and black-necked stilts (Figure 6.8).

Teratogenesis is a less sensitive measure of selenosis in birds than is the hatchability of their eggs when incubated to full term. Captive studies have been conducted in which mallards were fed diets containing various levels of Se (Heinz et al. 1987, 1989; Stanley et al. 1994, 1996; Heinz and Hoffman 1996, 1998). Depending on how the data in those studies were analyzed, it was calculated that a 10% hatch failure rate would correspond with egg Se concentrations (in dry weight) of 7.7 mg/kg (Beckon et al. 2008), 12 mg/kg (Adams et al. 2003), or 12.5 mg/kg (Ohlendorf 2003) (Table 6.4). Field data for balck-necked stilts also support the conclusion that egg inviability, which

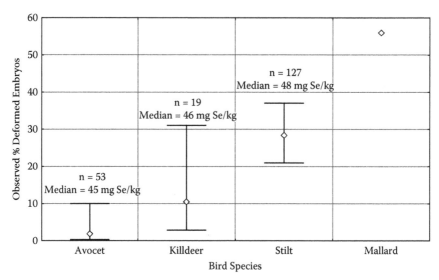

FIGURE 6.8 Comparative embryonic sensitivity of waterbird species to Se based on eggs containing 40 to 60 mg Se/kg. Whiskers are the binomial 95% confidence limits. Data for avocets, killdeer, and stilts are from Skorupa, unpublished field data; comparable field data for mallards in the 40 to 60 mg Se/ kg exposure range are unavailable, the point plotted here for mallards is from laboratory data in Heinz et al. 1989.

is expressed as clutch inviability if at least one egg in the clutch did not hatch, is a more sensitive endpoint that teratogenesis. Clutch-wise EC10s for black-necked stilts, again depending on how the data were analyzed, have been reported to correspond with egg Se concentrations (in dry weight) of 6 to 7 mg/kg (USDOI 1998) and 14 mg/kg (EC11.8; Lam et al. 2005). Adams et al. (2003) used the equation from Skorupa (1998b) to relate the probability of a given egg Se concentration resulting in an inviable egg or an inviable clutch. They then derived EC10 values ranging from 21 to 31 mg/kg dw for stilt egg inviability (Table 6.4). Because the studies of stilt and mallard egg hatchability are based on different sampling units (affected clutches for stilts and affected eggs for mallards), estimates of EC10 values across these two species for egg hatchability do not necessarily reflect relative sensitivity (Skorupa 1999). The apparent overlap in mallard egg inviability EC10s and stilt clutch-wise EC10s suggests that mallards and stilts may be similarly sensitive to Se; however, the analysis of Adams et al. (2003) suggests stilts may be less sensitive. Mallards and black-necked stilts are clearly more sensitive to Se than the American avocet, for which hatchability does not begin to decline until Se concentrations in eggs exceed 60 mg Se/kg (USDOI 1998; Table 6.4). In a captive-feeding study, hatchability of black-crowned night-heron (*Nycticorax nycticorax*) eggs was not adversely affected by dietary exposure to selenomethionine, resulting in a mean concentration of 16.5 mg Se/kg dw in eggs, suggesting that this species is less susceptible to reproductive impairment than mallards (Smith et al. 1988; Table 6.4). However, because an effect-threshold was not established in that study, heron susceptibility relative to that of more tolerant species like the American avocet remains unknown. In red-winged blackbirds (*Agelaius phoeniceus*), the estimated threshold

TABLE 6.4
Hatchability of Bird Eggs in Relation to Se Concentrations in Eggs

Species	Concentration (mg Se/kg dw)	Effect	Considerations	References
Mallard	7.7–15	Hatchability EC10	Based on analysis of results from 5 to 6 captive-feeding studies	Adams et al. 2003 Ohlendorf 2003, 2007 Beckon et al. 2008
Black-necked stilt	6–7	Threshold point for hatchability effects	Field study where eggs were randomly selected from each clutch and Se hatch success compared to that of group with a lower range of Se concentrations	USDOI 1998
	21–31	Hatchability EC10	Same field study as above but different data analysis approach used	Adams et al. 2003
Black-crowned night heron	16.5	NOAEL	Captive feeding study—mean egg Se concentration for group fed Se in their diet	Smith et al. 1988
American avocet	60	Low bound of a concentration range associated with reproductive impairment in 20% of clutches	Field study	USDOI 1998
Red-winged blackbird	22	Threshold for adverse effects	Field study examined hatchability of eggs incubated to full term	Harding 2008
Eastern screech owl	37	5% hatchability of incubated eggs (adjusted to hatchability of control eggs)	Captive study in which parent birds were fed diet containing 13.2 ppm Se (wet Wt)	Wiemeyer and Hoffman 1996
American kestrel	25	Hatchability NOAEL	Captive study in which parents were fed a diet containing 12 ppm Se (dry wt)	Santolo et al. 1999

for reproductive impairment was approximately 22 mg Se/kg dw egg (Harding 2008; Table 6.4). Thus, it appears that reproductive impairment at the EC10 level may begin in the range 10 to 20 mg Se/kg dw in eggs of several species. However, some species, such as the American avocet and possibly the black-crowned night-heron, remain unaffected at higher Se concentrations.

Mechanisms that might explain why a given concentration of Se is more likely to cause reproductive impairment in some species than in others have not yet been elucidated. Hoffman et al. (2002) examined sublethal effects and oxidative stress in stilts and avocets from two Se-contaminated sites and a reference site. Oxidative stress was greater in avocets from the most contaminated site than in those from the other sites, while oxidative stress in stilts was not noticeably higher at the most contaminated site. Selenium concentrations in avocet hatchlings at the most contaminated site averaged about 30 mg Se/kg, whereas stilt embryos averaged 21 mg Se/kg at that site. Despite the greater concentrations of Se in avocets than in stilts, it remains surprising that oxidative stress appeared less severe in the latter species given that it is about twice as sensitive as the former to the teratogenic effects of Se. Thus, in the case of these 2 species, there is no evidence that their relative susceptibility to Se-induced teratogenicity can be explained by inter-specific differences in the potential of Se to produce oxidative stress.

The documented inter-specific variability in sensitivity of reproductive impairment to Se exposure suggests that predicting the severity of toxic effects in an array of species in a field situation based on the results of laboratory toxicity tests using different "indicator" species may be fraught with uncertainty. Moreover, as evidenced from the disparity between stilts and avocets, even closely related species can be differentially susceptible to Se. The assessment of Se risk to aquatic-dependent species of birds would be improved by a better understanding of the physiological and biochemical mechanisms underlying embryotoxicity of Se in birds.

6.4.9 MAMMALS

Mammalian species that rely on water for their food such as mink (*Mustela vison*), otter (*Lutra canadensis*), muskrat (*Ondatra zibethicus*), raccoon (*Procyon lotor*), and beaver (*Castor canadensis*) also have the potential for significant Se accumulation via dietary exposure in aquatic settings. Studies conducted in both contaminated (Clark 1987; Clark et al. 1989) and non-contaminated (Wren 1984; Wren et al. 1986; Gamberg et al. 2005) settings clearly demonstrate the ability of these semiaquatic mammals to accumulate Se.

In the 1984 studies on Kesterson National Wildlife Refuge (California), Clark (1987) collected 332 mammalian organisms comprising 10 different species. Of those species, muskrats comprised 18 organisms split between Kesterson ($n = 11$) and a reference site (Volta Wildlife area, $n = 7$). Average liver Se concentrations for Kesterson organisms were as high as 32.1 mg/kg dw, while Volta liver concentrations were 1.5 mg/kg dw. By contrast, other rodent species such as the house mouse ($n = 18$; *Mus musculus*), western harvest mouse ($n = 45$; *Reithrodontomys megalotis*), and California ground squirrel ($n = 5$; *Spermophilus beecheyi*), whose diet consisted principally of vegetation not associated with pond water, had liver concentrations of 14.5, 15.3, and 1.82 mg Se/kg dw, respectively. However, other small mammals not often considered "aquatic-dependent," such as the California vole (*Microtus californicus*), deer mouse (*Peromyscus maniculatus*), and the ornate shrew (*Sorex ornatus*), also had elevated Se levels in liver with the shrew having the highest average concentration of 92.7 mg/kg dw ($n = 8$). After noting these elevated concentrations in small mammals, Clark et al. (1989) also collected raccoons from Kesterson ($n = 8$) and Volta ($n = 8$) to investigate accumulation and potential effects on a tertiary mammalian predator. Selenium concentrations in blood, liver, hair, and feces from animals trapped on

Kesterson averaged 2.61 mg/L, 19.9, 28.3, and 21.6 mg Se/kg dw, respectively, compared to concentrations of 0.27 mg/L, 1.69, 0.93, and 1.05 mg Se/kg dw in the same tissues collected from Volta raccoons. Gamberg et al. (2005) documented Se concentrations in the tissues of mink from noncontaminated areas of the Canadian Yukon. Average total Se concentrations ($n = 98$) found in kidney, liver, and brain were 7.45, 4.5, and 1.70 mg Se/kg (converted from wet weight concentrations using % moisture data provided in the paper). In a relatively pristine setting in an undisturbed watershed in south central Ontario (Canada), Wren (1984) reported similar low Se concentrations in the liver, kidney, intestine, and muscle of raccoon, beaver, and otter.

Regarding sensitivity of mammals, none of the documented incidences of elevated Se concentrations in tissues indicated any negative impacts on the organisms. The most quantitative studies conducted were the Clark (1987) and Clark et al. (1989) studies completed on Kesterson National Wildlife Refuge. In Clark (1987), 88 California voles were captured on Kesterson and 89 were captured on Volta. Sex ratios, condition, organ weights, and reproductive status were observed and were not significantly different for most of the endpoints. Interestingly, no pregnant females were found on Kesterson (0/50), whereas 12 out of 29 females were found pregnant on Volta. The author attributed this finding not to Se, but rather to other factors, such as diet, that resulted in different reproductive schedules between the two sites. In the raccoon study (Clark et al. 1989), body condition, blood parameters, histopathology, and evidence of pregnancy were investigated in the animals. No effects were noted, and one pregnant female from Kesterson was trapped. In ecological settings such as Kesterson, aquatic-dependent mammalian species appear unaffected by high Se exposure despite extirpation of fish populations and severe effects on avian waterfowl (Ohlendorf et al. 1986). In fact, studies (e.g., Wren et al. 1986; Khan and Wang 2009) suggest that Se in the tissues of wild mammals has beneficial effects in the binding of mercury, thereby reducing its impacts to the organism. It should be noted, however, that relative to the avian and fish literature, very little quantitative, robust data exist that rigorously examine the effects of Se in wild mammalian species; moreover, no controlled dosing experiments were found. From the available toxicity data for laboratory rodents, the most sensitive endpoint subsequent to Se exposure appears to be growth (USEPA 2007). This subtle endpoint would be difficult to document in field studies with wild species.

One likely explanation for the lower sensitivity of mammals compared to other vertebrates, such as fish and birds in these Se-contaminated settings, is the differences they have in the ratio of essentiality and toxicity compared to fish and birds. As stated at the beginning of this chapter, that ratio might range from 7 to 30 for fish and birds (but typically less than 10), while this ratio in mammals is greater. NRC (2006) reviewed nutritional adequacy and toxicity for several elements for laboratory rodents (Watson 1996) and small ruminants. For laboratory rodents, they recommended a diet with 0.15 mg Se/kg as adequate, whereas a diet containing 5.0 mg Se/kg may lead to effects on growth in weanling pups (i.e., 33-fold difference). The recommendation for cervids and other ruminants for adequate Se in the diet is 0.25 mg/kg^{-1}, whereas 5.0 mg/kg^{-1} in food is the maximum tolerable level (20×). No data exist for wild aquatic-dependent mammalian species, but of these two ratios, the species of most potential concern (i.e., small to midsize mammals) the ratio of 33-fold difference between essentiality and toxicity is likely the most pertinent.

6.4.10 COMPARATIVE SENSITIVITY SUMMARY AND IMPLICATIONS FOR SE THRESHOLD DEVELOPMENT

Fish and birds are generally the most sensitive taxa to Se in aquatic systems; the sensitivity of reptiles and amphibians is less understood. There are several options for evaluating whether environmental concentrations of Se at a site are potentially toxic. For example, Se may be measured in environmental media, including water, sediment, fish, or bird diets, fish tissue, and bird tissue. Further, different tissues may be analyzed. Tissues commonly analyzed in fish are whole body, muscle, liver, ovaries, and eggs. Overall, there is a clear consensus that tissue Se is the most reliable indicator of toxic effects in the field (Chapter 7). Previous sections have discussed the many site-specific factors that influence Se speciation and bioaccumulation, such as variable intake of Se-rich foods, factors that ultimately dictate whether the Se in an aquatic system is toxic for fish or birds. However, by measuring Se in fish or bird tissues, site- and species-specific variation in Se bioaccumulation can be determined. Over the past 15 years, several studies have recommended tissue-based Se benchmarks for fish and birds (Lemly 1993b, 1996a; USDOI 1998; DeForest et al. 1999; Hamilton 2002; Ohlendorf 2003; Adams et al. 2003; Chapman 2007). Moreover, the USEPA (2004) has developed a draft fish tissue-based Se criterion. Although there is not always consensus on the benchmarks recommended, there is consensus that tissue-based Se benchmarks are the most appropriate medium for linking Se concentrations to toxicity. The next step is to develop toxicity studies that directly relate Se toxicity to the internal Se concentration in the organism.

For fish and birds, the most appropriate tissue for linking Se concentrations with toxicity is still under debate. As discussed above, for both fish and birds, the critical exposure route and endpoint for Se toxicity is maternal transfer of Se to the eggs and subsequent effects on either the developing larvae (fish) or embryo (birds). As such, egg Se appears to be the most appropriate tissue for linking fish or bird Se exposure to toxicity. For fish, likely Se benchmarks are eggs and ovaries because interspecific egg- and ovary-based Se EC10 values are reasonably consistent across species tested to date (Figure 6.4). For birds, use of egg Se rather than dietary Se is beneficial because many birds are highly mobile, and egg Se is a direct reflection of Se exposure during the critical reproductive period. The use of these toxicity thresholds in site-specific evaluations is discussed further in Section 6.7.1.

6.5 FACTORS THAT MODIFY Se TOXICITY

6.5.1 INTERACTIONS

Table 6.5 summarizes the known interactions of Se with other factors (e.g., metals/metalloids, biotic and abiotic stressors). Interactions with other trace elements (e.g., As, Cd, Cu, Pb, Hg, S, Tl, Sb, Pb, Zn) are typically antagonistic, although there are three exceptions, as noted below. At extremely high concentrations, these antagonistic effects may be reduced or nonexistent (Stanley et al. 1994).

The most well-known antagonistic reactions occur between organic Hg (methyl Hg [meHg]) and organic Se; both reduce the bioavailability and thus the toxicity of

TABLE 6.5

Selenium Interactions with Other Factors

Factor	Organism	Interaction	Reference
Arsenic (As)	Chicken	As reduced Se toxicity (e.g., effects on growth, egg production and weight, hatching success)	Hill 1975; Howell and Hill 1978; Thapar et al. 1969
	Mallard	As improved hatching success and reduced embryo mortality due to Se	Stanley et al. 1994
		As reduced duckling mortality, growth, and hepatic lesions due to Se; Se similarly reduced As toxicity	Hoffman et al. 1992a
	As plant hyperaccumulator	Se alleviated As oxidative stress and improved As uptake essential to this plant	Srivastava et al. 2009
	Rat	As reduced Se toxicity	Moxon 1938; Dubois et al. 1940; Palmer et al. 1983
		Synergistic toxicity of As and organic Se	Kraus and Ganther 1989
	Dog	As prevented Se poisoning in dogs	Rhian and Moxon 1943
Boron (B)	Mallard	Bo and Se synergistically suppressed growth, altered blood protein and, with restricted dietary protein, decreased survival	Hoffman et al. 1991b
		No interaction with adult health, reproductive success, duckling growth and survival, tissue residues	Stanley et al. 1996
Cadmium (Cd)	Chicken	Cd reduced Se toxicity	Hill 1975; Howell and Hill 1978
	Rat	Se reduced liver damage, more so in combination with Zn	Jihen et al. 2008, 2009
	Cabbage and Lettuce	Se together with Zn reduced Cd absorption by roots	He et al. 2004
Copper (Cu)	Chicken	Cu reduced Se toxicity	Hill 1975; Howell and Hill 1978
Lead (Pb)	Cabbage and Lettuce	Se together with Zn reduced Cd absorption by roots	He et al. 2004
Mercury (Hg)	Chicken	Hg reduced Se toxicity	Hill 1975; Howell and Hill 1978
	Cricket	Se increased survival and growth	Ralston et al. 2006, 2008
	Aquatic oligochaete	Se reduced meHg bioaccumulation	Nuutinen and Kukkonen 1998

(continued)

TABLE 6.5 (CONTINUED)
Selenium Interactions with Other Factors

Factor	Organism	Interaction	Reference
	Four waterbird species	Se may act as a binding site for demethylated Hg in bird livers, reducing the potential for secondary toxicity	Eagles-Smith et al. 2009
	Fish, birds, other fauna	Se reduced meHg toxicity	Cuvin-Aralar and Furness 1991; Belzile et al. 2006; Yang et al. 2008
	Fish	Se reduced bioaccumulation of meHg in fish	Paulsson and Lindbergh 1989; Southworth et al. 1994, 2000; Chen et al. 2001; Peterson et al. 2009
	Mallard	Antagonistic effects in adults, but additive or synergistic in embryos	Hoffman and Heinz 1998; Heinz and Hoffman 1998
Sulfur (S)	Fish, invertebrates	S reduces Se bioaccumulation	Hansen et al. 1993; Bailey et al. 1995; Ogle and Knight 1996; Riedel and Sanders 1996
Tellurium (Te)	Chicken	Tl reduced Se toxicity	Hill 1975; Howell and Hill 1978
Tin (Sb)	Chicken	Sb reduced Se toxicity	Hill 1975; Howell and Hill 1978
Lead (Pb)	Chicken	Pb reduced Se toxicity	Hill 1975; Howell and Hill 1978
Zinc (Zn)	Rat	Zn and Se together provide more protection against Cd-induced liver damage than either alone	Jihen et al. 2008, 2009
Cyanobacterial algal bloom	Tilapia	Se reduced algal toxicity (oxidative stress and histological lesions)	Atencio et al. 2009
UV radiation	Plant	Se protects against UV-reduced plant growth	Hartikainen and Xue 1999
Oxidative stress	Chinook salmon	Dietary α-tocopherol + ascorbic acid decreased oxidative stress, but Se and Fe did not	Welker and Congleton 2009

Note: That both organic and inorganic Se interactions are considered.

the other, possibly via the formation of metabolically inert mercury selenides (HgSe; Ralston et al. 2006; Yang et al. 2008; Khan and Wang 2009; Peterson et al. 2009). The one exception of additive or synergistic effects in mallard embryos (Heinz and Hoffman 1998) despite antagonistic effects in adult mallards (Hoffman and Heinz 1998) remains unexplained. For other trace elements, antagonistic reactions between As and Se are also well documented. The single exception (Kraus and Ganther 1989) also remains unexplained. Two studies examined the interactive effects of Se and B on mallards and found opposing results: Hoffman et al. (1991b) reported synergistic toxicity, while Stanley et al. (1996) did not. Selenium decreased toxicity to tilapia due to an algal bloom (Atencio et al. 2009) and toxicity to a plant due to UV-exposure (Hartikainen and Xue 1999). However, it had no effect on oxidative stress in chinook salmon (Welker and Congleton 2009).

With the possible exception of interactions between organic Se and methyl Hg, the mechanisms and extent of antagonistic reactions between Se and other factors (e.g., other elements, and biotic and abiotic stressors) are unclear. Further, there are a few studies showing synergistic, not antagonistic, interactions that remain unexplained.

6.5.2 NUTRITIONAL FACTORS

Interactions of Se with other stressors has been shown to result in increased or decreased toxicity, depending on the particular stressors involved (Section 6.5.1; Table 6.5). Nutritional factors can also significantly influence Se toxicity. Selenium appears to enhance the nutritional uptake of other essential elements by at least some plants. For example, He et al. (2004) observed that the addition of Se and Zn to soils increased the uptake of the essential elements Fe, Mn, Cu, Ca, and Mg by Chinese cabbage and lettuce, with positive effects on growth. Reduced dietary protein inter-acts with increased Se exposure to increase Se toxicity to mallards (Hoffman et al. 1991b, 1992a,b). Similarly, dietary restriction of protein increases Se toxicity in growing chickens and mammals (Combs and Combs 1986). Conversely, increased dietary protein reduced Se toxicity in rats (Gortner 1940). Increased dietary protein reduced Se toxicity in chickens in one study (El-Begearmi and Combs 1982) but in another study showed no effect (Hill 1979), possibly because different types of pro-tein impart different levels of protection against Se toxicity (Levander and Morris 1970). Variations in available dietary protein and types of protein may therefore influence the toxicity of Se to water birds and mammals and likely also influence the toxicity of Se to other vertebrates.

In contrast to the beneficial effects of increased dietary protein, excess dietary carbohydrate enhanced dietary Se toxicity in rainbow trout (Hilton and Hodson 1983). The mechanisms of increased toxicity due to carbohydrates and of decreased toxicity due to protein are unknown, as are the extent of these demonstrated effects to different organisms than those tested.

6.5.3 TOLERANCE

Chapman (2008) summarizes the extensive research in the literature pertaining to metals tolerance, including both heritable genetic adaptation (i.e., modifications of tolerance by changes in heritable genetic material) and physiological acclimation (i.e., shifting of tolerance within genetically defined limits by up-regulating existing processes). Selection for metals-resistant populations, resulting in inheritable genetic adaptations, can occur following long-term exposure to elevated metals concentra-tions. Shorter-term exposures can result in reduced metals uptake and increased detoxification mechanisms to deal with metal exposure without selection for a metal-tolerant population. Tolerance is largely metal specific, and resistance to one stressor does not necessarily confer resistance to other stressors. In addition, there are energetic costs to acclimation; reduced energy available for growth or reproduction may also result in some level of reduced fitness. Genetic adaptation can involve reduced genetic diversity via selective pressure eliminating the least fit (i.e., most sensitive) organisms but may or may not involve energetic costs.

The possibility of Se tolerance in aquatic organisms has been suggested (e.g., for westslope cutthroat trout [*Salmo clarkii*] by Kennedy et al. 2000), but remains to be convincingly demonstrated. However, Se tolerance has been demonstrated in terrestrial organisms. For example, plants showing hypertolerance to Se (i.e., Se hyperaccumulators) are the result of long-term evolutionary selective pressures in naturally Se-enriched environments (Chiang et al. 2006), which protects them from herbivory and pathogens but has also led to the evolution of tolerant herbivores (Freeman et al. 2006; Quinn et al. 2007). Acclimation and adaptation are normal responses of organisms to adjust the boundaries of their ecological niches in order to maximize their chances to survive and reproduce.

The potential for Se tolerance in aquatic biota such as fish is an important research need. Such research could involve, for instance, parallel toxicity tests of suspected "tolerant" species (e.g., collected from areas with highly elevated Se concentrations) and intolerant species (e.g., collected from areas with relatively low Se concentrations typical of background conditions). Undertaking such tests, particularly interspecific experimental designs, would be powerful for testing site-specific adaptation versus acclimation of populations.

6.5.4 MARINE VERSUS FRESHWATER ENVIRONMENTS

Some marine animals bioaccumulate much greater concentrations of Se than freshwater species without teratogenic effects on offspring (Muir et al. 1999). Higher concentrations of Se have been observed across many taxa from bacteria to brine shrimp (*Artemia* spp.) to marine mammals to seabirds (Dietz et al. 2000; Brix et al. 2004; Oremland et al. 2004). An important exception appears to be fish, which apparently do not bioaccumulate greatly higher concentrations of Se in uncontaminated waters than do freshwater fish (Stewart et al. 2004; Campbell et al. 2005; Kelly et al. 2008; Burger et al. 2007; McMeans et al. 2007). While accumulations to elevated levels can occur in long-lived, piscivorous fish such as tunas, swordfish, and marlins (Nigro and Leonzio 1996; Eisler 2000; Table 6.6), it is unclear whether Se is impacting these wild populations.

Se accumulation to elevated levels in the tissues of many marine species without apparent ill effect is an interesting observation, because it suggests that there are fundamental distinctions in the essentiality and toxicity of Se in animals adapted to seawater and to hyper- rather than hypo-osmotic waters. High Se uptake by primary producers and high assimilation efficiency or feeding rates as discussed in Chapter 5 explain higher tissue concentrations of Se in marine species, but do not clarify the exact mechanisms that mitigate toxicity. Potential explanations for greater Se tolerance include complexation with, and detoxification of, Hg (Koeman et al. 1973; Eagles-Smith et al. 2009). Given that Se is central to antioxidant mediation, it is not surprising that Se requirements could be greater in marine species; however, mechanisms, and the extent and interaction with other stressors such as UV radiation (which poses a greater

TABLE 6.6
Selenium Tissue Concentrations in Wild Marine and Freshwater Species

Organism	n	Location	[Se] mg/kg dw (Mean ± SEM)	Reference
		Seabirds		
Common eiders (*Somateria mollissima*), nesting females	40	North shore and islands of Beaufort Sea, Alaska	36.1 ± 1.7 B	Franson et al. 2004
Common eiders	20	North shore and islands of Beaufort Sea, Alaska	2.28 ± 0.09 E	Franson et al. 2004
Common eiders, males	30	Yukon-Kuskokwim Delta (Y-K Delta), Aleutian Islands, Saint Lawrence Island, Alaska	9.29[a] (2.00–21.2) L	Stout et al. 2002
Common eiders, females	21	Y-K Delta, Aleutian Islands, Saint Lawrence Island	7.85[a] (2.50–44.0) L	Stout et al. 2002
King eiders (*Somateria spectabilis*), males	33	Barrow, AK, USA	34.5[a] (14.3–93.0) L	Stout et al. 2002
King eiders, females	21	Barrow, AK, USA	27.6[a] (9.60–63.1) L	Stout et al. 2002
Steller's eiders (*Polysticta stelleri*), males	4	Barrow, Togiak, and Kotzebue, AK, USA; Russia	25.6 (13.0–56.8) L	Stout et al. 2002
Steller's eiders, females	6	Barrow, Togiak, and Kotzebue, AK, USA; Russia	17.6[a] (8.18–31.4) L	Stout et al. 2002
Spectacled eiders (*Somateria fischeri*), males	28	Y-K Delta, Barrow, and Saint Lawrence Island, AK, USA; Russia	124[a] (35.4–401) L	Stout et al. 2002
Spectacled eiders, females	10	Y-K Delta, Barrow, and Saint Lawrence Island, AK, USA; Russia	43.5[a] (4.98–235) L	Stout et al. 2002
Spectacled eiders, males	2	Saint Lawrence Island and Togiak, AK, USA	76.0[a] (75 to 77) L	Henny et al. 1995
Spectacled eiders, females	1	Saint Lawrence Island, AK, USA	35.0 L	Henny et al. 1995
White-winged scoters (*Melanitta fuscal*), males	8	Cape Yakataga, Cape Suckling, AK, USA	21.7[a] (12 to 75) L	Henny et al. 1995
Black-legged kittiwakes (*Rissa tridactyla*)	63 (21)	Prince Leopold Island, Canada	3.52[a] ± 0.26 E	Braune 2007
Black-legged kittiwakes	10 (1)	Prince Leopold Island, Canada	36.2 L	Braune and Scheuhammer 2008

(continued)

TABLE 6.6 (CONTINUED)
Selenium Tissue Concentrations in Wild Marine and Freshwater Species

Organism	n	Location	[Se] mg/kg dw (Mean ± SEM)	Reference
Northern fulmars (*Fulmarus glacialis*)	93 (31)	Prince Leopold Island, Canada	4.12[a] ± 0.15 E	Braune 2007
Northern fulmars	10 (1)	Prince Leopold Island, Canada	34.4 L	Braune and Scheuhammer 2008
Thick-billed murres (*Uria lomvia*)	90 (30)	Prince Leopold Island, Canada	2.65[a] ± 0.10 E	Braune 2007
Thick-billed murres	29 (4)	Ivujivik, Salluit, Coats Island, Prince Leopold Island, Canada	5.68 ± 0.17 L	Braune and Scheuhammer 2008
Black guillemots (*Ceppus grylle*)	15 (2)	Ivujivik, Prince Leopold Island, Canada	9.95 ± 0.85 L	Braune and Scheuhammer 2008
Black guillemots	10	North water polynya	15.85 ± 5.96 L	Campbell et al. 2005
Black-winged scoters (*Melanitta perspicillata*), females	3	Cape Yakataga, AK, USA	22.1[a] (14 to 32) L	Henny et al. 1995
Glaucous gulls (*Larus hyperboreus*)	7 (2)	Akpatok Island, Coats Island, Canada	14.40 ± 5.20 L	Braune and Scheuhammer 2008
Glaucous gulls	9	North water polynya	12.65 ± 1.90 L	Campbell et al. 2005
Thayer's gull (*Larus thayeri*)	1	North water polynya	15.78 ± nr L	Campbell et al. 2005
Marine Mammals				
Porpoise: *Phocoena phocoena*	3	Dutch coast of North Sea		
Dolphins:				
Tursiops truncates	3	A dolphinarium		
Delphinus delphis	2	New Zealand	24.52[a]	
Lagenorhynchus obscures (species pooled)	1	New Zealand	(2.0 to 266.4) L	
Sotalia guianensis	2 (1)	Surinam		Koeman et al. 1973
Common seal (*Phoca vitulina*)	16	Wadden Sea, Dutch coast of North Sea	46.71[a] (2.3 to 416.3) L	Koeman et al. 1973
Ringed seal (*Phoca hispida*)	9	North water polynya	33.93 ± 22.38 L	Campbell et al. 2005
Polar bears (*Ursus maritimus*) M:F = 5:7	13	East Baffin Island, Canada	30.50 ± 12.49[b] L	Rush et al. 2008
M:F = 3:1	13	Lancaster Sound, Canada	47.89 ± 12.49[b] L	Rush et al. 2008

TABLE 6.6 (CONTINUED)
Selenium Tissue Concentrations in Wild Marine and Freshwater Species

Organism	n	Location	[Se] mg/kg dw (Mean ± SEM)	Reference
M:F = 7:5	12	Northern Baffin Island, Canada	38.73 ± 12.65[b] L	Rush et al. 2008
M:F = 8:3	11	Southeast Beaufort Sea, Canada	72.66 ± 12.82[b] L	Rush et al. 2008
M:F = 5:6	11	Southeast Hudson Bay, Canada	10.99 ± 12.82[b] L	Rush et al. 2008
1994 to 1999 M:F = 0:6	6	Chukchi Sea, AK, USA	16.58 ± 10.29[b] L	Rush et al. 2008
1983 to 2000 M:F = 19:27	46	Avanersuaq, Greenland	18.65 ± 12.22[b] L	Rush et al. 2008
1983 to 2000 M:F = 40:42	82	Ittoqqortoormiit, Greenland	16.15 ± 11.22[b] L	Rush et al. 2008
Marine Fish				
Leopard shark (*Triakis semifasciata*)	2	North San Francisco Bay, CA, USA	4, 10 L	Stewart et al. 2004
Pacific sleeper shark (*Somniosus pacificus*) M:F = 7:7	14	Prince William Sound, AK, USA	1.81 ± 0.09 L	McMeans et al. 2007
Greenland shark (*Somniosus microcephalus*) M:F = 14:10	24	Cumberland Sound, Greenland	1.72 ± 0.10 L	McMeans et al. 2007
Starry flounder (*Platichthys stellatus*)	3	North San Francisco Bay, CA, USA	7[c] (3 to 13) L	Stewart et al. 2004
Yellowfin goby (*Acanthogobius flavimanus*)	12 (3)	North San Francisco Bay, CA, USA	2[c] (1 to 7) L	Stewart et al. 2004
Arctic cod (*Boreogadus saida*)	2	North water polynya	5.13 ± nr L	Campbell et al. 2005
Pacific cod (*Gadus macrocephalus*)	16	Nikolski (Umnak I), AK, USA	4.63[a] (2.78 to 7.62) L	Burger et al. 2007
Pacific cod	6	Adak Island, AK, USA	5.29[a] (3.30 to 8.49) L	Burger et al. 2007
Pacific cod	77	Amchitka Island, AK, USA	4.66[a] (0.94 to 13.32) L	Burger et al. 2007
Pacific cod	42	Kiska Island, AK, USA	4.33[a] (0.23 to 12.42) L	Burger et al. 2007
Japanese tunas, 4 species	nr	Japan	33.3 to 50.0 L	Eisler 2000
Tuna (*Thunnus thynnus*)	nr	Southern Tyrrhenian Sea, Italy	10 L	Nigro and Leonzio 1996

(continued)

TABLE 6.6 (CONTINUED)
Selenium Tissue Concentrations in Wild Marine and Freshwater Species

Organism	n	Location	[Se] mg/kg dw (Mean ± SEM)	Reference
Marine Fish				
Swordfish (*Xiphias gladius*)	nr	Southern Tyrrhenian Sea, Italy	19 L	Nigro and Leonzio 1996
Black marlin (*Makaira indica*)	nr	nr	4.7 to 45.0 L	Eisler 2000
Striped bass (*Morone saxitilis*)	nr	nr	1.0 to 4.3 L	Eisler 2000
Anadromous Fish				
Striped bass, adult (*Morone saxatilis*)	15	North San Francisco Bay, CA, USA	12[c] (8 to 14) L	Stewart et al. 2004
Striped bass, juvenile	16	North San Francisco Bay, CA, USA	13[c] (12 to 14) L	Stewart et al. 2004
Atlantic salmon (*Salmo salar*) chinook salmon (*Oncorhynchus tshawytscha*) coho salmon (*Oncorhynchus kisutch*) Wild chinook, coho, chum (*Oncorhynchus keta*) sockeye (*Oncorhynchus nerka*) chum (*Oncorhynchus keta*)	110	8 British Columbia (Canada) salmon farms	0.67 ± 0.01[c] M	Kelly et al. 2008
pink (*Oncorhynchus gorbuscha*)	91	Coastal British Columbia, Canada	0.57 ± 0.02[c] M	Kelly et al. 2008
Freshwater Fish				
White sturgeon (*Acipenser transmontanus*)	15	North San Francisco Bay, CA, USA	25[c] (13 to 32) L	Stewart et al. 2004
Sacramento splittail (*Pogonichthys macrolepidotus*)	12 (3)	North San Francisco Bay, CA, USA	13[c] (7 to 20) L	Stewart et al. 2004
Bluegill sunfish (*Lepomis macrochirus*)	3 to 5 (1)	San Joaquin River Basin, CA, USA	1.85[a] (0.46 to 5.28) W	Saiki et al. 1992
Common carp (*Cyprinus carpio*)	3 to 5 (1)	San Joaquin River Basin, CA, USA	2.05[a] (0.57 to 4.28) W	Saiki et al. 1992
Mosquito fish (*Gambusia affinis*)	3 to 5 (1)	San Joaquin River Basin, CA, USA	1.63[a] (0.53 to 3.44) W	Saiki et al. 1992

TABLE 6.6 (CONTINUED)
Selenium Tissue Concentrations in Wild Marine and Freshwater Species

Organism	n	Location	[Se] mg/kg dw (Mean ± SEM)	Reference
		Freshwater Fish		
Largemouth bass (*Micropterus salmoides*)	3 to 5 (1)	San Joaquin River Basin, CA, USA	2.31[a] (0.92 to 5.25) W	Saiki et al. 1992

[a] Geometric mean (range).
[b] Converted from wet weight, assuming 70% moisture content.
[c] Median and (25th to 75th percentile).

Note: nr = not reported. WB = whole body, B = blood, E = egg, L = liver, dw = dry weight. If samples were pooled, the number of pools is shown in parentheses. Unless specified, gender or sex ratios were not reported. Data are representative rather than all inclusive.

threat in marine waters with low spectral absorbance compared to freshwater systems; Johannessen et al. 2003) warrant further investigation in oceanic ecosystems.

Selenium plays an important role in salt tolerance. It is well known that biota, from bacteria to mammals, use organic compounds to osmoregulate in highly saline environments. Such compounds are called "compatible solutes" because they do not inhibit cell macromolecules or function. Some organic osmolytes contain Se. The most extreme example occurs in the salt-tolerant plant, *Astragalus bisulcatus,* which has a nutritional requirement for Se and has been observed to accumulate tissue concentrations of Se > 500 mg/g. *Astragalus* accumulates approximately 80% of its Se as methylselenocysteine (Brown and Shrift 1982). This is the same presumed osmolyte that has been identified in the Se-tolerant diamondback moth (*Plutella xylostella*) and its parasite, the Se-tolerant wasp (*Diadegma insulare*) (Freeman et al. 2006). In fish, most osmolytes are neutral free amino acids such as taurine and glycine, small carbohydrates, such as myo-inositol, and methylamines, such as trimethylamine oxide (Fiess et al. 2007). Interestingly, bivalves surrounding deep-sea thermal vents contain elevated concentrations of thiotaurine but only in tissues containing sulfur-fixing endosymbionts (Brand et al. 2007), suggesting that Se could serve in the place of sulfur. However, to our knowledge, it has not been demonstrated that bivalves or marine vertebrates enzymatically fix Se as components of organic osmolytes. As part of their osmoregulatory physiology, many marine animals have salt glands, which provide them with an additional excretory pathway to maintain intracellular homeostasis. Higher concentrations of metals and Se in the tissues of salt glands suggest that this organ is also an excretory pathway for Se (Burger et al. 2000). Future research needs include manipulative testing of this correlative relationship as well as investigation of whether bivalves or marine vertebrates enzymatically fix Se as a component of organic osmolytes.

Marine birds differ from freshwater birds in the way in which they partition Se among different tissues. This difference may have consequences for the relative susceptibilities of marine and freshwater birds to Se-mediated reproductive toxicity. Some marine birds accumulate relatively high concentrations of Se in their livers, but Se concentrations in their muscles, blood, and eggs are similar to those in freshwater birds. For example, hepatic Se concentrations in several species of marine birds that spend all, or most, of their lives, at sea averaged 37 and 44 mg/kg dw, respectively, whereas hepatic Se concentrations in several species of freshwater birds averaged only 11 mg/kg dw. Moreover, apparently natural Se concentrations as high as 300 mg/kg dw have been recorded in some marine birds. Conversely, Se concentrations in the eggs of marine and freshwater bird species are comparable; marine species average about 4 mg Se/kg dw in eggs, whereas freshwater species average just under 3 mg Se/kg dw (Ohlendorf and Harrison 1986; Bischoff et al. 2002; Braune and Simon 2004; Braune and Scheuhammer 2008; Braune et al. 2002; DeVink et al. 2008b; Grand et al. 2002; Harding et al. 2005; Scheuhammer et al. 1998, 2001; Skorupa and Ohlendorf 1991; Trust et al. 2000). Franson et al. (2007) fed common eiders (a marine bird) a commercial diet containing either 20 or 60 mg Se/kg dw as selenomethionine. Mean Se concentrations in liver and muscle and peak levels in blood of birds fed the low and high Se diets were as follows (mg/kg dw): 351 and 735, 85 and 88, and 14 and 17, respectively. In comparison, when mallards, a species found most often in freshwater ecosystems, were fed a commercial diet containing either 25 or 60 mg Se/kg dw as selenomethionine, their livers contained Se concentrations only about 25% as high as those in eiders (98 and 200 mg/kg on the low- and high-Se diets, respectively), whereas concentrations in their muscles and blood were similar to those in eiders (O'Toole and Raisbeck 1997). It remains unknown how seabirds appear to preferentially store Se in their livers when compared to freshwater birds. Nevertheless, this preferential storage of Se in livers by marine birds compared to freshwater birds coupled with the similarity between the two groups of birds in Se levels in blood may be one way in which marine birds appear able to avoid excessive in ovo exposure to Se, even when consuming diets with relatively high naturally occurring Se concentrations. Although marine birds may differ from freshwater birds in tissue partitioning of dietary Se to liver and eggs, thus affording a greater degree of protection to the former group from reproductive problems associated with excessive exposure to dietary Se, the higher concentrations of Se in livers of marine birds may have consequences for toxic effects in adults. The frequency of occurrence and severity of liver and feather lesions were similar between common eiders fed diets containing 20 and 60 mg Se/kg dw and mallards fed diets containing 25 and 60 mg Se/kg dw as selenomethionine (O'Toole and Raisbeck 1997; Franson et al. 2007).

6.5.5 WINTER STRESS SYNDROME

Fish species that experience prolonged winter periods are at risk of overwinter mortality due to many natural factors such as low temperature (i.e., thermal stress), low oxygen (i.e., hypoxia/anoxia), reduced food availability (i.e., starvation), increased predation, disease, and parasitism (reviewed in Hurst 2007). Most of these factors act more severely in smaller fish due to size-dependent changes in surface: volume relationships (Post and Parkinson 2001). Since survival of fish beyond the first year of life is a critical determinant of future year class strength, significant overwinter mortality of early life stages (e.g., young-of-the-year) can cause a "recruitment bottleneck" and negatively impact fish population dynamics (Post and Parkinson 2001; Hurst 2007). Thus, factors that increase the frequency of overwinter mortality of fish can negatively influence the sustainability of fish populations. In aquatic ecotoxicology, the term *winter stress syndrome* has been proposed to describe the potential for contaminants to potentiate overwinter mortality (Lemly 1996b). The three conditions noted by Lemly (1996b) for winter stress syndrome to occur are 1) the presence of a metabolic stressor (natural or anthropogenic), 2) cold water temperatures, and 3) the response of fish to cold with reduced activity and foraging. Importantly, the potential that a given stressor will lead to winter stress syndrome depends on its propensity to increase metabolism. Metabolic stressors can include such variables as exposure to inorganic or organic chemicals, parasites, altered pH, or elevated concentrations of suspended sediment. The presence of multiple metabolic stressors increases the probability of occurrence of winter stress syndrome (Lemly 1993c).

The winter stress syndrome hypothesis is based on a laboratory study in which juvenile bluegill sunfish were exposed for 180 days to dietary and waterborne Se under either summer or winter conditions (Lemly 1993c). Winter conditions of low water temperature (4 °C) exacerbated the toxicity of Se, indicated by increased mortality, decreased condition factor (weight-at-length) and decreased energy (lipid) stores. Lemly's (1993c) laboratory study has recently been replicated except for photoperiod, with similar results overall (McIntyre et al. 2008). The juvenile bluegill sunfish studies (Lemly 1993c; McIntyre et al. 2008) indicated that Se likely causes increased metabolism, which may occur, in part, through oxidative stress (Spallholz and Hoffman 2002; Palace et al. 2004). Thus, Se has the potential to cause winter stress syndrome in field settings.

Although the concept of winter stress syndrome is a scientifically sound hypothesis, it has rarely been tested explicitly under field conditions of elevated Se, or other chemical, exposures (Janz 2008). Higher lipid metabolism and lower triglyceride levels in fish experiencing chronic metal exposure in the fall relative to fish inhabiting uncontaminated lakes have been reported (Levesque et al. 2002). Other studies investigated aspects of the winter stress syndrome hypothesis in several native fish species inhabiting areas receiving complex metal mine effluents containing elevated Se (Bennett and Janz 2007a,b; Kelly and Janz 2008, 2009; Weber et al. 2008; Driedger et al. 2009). In these studies, juvenile fish were collected just prior to ice-on and immediately following ice-off from lakes and creeks receiving discharges from uranium mining (Bennett and Janz 2007a,b) and base metal (copper or nickel) mining (Driedger et al. 2009) operations in northern Canada. Measures and indicators of growth (length, weight, condition factor, muscle RNA:DNA ratio, muscle proteins) and energy storage (whole-body lipids, whole-body triglycerides, liver triglycerides, and liver glycogen) were determined in fish collected along gradients of exposure and from reference sites. Based on the winter stress syndrome concept, it was hypothesized that fish collected in spring from all sites (both exposure and reference) would exhibit decreases in growth and energy storage measures compared to the previous autumn and that these measures would be decreased to a greater extent in juvenile fish collected from exposure sites. In contrast to these hypotheses, juvenile northern pike, burbot (*Lota lota*), fathead minnow (*Pimephales promelas*), creek chub (*Semotilus atromaculatus*), and white sucker (*Catostomus commersoni*) collected from exposure sites generally exhibited similar or greater growth and energy stores in spring when compared to the previous autumn, in comparison with reference sites (Bennett and Janz 2007a,b; Driedger et al. 2009). Only slimy sculpin (*Cottus cognatus*) exhibited changes in energy stores (whole-body triglycerides) that were consistent with the winter stress syndrome hypothesis (Bennett and Janz 2007b). In contrast, the majority of these species collected from reference sites exhibited overwinter decreases in energy stores that were consistent with the overwinter fish biology literature (Hurst 2007). In these studies, Se residues were also measured in selected species. Whole-body Se concentrations in juvenile fathead minnows and white sucker ranged from 11 to 43 mg/kg dw (Driedger et al. 2009) and in juvenile northern pike muscle ranged from 17 to 23 mg/kg dw at exposure sites (Kelly and Janz 2009). At both study locations, Se was the predominant element consistently elevated in fish tissues.

Further work investigating whether winter stress syndrome occurs under field conditions of Se exposure is needed to fully assess the hypothesis (Janz 2008). As Lemly (1993c) notes, basic knowledge of life history characteristics and feeding ecology, particularly for juvenile fish, would allow identification of potentially vulnerable fish species in temperate regions of the world. Unfortunately, there are few studies with direct observations and concrete conclusions regarding the feeding ecology of juvenile fishes. It is possible that winter stress syndrome is most important in species at the northern limit of their ranges, and future studies should focus on this aspect. Knowledge of local fish community ecology is essential when assessing the potential importance of overwinter mortality in aquatic ecotoxicological investigations of Se.

Limited evidence suggests that winter stress may occur in avian species as well (e.g., chickens [Tully and Franke 1935]). Heinz and Fitzgerald (1993) exposed mallard ducks ($n = 5$) to diets supplemented with 0, 10, 20, 40, and 80 mg Se/kg dw for 16 weeks (November 16–March 7) in outdoor pens. By week 8, all the ducks in the 80 mg Se/kg dose level died. Ninety-five percent of the ducks exposed to 40 mg Se/kg died after 11 weeks. After 16 weeks, 75% of the birds exposed to the 20 mg Se/kg diet survived, while none died at the 0 and 10 mg Se/kg dose levels. The authors observed the majority of mortalities between December 26 and January 11, when temperatures were consistently below freezing. Following necropsy, these animals were extremely emaciated with no body fat. In similar studies (Heinz et al. 1987, 1989, 1990) conducted during spring and summer, little to no mortality occurred in mallards exposed to diets supplemented with 10 to 32 mg Se/kg. The authors noted that in this and other studies birds fed the elevated Se diets became sick and did not eat as much food, which ultimately resulted in emaciation and death. However, they argued that the increased energy demands of cold weather probably forced the ducks to eat more Se-treated food, leading to effects faster than would occur in warm weather.

6.5.6 Comparison of Fish Early Life Stages, Juveniles, and Adults

The larval life stage of an anadromous fish species may not receive a large Se dose via maternal transfer, but it could begin to feed on an elevated Se diet if rearing occurs in an area with elevated Se concentrations. Therefore, although maternal transfer is often considered to be the classic pathway by which early life stages are exposed to Se, dietary Se exposures alone by young juvenile fish are important in some situations. Hamilton et al. (1990), for example, hypothesized that juvenile chinook salmon feeding in nursery areas of the Sacramento–San Joaquin Delta (California), an area with elevated food-chain Se due to irrigation in the San Joaquin river system, could be adversely affected by Se when they undergo parr-smolt transformation prior to migration to the sea. To test this, the authors conducted 90-d dietary organic Se exposures and observed a dose-dependent reduction in juvenile survival and growth. In a 10-d seawater challenge test following 120-d organic Se exposures in brackish waters, they observed a dose-dependent reduction in smolt survival. Elevated Se levels apparently caused a physiological imbalance that impaired performance during the seawater challenge test (Hamilton et al. 1990), but the precise mechanism for this effect is unclear.

In another juvenile fish study, Vidal et al. (2005) exposed juvenile rainbow trout to dietary selenomethionine for 90 days. In addition to growth, they measured the reduced and oxidized glutathione and thiobarbituric acid reactive substance levels in the livers of the trout to assess oxidative damage caused by Se. Lipid peroxidation and GSH:GSSG ratios were unchanged by all Se treatments (dietary exposures up to 18 mg Se/kg dw), suggesting that mechanisms other than oxidative stress caused the observed toxicity.

6.5.7 POPULATION-LEVEL LINKAGES (FISH AND BIRDS)

In practice, establishing cause–effect linkages between toxic effects at the individual organism level and adverse impacts on populations is challenging and has rarely been demonstrated in the field of ecotoxicology. However, as discussed in Chapter 3, there have been several examples of significant declines of fish (e.g., the Belews Lake incident) and water bird (e.g., the Kesterson Reservoir incident) populations caused by Se contamination of aquatic ecosystems. These impacts were primarily due to reproductive failure resulting from embryo- and early-life-stage mortality of fish (mainly due to deformities) and birds (mainly due to failed hatchability). Based on extensive data available for the Belews Lake fish population collapses (Appendix A), Lemly (1997a) proposed a "teratogenicity index" that relates the proportion of deformed fish to an adverse population impact. To our knowledge, Lemly's (1997a) deformity index has not been applied to other aquatic ecosystems contaminated with Se. Nevertheless, it has provided one of the clearest linkages to date between individual and population level effects in ecotoxicology and illustrates the significant potential hazard posed by Se in aquatic ecosystems. Linkages between organic Se toxicity and population-level impacts are discussed further in Section 6.6.

6.6 LINKAGES BETWEEN Se TOXICITY AT SUBORGANISMAL AND ORGANISMAL LEVELS TO POPULATION- AND COMMUNITY-LEVEL IMPACTS

6.6.1 POTENTIAL POPULATION AND COMMUNITY-LEVEL IMPACTS

Selenium, when in excess of nutritional levels, can be toxic at multiple levels of ecological organization. Lower trophic level organisms (primary producers and most primary consumers) are less sensitive to Se toxicity than higher trophic level organisms,

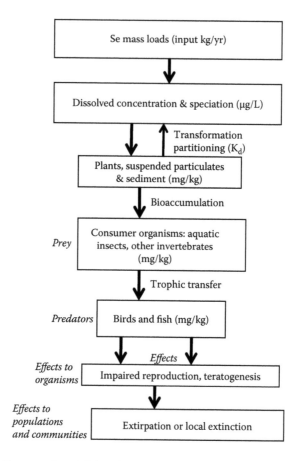

FIGURE 6.9   Conceptual model describing linked factors that determine the effects of Se on ecosystems. (After Presser and Luoma 2006.)

although as noted in Section 6.5, this has not been fully established. Thus, since lower trophic-level organisms are relatively insensitive to Se toxicity, they act as vectors transferring Se to more vulnerable organisms such as fish, birds, and potentially also to oviparous amphibians and reptiles. Selenium has the potential to adversely impact ecologically and toxicologically vulnerable fish, waterbird, amphibian, and reptile populations and cause profound ecosystem changes (Figure 6.9). For example, Se enrichment of reservoir environments (Chapter 3; Appendix A) provide classic examples of adverse effects through different levels of biological organization; they effectively comprise integrated whole-ecosystem "experiments" of trophic transfer, resulting in a cascade of population and community impacts (e.g., Gillespie and Baumann 1986; Sorensen 1991; Lemly 2002). Recovery from these impacts has also been documented once Se sources were eliminated (Lemly 1997b; Crutchfield 2000; Finley and Garrett 2007).

6.6.1.1 Fish

In contrast to reservoirs or other contained lentic environments, some lotic environments have few or no adverse effects, despite elevated Se concentrations. This is the case even when the species exposed in the field are the same or closely related to those demonstrating adverse effects in laboratory settings at lower Se tissue concentrations. Appendix B illustrates an extreme example in streams from the Great Plains region of southeast Colorado. Those streams had greatly elevated Se concentrations in water and biota, relative to effects-thresholds for fish. The elevated Se was apparently long-standing and predominantly of natural origin via the regional geology and groundwater. One minnow species, the central stoneroller (*Campostoma anomalum*), was common in a stream with a mean Se concentration of 418 µg/L and with whole-body tissue concentrations of 20 to 50 mg Se/kg dry weight. Appendix B includes speculation on several plausible factors that might explain this species' persistence in a stream with those extremely high Se concentrations, including 1) whether the low organic carbon and high sulfate in sediments limited the accumulation of Se in sediments and subsequent trophic transfer to fish, 2) whether central stonerollers were inherently less sensitive to Se than better studied species, and/or 3) whether the reproductive strategy of central stonerollers and perhaps other small-bodied Great Plains fishes that have evolved under harsh conditions makes their populations more resilient than less fecund, longer-lived fish species (Matthews 1987).

Data from Thompson Creek (Idaho) illustrate the difficulty of detecting responses in stream fish communities exposed to moderately elevated Se concentrations. Thompson Creek is a cold-water, mountain stream with elevated Se concentrations from groundwater input originating in waste dumps for overburden from a large, open-pit molybdenum mine. Species richness and abundance of fish and benthic communities in the stream have been monitored for about 20 years at fixed monitoring points, using consistent methods. Since Se contamination was detected in the late 1990s, Se residues in sediments, organic detritus, invertebrates, and the dominant fish species, shorthead sculpin (*Cottus confusus*) and cutthroat/rainbow trout hybrids ("trout") were measured annually for 7 years (CEC 2004; GEI 2008b).

At Thompson Creek, trout whole-body Se residues ranging from 4 to 14 mg Se/kg dw had no apparent relation to trout density. Sculpin densities were negatively related to whole-body Se residues over a range of 9 to 18 mg Se/kg dw, although the relationship was not statistically significant ($p = 0.12$) (Figure 6.10). In contrast, in a tributary and adjacent pond with water concentrations ranging from 30 to 35 µg Se/L, trout collected in 2000 had average whole-body concentrations of about 13 mg Se/kg dw and, when the site was resurveyed in 2003, no trout could be found (Mebane 2000; CEC 2004). These limited observations suggest that in a small, closed population, whole-body tissue residues approximately 2-fold higher than laboratory thresholds of 6 mg Se/kg dw (DeForest et al. 1999) could contribute to a local extirpation.

The Thompson Creek long-term monitoring record also illustrates how natural variability in fish densities can make all but persistent population declines difficult to detect through field monitoring. Over the 20-year period of record at the upstream reference site along Thompson Creek, the median year-to-year variation

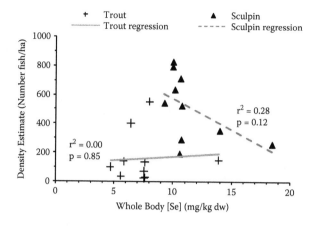

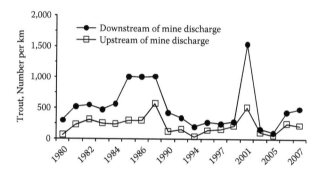

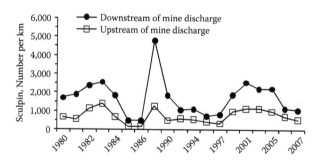

FIGURE 6.10 Whole-body Se residues (dw; A) and variability in density estimates for trout (B) and sculpin (C) in Thompson Creek, Idaho. (Data from GEI Consultants 2008b, 2008c; Canton and Baker 2008.)

in trout densities was 36% (mean 76%) and ranged from 1 to 260%. With sculpin, the median year-to-year variation in densities was only 22% (mean 70%) but ranged from 1 to 590% (Figure 6.10B,C; GEI 2008b).

6.6.1.2   Birds

For birds, an avian population's ability to tolerate even low rates of reproductive impairment from evolutionarily novel sources, such as anthropogenically mobilized Se, is strongly contingent on how well the population has adjusted to historic evolutionary pressures such as nest predation (Terborgh 1989). Evaluating the linkage between Se-induced individual toxicity and population-level implications requires demographic modeling or long-term field monitoring. Detecting a Se signal is most feasible for closed populations such as fish populations in reservoirs, lakes, and ponds. Detecting a Se signal in bird populations is challenging because they are highly mobile, often accessing resources across entire continents, and their populations are almost always demographically open. Consequently, modeling and field investigation of population level effects are most appropriate at the landscape scale and require meta-population analyses.

The mere perpetual presence from year-to-year of normal densities of breeding water birds at a Se-contaminated study site is not evidence for lack of population-level effect. For example, the abundance of breeding water birds at Kesterson Reservoir did not crash (or even notably decline) despite the complete reproductive failure of several species (Ohlendorf 1989). Because the bird populations at Kesterson were demographically open, even though Kesterson was indisputably a demographic sink, immigration from unaffected source populations substituted for the lost productivity at Kesterson. In this respect, Se exposure of breeding birds may often be more forgiving at the population level than for fish because flight links them to a much larger meta-population and a larger effective demographic landscape. However, the landscape-level relationship between demographic source and demographic sink subpopulations can be complex and can lead to counter-intuitive outcomes. In other words, populations are structured by actual habitat quality (e.g., low value of the invertebrates as food due to selenosis) rather than perceived habitat quality (e.g., high level of aquatic invertebrate productivity). In such cases, a seemingly attractive sink site (such as Kesterson was) can ultimately drain source populations dry. The critical landscape level impact is then geographically removed from the contaminated site itself.

Basic demographic modeling of potential population-level impacts from Se exposure in wild birds requires data relating tissue levels of Se to overwinter survivorship, and posthatch mortality. These data are meager. Winter stress syndrome has been reported for Se-exposed birds in two studies (Tully and Franke 1935; Heinz and Fitzgerald 1993). Similarly, there has been only one rigorous (i.e., radio telemetry) field study of Se effects on posthatch survivorship of water bird chicks (Marn 2003). Posthatch survivorship might be critical because experimental (Heinz 1996; Fairbrother et al. 2004) and nontelemetered observational field studies (Williams et al. 1989) suggest that posthatch mortality can be a larger component of overall Se-induced reproductive impairment than embryo mortality (see discussion in Section 6.2.2.2.3 and in Seiler et al. 2003). For example, Marn (2003) studied the least Se-sensitive species of waterbird, the American avocet, according to current data, and yet clearly documented a Se effect on posthatch survivorship of chicks. Until more exposure–response information is produced for the endpoints of over-winter survivorship and posthatch chick survivorship, the basis for demographic modeling of potential population-level effects in birds will be weak.

6.6.2 POPULATION RECOVERY

Recovery of populations and communities from Se stress is a counterpart to the pre-ceding examples of case studies on studying population- or community-level effects. The following discussion focuses on lentic fish, since population-level effects with birds are difficult to interpret because of the opportunity for immigration from other populations, as discussed above. The case study of Belews Lake not only provides an example of Se impacts on fish populations, it also provides data for evaluating the recovery of a severely impacted system after Se discharges to the lake were elimi-nated in 1985. Selenium-contaminated water was first discharged into Belews Lake in fall 1975, and in 1976 no centrarchids in the 0–64 mm size range were found and only 8 black and flat bullheads were found in the projected young-of-the-year size range (Cumbie and Van Horn 1978). Based on fish sampling conducted in 1977, 1980, and 1981, diversity ranged from 7 to 13 taxa and estimated biomass ranged from 5.67 to 15.02 kg/ha (Barwick and Harrell 1997). From 1984 through 1994 diversity ranged from 14 to 22 taxa and biomass ranged from 36.39 to 79.66 kg/ha (Barwick and Harrell 1997). In 1984 fish taxa and biomass were dominated by Se-tolerant species (e.g., green sunfish, common carp), but by 1994 fish biomass was dominated by more Se-sensitive species, such as bluegill sunfish and largemouth bass (Barwick and Harrell 1997). Overall, based on taxa diversity and standing stock estimates, fish populations in Belews Lake had generally recovered by the mid-1990s (Finley and Garrett 2007).

Correspondingly, Se concentrations in fish tissue have declined as fish popula-tions in Belews Lake recovered. In 1976, Se concentrations in composite muscle samples collected from areas of the lake with elevated aqueous Se were on the order of 7.96 to 22.3 mg Se/kg ww (31.8 to 89.2 mg Se/kg dw assuming 75% moisture) and in 1977 Se concentrations in composite muscle samples ranged from 6.32 to 54.6 mg Se/kg ww (25.3 to 218 mg Se/kg dw assuming 75% mois-ture) (Cumbie and Van Horn 1978). By 1992, maximum muscle Se concentrations were 2.6, 3.8, and 3.2 mg Se/kg ww (10.4, 15.2, and 12.8 mg Se/kg dw assum-ing 75% moisture) for catfish, green sunfish, and bluegill sunfish, respectively (Barwick and Harrell 1997). Similarly, Finley and Garret (2007) reported that median estimated whole-body Se concentrations in carp, redear sunfish, and crap-pie were approximately 22, 17, and 18 mg Se/kg dw, respectively, in 1994–1996, and approximately 9, 10, and 9 mg Se/kg dw in 2004–2006. Lemly (1997b) also reported substantial reductions in fish tissue Se concentrations from pre-1986 compared to 1996. For example, Se concentrations in bluegill eggs ranged from 40–133 mg Se/kg dw prior to 1986 and from 3–20 mg Se/kg dw in 1996. Similar reductions were observed for other fish species. The reductions in Se concentra-tions in fish tissue at Belews Lake and the corresponding recovery of fish popula-tions supports the hypothesis that Se toxicity at the organism level has a direct relationship to population-level impacts. In the late 1970s and early 1980s, Se concentrations in fish tissues exceeded all known levels demonstrated to cause toxicity, while by the mid-1990s, Se concentrations in fish tissues approached or were below recommended toxicity thresholds.

However, reductions in environmental Se concentrations or persistent residual contamination in tissues do not necessarily indicate biological recovery or the lack thereof.

Differing recovery trajectories have been demonstrated for different ecosystem components in the reservoir studies. In both Belews Lake and Hyco Lake, recovery was relatively fast (~2 to 3 years) for overall fish assemblage biomass as well as recolonization by fish species that were previously extirpated from the reservoir. However, the relative composition of the fish assemblages was markedly different from that pre-exposure or reference areas (Lemly 1997b; Crutchfield 2000). Some 20 years after major reductions in Se loading were implemented in Belews Lake, the fish community composition was largely stabilized, approaching a new equilibrium (Finley and Garrett 2007). This emphasizes the general challenge of defining recovery and the limitations of the concept of ecosystem "equilibrium."

6.6.3 FACTORS THAT MAY CONFOUND LINKING SE FISH TISSUE RESIDUES TO POPULATION-LEVEL IMPACTS

As discussed in the previous section, effects thresholds derived from fish tissues are logically more relevant to predicting effects in aquatic ecosystems than thresholds based on abiotic media (water, sediment) because tissue-based thresholds bypass the confounding influences of variable exposures and bioavailability to the organism (Chapter 7). However, factors such as density compensation and the role of refugia may make population impacts difficult to detect.

6.6.3.1 Density Compensation

Density compensation in this context refers to patterns where densities of organisms increase, food and space become limiting, and individuals begin to suffer from lack of resources or from competition and interference. When habitats are overcrowded, competition for food and space intensifies, causing individuals to expend more energy foraging or defending territories, which may result in lower growth or displacement to suboptimal habitats. Such shifts in energy allocation can reduce overwinter survival and increase the risk of being captured and eaten. In a declining population, as densities thin, overcrowding is relaxed, resulting in compensatory increases in growth, survival, and the overall chance of reproductive success. Thus, density compensation for overcrowding will tend to stabilize populations and lessen extinction risks as populations decline. In contrast to these compensatory effects, when habitats are undercrowded, population declines may lead to further declines in population growth (also referred to as Allee or depensation effects). For example, when spawning adults are scarce, they may not find each other or expend more energy finding each other and reduce the probability of fertilization (Chapman 1966; Liermann and Hilborn 2001; Rose et al. 2001). Thus, depending on the compensatory capacity, loss of individual organisms may not result in commensurate effects to the overall population abundance.

Modeling of cutthroat trout populations in tributaries of the upper Snake River (Idaho) has demonstrated the theoretical potential for density dependence to compensate for substantial juvenile mortality. Using functions fit to toxicity test data from the literature, Van Kirk and Hill (2007) projected decreased prewinter survival and growth of juveniles due to Se exposure. Population-level effects of Se were simulated by relating growth reductions to reduced fecundity and by extrapolating the Se-survival to reduced population size. However, because juvenile survival in trout is highly density-dependent, particularly during winter when individuals compete for limited concealment cover, trout populations may compensate for increased juvenile mortality via reduced density-dependent effects. Van Kirk and Hill (2007) suggested that population-level effects would be lower than individual-level effects until juvenile mortality rates exceeded 80%.

In another example of fish populations compensating for large juvenile mortalities, a toxic algal bloom along the coast of southern Norway killed an average of 60% of age-0 cod (*Gadus morhua*) populations with no effects on the cod population persisting beyond 3 years (Chan et al. 2003). If fish populations compensated for additive mortality to early life stages at rates close to those estimated by Van Kirk and Hill (2007), the release from density-dependent inhibition could mitigate fatal early life stage deformities caused by Se. However, density dependence is controversial because it is notoriously difficult to reliably estimate from even well-studied populations, and unreliable estimates can seriously underestimate the risks of population decline or extinction (Barnthouse et al. 1984; Rose et al. 2001). Indeed, Ginzberg et al. (1990) caution that even when working with plausible estimates of density dependence, "by choosing the model of density dependence carefully, one can achieve any quasi extinction risk desired." A prime example of the result of overgeneralizing the mitigating effects of density compensation, are marine fisheries that have become depleted under management plans that recommended annual harvests of approximately 25% of the population (Myers et al. 1997).

Further, other trout stream populations may have substantially less compensatory reserve than that estimated for the Snake River cutthroat trout populations. The cutthroat trout populations modeled by Van Kirk and Hill (2007) compensated for Se-induced mortality because this additional mortality occurred in winter before the density-dependent effects. If density-dependent mortality occurred in juvenile trout not in winter but rather during spring or summer when low flows and high temperatures were a limiting factor, this compensatory ability could be greatly reduced, if not lost completely (Elliott 1989; Van Kirk and Hill 2007).

6.6.3.2 Role of Refugia

Hydrologically intact stream networks provide important habitat linkages, reduce extinction risks through metapopulations, and reduce time to recovery from disturbances among many other important stream and landscape ecological functions. Stream networks also provide important access to refugia from naturally occurring or anthropogenic hazardous conditions such as high summer water temperatures or low flows (Sedell et al. 1990; Poole et al. 2004). These connections undoubtedly add resilience to some fish populations. With birds, as discussed in Section 6.6.1.2, immigration from unaffected demographic source populations would likewise mitigate declines in abundance at smaller scales.

Palace et al. (2007) have shown characteristic deformities related to elevated Se concentrations in larval rainbow trout derived from adults that were captured downstream from a coal mining operation in Alberta (Canada). Deformities such as these might be expected to reduce recruitment in the population and overall population numbers, but adult rainbow trout are still present in the affected streams. One suggestion for this apparent contradiction is that adult fish migrate to the affected system from nearby areas with background Se concentrations. Analyses of concentrations of Se in annual growth zones of otoliths (calcified structures in the inner ear of fish) suggested that fish from the mine-impacted system are recent immigrants from nearby reference streams not receiving Se-bearing effluent (Palace et al. 2007). In general, while such movements may allow stream fish to reduce risks from elevated Se, they complicate efforts to relate experimental effects of elevated Se to field populations. Some larger-bodied fishes such as suckers and the larger salmonids may move over 100 km seasonally. Smaller stream-resident salmonids or centrarchids commonly move on the order of 100 m to several km, whereas some small-bodied fish may only move on the order of 10 m (Munther 1970; Hill and Grossman 1987; Gibbons et al. 1998; Munkittrick et al. 2001; Baxter 2002; Schmetterling and Adams 2004). It follows that small-bodied fishes such as sculpin and small cyprinids or percids may be more locally vulnerable to elevated Se in streams and more indicative of local effects than more motile species.

The challenge and difficulty of reliably extrapolating effects from laboratory experiments to populations or ecosystems has been long recognized and is by no means unique to Se (Chapman 1983; Suter et al. 1985). Unambiguous identification of ecosystem level effects due to Se is rare in ecotoxicology. The Hyco and Belews Lake (North Carolina) power plant cooling reservoirs had been operated with a history of systematic baseline and ongoing chemical and biological monitoring with testing facilities and staff scientists onsite (Chapter 3; Appendix A). These Experimental Reservoir Area settings essentially provided before-after-control-impact and recovery study designs. The reservoirs were closed, dammed systems with limited opportunity for emigration or migration to refugia. In these settings, profound Se-related, population-level impacts were clearly apparent. In contrast, in interconnected streams motile fish species may freely migrate in and out of areas with Se enrichment. Large, readily captured species common to streams in western North America such as suckers, fluvial trout, and mountain whitefish (*Prosopium williamsoni*) often have annual ranges of hundreds of kilometers (Baxter 2002). Further, the hydrology of streams is usually more variable than that of power plant cooling reservoirs. Spates and drought features of stream environments often contribute to highly variable stream fish populations. This combination of high natural variability, compensating factors, and the generally less efficient trophic transfer of Se in lotic versus lentic systems (Orr et al. 2006) makes population- or community-level effects of Se difficult to detect.

6.7 SITE-SPECIFIC STUDIES FOR DETECTING PRESENCE OR ABSENCE OF Se EFFECTS

A general challenge in evaluating risks from elevated Se in aquatic environments is the difficulty of extrapolating effects from laboratory to field settings and generalizing between different field settings (Lemly and Skorupa 2007; McDonald and Chapman 2007; Ohlendorf et al. 2008). A tiered assessment approach is recommended, starting with relatively low-cost monitoring of exposure and comparison to screening benchmarks (e.g., water or tissue concentrations). Lemly and Skorupa (2007) recommended proceeding to a Se management plan that includes loading reductions if Se tissue benchmarks are exceeded, without necessarily investing time and resources into site-specific toxicity testing or population assessment. Ohlendorf et al. (2008) focus on best practices for site-specific assessment of bioaccumulation and trophic transfer of Se in aquatic ecosystems. McDonald and Chapman (2007) recommend that, if tissue residue benchmarks are exceeded, more definitive risks should be evaluated through reproductive toxicity testing of fish collected from the site of concern, and/or assessment of fish populations in the area of interest. This section assesses the use of reproductive toxicity testing and/or resident population assessment at higher tiers of a site-specific evaluation relative to the appropriate toxicity endpoint, using fish as an example.

6.7.1 SELENIUM EXPOSURE MEASUREMENT

To evaluate whether environmental concentrations of Se at a site are at potentially toxic levels, Se can be measured in a variety of abiotic (sediments, water) and biotic (diet, tissue) media. However, as noted herein (Section 6.5.10), tissue Se is the most reliable predictor of toxic effects (Lemly 1993b, 1996a; USDOI 1998; DeForest et al. 1999; Hamilton 2002; Ohlendorf 2003; Adams et al. 2003; USEPA 2004; Chapman 2007). Further, estimates of risk with the lowest uncertainty are derived from measurements of Se in ovaries and/or eggs (Chapter 7).

6.7.2 REPRODUCTIVE EFFECTS TESTING

McDonald and Chapman (2007) and Janz and Muscatello (2008) recommend reproductive toxicity testing conducted by capturing spawning fish from exposure and reference sites, collecting gametes for fertilization, rearing the fertilized eggs to the swim-up fry stage, and examining the fry for prevalence of deformities. This approach provides highly relevant site-specific information regarding Se effects. While relatively few species have been tested in this manner ($n = 8$; Figure 6.4), thresholds of effects for mortality or deformities of early life stage fish have been remarkably similar across those species when expressed as a factor of egg or ovary concentrations. Effects thresholds for 8 species in Figure 6.4, as egg or ovary Se concentration, ranged over a factor of 1.4, from 17 to 24 mg/kg dw. If the species of interest at a site have previously been tested, obtaining site-specific data on concentrations of Se in ovaries or eggs can provide useful estimates of risk without undertaking site-specific reproductive toxicity testing. If similar species have been tested, uncertainty is probably within about a factor of 2 (i.e., not unreasonable for initial screening). However, the number of species tested to date remains small, thus it is possible that untested species may have lower or higher thresholds than the range shown in Figure 6.4. Clearly, properly designed and executed species-specific studies are the best way to reduce uncertainties in risk estimates if one is not sure of the relative sensitivities of the species of interest. As discussed below in Section 6.7.3, additional considerations are required to assess the possible population impact of any reproductive toxicity noted in fish species sampled from natural systems.

Other than the fathead minnow, we are not aware of reproductive toxicity testing with small-bodied fish (Figure 6.4). Small-bodied fish would be expected to experience greater exposures to locally elevated contaminant concentrations than more motile, larger-bodied fish species with lower site fidelity such as salmonids or catostomids (Gibbons et al. 1998; Munkittrick et al. 2001). For instance, sculpin may only move tens of meters or less over their lifetimes, while even "stream resident" salmonids typically travel hundreds of meters to tens of kilometers, and larger salmonids move much greater distances (Schmetterling and Adams 2004).

Thus, reproductive testing of small-bodied species is recommended, but may require pilot studies to determine husbandry requirements. Geckler et al. (1976) successfully conducted chronic toxicity tests with progeny from field-collected darter (Percidae) and minnow (Cyprinidae) species. Besser et al. (2007) were able to induce laboratory spawning and conduct early-life stage tests with mottled sculpin (*Cottus bairdi*) that were collected shortly before their normal spawning period and held for a short period of time under conditions similar to those in their native streams. However, they found that Ozark sculpin (*C. hypselurus*), which were similarly handled, failed to produce viable eggs, as did shorthead sculpin, which were collected before their spawning season and held overwinter in the laboratory. These studies suggest that some, but not necessarily all, field-collected small-bodied fish may be amenable to laboratory spawning and testing, particularly if sexually mature brood stock are collected shortly before their normal spawning period and held for a short period of time under conditions similar to those in their native streams. Because of their higher site fidelity, shorter life spans, and usual abundance, small-bodied fish may be useful sentinel species for monitoring effects of Se in streams and rivers, and the available data set on reproductive effects of Se should be expanded to these species.

6.7.3 MONITORING AND ASSESSING FISH POPULATIONS

Field assessments of fish (and other relevant, potentially Se-affected) populations are an intuitive approach to evaluating whether elevated Se concentrations could be linked to community-level impacts in aquatic ecosystems. Technical details relevant to the design of monitoring programs for detecting effects of Se are available elsewhere (Environment Canada 2002; Guy and Brown 2007; Johnson et al. 2007; Ohlendorf et al. 2008). This subsection focuses on general principles, using fish as an example.

Linking effects observed during field monitoring of fish populations to causal factors can be challenging. Selenium toxicity to early life stages may not be reflected at the population level because of density compensation or immigration from other source populations (Section 6.6). There are practical limitations in using field monitoring to reliably detect and link apparent population-level effects to elevated Se or other stressors. Further, monitoring programs and statistical tests are not customarily designed to detect noneffects. However, such biomonitoring can be optimized to deal with Se exposure. Monitoring and accompanying statistics are customarily intended to detect effects with a given likelihood of falsely detecting an effect when no true effect is present (Type I error, α). Natural variability and measurement error inherent to biomonitoring programs make statistically based comparative procedures necessary to detect adverse effects; traditionally, these have been significance or null hypothesis tests. This approach has been repeatedly criticized across disciplines on logical grounds, but the practice has endured (e.g., Berkson 1942; McCloskey 1995; Johnson 1999; Newman 2008). Alternative approaches for evaluating trends or comparing sites that should be considered include comparing confidence intervals for nonequivalence or testing if linear trends are near zero (Parkhurst 2001; Dixon and Pechmann 2005; McGarvey 2007).

The companion problem of Type II errors (β, failing to detect adverse effects that are present) has been ignored in some monitoring studies. However, the Environment Canada (2002) Environmental Effects Monitoring program seeks to balance risks of misguided remediation or unnecessary expenses resulting from spurious results (Type I error) with risks of undetected ecosystem degradation (Type II error) by setting $\alpha = \beta$, with both at 0.1 or less. In practice, tests that incorporate statistical power must also specify a priori the size of effect that they are trying to detect.

For a monitoring variable, selecting the magnitude of difference that distinguishes between a biologically important or negligible effect is an important aspect of monitoring since effect size allows ranking of the importance of impacts or alternative outcomes as well as hypotheses testing. However, selecting the effects magnitude of interest for a monitoring variable is not a trivial problem. The stipulation of an effect size threshold is a judgment about biology, not simply a statistical or procedural decision, and relies on many underlying explicit or implicit judgments about the biological importance of an effect of a nominated magnitude. Because of the difficulty in selecting broadly applicable criteria for what constitutes a biologically significant effect among different species and populations, some authors have argued that the selection of "critical" sizes for effects monitoring may need to be made by subjective consensus of those with relevant expertise (Mapstone 1995; Reed and Blaustein 1997; Munkittrick et al. 2009).

In the absence of a regulatory definition that designates critical effect sizes for interpreting monitoring results, three general approaches have been suggested (Munkittrick et al. 2001, 2009):

1) Select an arbitrary difference from reference conditions, such as two standard deviations (SD) from the mean, or other statistical extremes of the reference condition, such as the 5th or 10th percentile.
2) Use a predetermined difference that constitutes a change of significant magnitude to cause concern for the endpoint (e.g., a 25% decline).
3) Attempt to define statistically significant differences of smaller magnitudes.

The first approach selects effect sizes by quantifying natural variability under reference conditions, and by basing the effect size on exceedance of some statistic such as two standard deviations from the mean or the region of data outside of 90% to 95% of the possible observations under reference conditions (Kilgour et al. 2007; Munkittrick et al. 2009). Reference conditions may be defined more or less objectively, although the rules and criteria for defining reference conditions such as "natural," "least-disturbed," and "best attainable" are subjective and debatable (Mebane and Essig 2003; Stoddard et al. 2006). Documenting environmental variability under reference conditions is fundamental because the higher the variability, the less likely detection of trends becomes. Environmental variability is not simply stochasticity or measurement error, but often includes dynamic stabilities, that is, properties that vary in a repeated, reasonably predictable fashion. Examples include year-to-year differences in stream flows, temperatures, and thus, habitat suitability, seasonal differences, and spatial variability in abiotic factors that may influence fish populations (Luoma et al. 2001).

Variability introduced by flow regimes is universally important for design and interpretation of an effective monitoring program to detect effects of Se exposures in lotic water bodies. In all of the data sets reviewed in Table 6.7, stream flow variability had a major influence on the variability of fish populations. For instance, over the monitoring record for Thompson Creek (Idaho), stream flows ranged from 0.05 to 10 m^3/s (a factor of 200 difference) and annually ranged from 0.1 to 2 m^3/s (a factor of 20 difference) (GEI 2008b). Flood flows may scour fish habitats, and low flows may reduce habitat areas and cause displacement or direct mortality due to temperature extremes and reduced overwinter (and other) habitat. Depending on factors such as spawning timing and emergence, and resident or migratory life history, stream flow extremes may affect species that occupy the same streams during summer differently (e.g., Cunjak 1996; Waters 1999; Lobón-Cerviá 2009).

Because the prevalent adverse effect of Se in laboratory toxicity tests with fish is reproductive failure due to deformities in early life stage fish, monitoring relevant characteristics of fish populations is recommended. These characteristics include changes or differences in the age distribution and relative abundance of different age classes over time or from reference conditions. Young-of-year (age-0) fish would be the most directly relevant age class to target to detect reproductive failure. However, abundance estimates of age-0 fish are often more variable than those of older and larger fish (Table 6.7), and are likely influenced by high measurement error from variability in emergence timing and low capture efficiency. This may limit the effectiveness of detecting trends in the relative abundances of age-0 fish between sites or over time using routine methods (e.g., electrofishing or direct observation). Instead, adaptation of nonroutine methods that are specifically targeted for detecting trends in survival to emergence of early life stage fish such as fry emergence studies may be needed (Curry and MacNeill 2004).

Thus, for Se, detecting an effect requires monitoring of recruitment failure and, in some instances, species richness and composition. Recruitment failure is the logical population-level consequence of reproductive impairment. The general indication of recruitment failure in fish populations is a shift in the age distribution toward older and fewer fish. Recruitment failure may also be characterized by an increased growth rate in response to a decreased population size that lowers resource competition (Munkittrick

TABLE 6.7
Examples of Year-to-Year Variability of Fish Abundances from Long-Term Records of Reference Sites

Monitoring Endpoint	Year-to-Year Variability, as Median Coefficients of Variability, CV (Range of y-to-y CVs)	Average Fish Densities or Abundances (±SD)	The 2 SD Decline from the Mean Required to Detect Effects and Trigger Changes in Management Practices
Densities of age-1 cutthroat trout in an isolated population with no fishing or other observed disturbances, average year-to-year difference, 11-yr record, Dead Horse Canyon Creek, Oregon	30 (0–72)% (House 1995)	25 (± 6) fish/100 m²	46%
Densities of age-0 cutthroat trout, average year-to-year difference, 11-yr record, Dead Horse Canyon Creek, Oregon	36 (7–111)% (House 1995)	16 (± 8) fish/100 m²	100%
Abundance of age-1 brook trout in an open population with limited fishing and few other disturbances, average year-to-year difference, 14-yr record, Hunt Creek, Michigan	10 (5–48)% (McFadden et al. 1967)	1996 (± 317), number of individuals	32%
Densities of age-0 brook trout, average year-to-year difference, 14-yr record, Hunt Creek, Michigan	16 (7–80)% (McFadden et al. 1967)	4813 (± 983), number of individuals	41%
Densities of mottled sculpin, all ages, year-to-year difference, 12-yr record, Coweeta Creek, Georgia	22 (7–173)% (Grossman et al. 2006)	31 (± 13) fish/100 m²	84%
Densities of mottled sculpin, all ages, year-to-year difference, 12-yr record, Ball Creek-A, Georgia	33 (0–61)% (Grossman et al. 2006)	34 (± 8) fish/100 m²	46%
Densities of mottled sculpin, all ages, year-to-year difference, 12-yr record, Ball Creek-B, Georgia	50 (2–134)% (Grossman et al. 2006)	49 (± 18) fish/100 m²	74%
Densities of shorthead sculpin, all ages, year-to-year difference, 19-yr record, Thompson Creek, Idaho, upstream of mining effluent	22 (0–590)% (Fig. 6.10; GEI 2008b)	737 (± 362) fish/km	98%
Densities of cutthroat/rainbow trout hybrids, all ages, year-to-year difference, 19-yr record, Thompson Creek, Idaho, upstream of mining effluent	35 (0–263)% (Fig. 6.10; GEI 2008b)	229 (±139) fish/km	100%

and Dixon 1989). Species richness and composition are effective measures for detecting community-level impacts of elevated Se, but only in locations where fish diversity is high (e.g., 20 to 30 species such as in reservoirs of the southeastern United States (Crutchfield 2000; Chapter 3; Appendix A)). However, this is not the case for many areas where elevated Se concentrations are a concern. For instance, in the Great Plains or in cold, temperate regions of North America, fish assemblages are depauperate and may have few native species at reference sites. In cold-water, mountainous streams, this limited fish assemblage is often dominated by sculpins and trout (Bramblett and Fausch 1991; Mebane et al. 2003; Bramblett et al. 2005).

Monitoring age distribution and abundance of fish populations may avoid diffusing resources on the plethora of measurement endpoints often collected in monitoring programs for nonspecific causes, or monitoring programs designed for other stressors such as urban wastewater or pulp mill effluents. Because adult fish can survive and appear healthy under chronic Se stress (Coyle et al. 1993; Lemly 2002), and the responses of macroinvertebrate communities to chronic Se stress are equivocal (Section 6.4), some measurement endpoints that are commonly used in other environmental settings may not necessarily be sensitive to effects of moderately elevated Se. Examples of such endpoints that thus may be of limited utility for detecting chronic Se stress include calculation of routine macroinvertebrate bioassessment metrics or biotic integrity indexes, and some common fish health measures such as fecundity, egg size, condition factors of adult fish, lipid content, liver size, gonadosomatic index, and gonad size.

In addition to determining the effect size and the likelihood of false positives (α, Type I errors), we must consider data variability that promotes Type II errors even when we monitor the correct aspects of the system. Data variability undermines the power of monitoring programs to detect changes. The variability in age-structured fish populations for several long-term studies with trout or sculpin in lotic environments are summarized in Table 6.7. In these studies, the median year-to-year variability ranged from 10% to 50%, and coefficients of variation (CVs) ranged from 0% to 590%. The "2 SD from the mean" approach to defining departure from reference conditions (Environment Canada 2002; Kilgour et al. 2007) would correspond to declines in abundances of about 32% to 100% (Table 6.7). These declines are large and variable enough to suggest avoiding using the "2 SD from the mean" approach to setting monitoring trigger effect sizes for assessing the status of field fish populations.

Ham and Pearsons (2000) evaluated the ability to detect change in eight salmonid populations based on annual abundance estimates over 9 to 15 years, using the equal error power scheme of setting $\alpha = \beta$, with both at 0.1. They found that, after 5 years, detectable effect sizes ranged from decreases of 19% to 79%. The smaller detectable effects occurred for the more abundant species, and the poorest trend detections occurred for the rare, but highly valued, species. Dauwalter et al. (2009) examined trends over time in inland trout populations in relation to temporal variability, effect size, error rates, and number of sampling sites. They found that, using the traditional error rate (α) of 0.05 at a single site with an average CV of 49%, it would take about 20 years to detect a 5% annual decline (i.e., an absolute decline of about 62% from initial abundance) with good power (1-β of 0.8). Using the median CV from Table 6.7 and relaxing α to 0.1, the power to detect an annual decline of 5% in 10 years (37% decline from initial abundance) would be about 0.55. Working with bream (*Abramis brama*, a cyprinid), Nagelkerke and van Densen (2007) found that, in most cases, more than 6 years of monitoring would be required to detect a population decline of 15% per year, which roughly corresponded to a halving of the population size over 6 years. Working with data for many species from the English North Sea groundfish survey, Maxwell and Jennings (2005) also found that the power to detect declines in abundant species was much higher than for rare and vulnerable species and they often failed to detect declines of the magnitude that would lead to species being listed as endangered. Field et al. (2007) used Australian woodland bird census data sets to evaluate ability to detect a change of conservation status from "Least Concern" to "Vulnerable" (i.e., a decline in abundance of ≥ 30% over 10 years). They found that, although initially there was very low power to detect change for most species (<0.5), by the 10th year 4 species had reached their target power level of 0.9. For some of the less prevalent and more difficult to detect species, power to detect change started to rise more rapidly as time passed.

Examples of critical effect sizes used to determine population-level impacts in monitoring programs have ranged from 20% to the complete loss of dominant fish species (Table 6.8). However, several effect sizes were in the range of 20%–30%

TABLE 6.8
Examples of the Magnitude of Critical Effect Sizes Detected in or Used to Evaluate Fish Population Monitoring and Assessment Endpoints

Monitoring Endpoint	Effect Size and Notes
Wide variety of endpoints	25% less than reference conditions (Munkittrick et al. 2009)
Wide variety of endpoints	Values outside of the range of most reference conditions, such as values below the 5th or 10th percentile of the reference condition (Munkittrick et al. 2009)
Fish community or populations	20% reduction in fish species richness or abundance measured in the field (Suter et al. 1999)
Sentinel fish reproductive performance	50% decline in proportion of young-of-year fish from reference sites (Gray et al. 2002)
Fish condition factor	>10% change from reference condition (Kilgour et al. 2005)
Fish community	Loss of any dominant or nonrare species (Kilgour et al. 2007)
Fish abundance	15% decline/year over 6 years (Nagelkerke and van Densen 2007)
Species abundance	30–50% decline over 10 years or 3 generations specieswide could trigger a Red List "vulnerable" listing (IUCN 2006)
Species abundance	50–70% decline over 10 years or 3 generations specieswide would trigger a Red List "endangered" listing (IUCN 2006)
Species abundance	80–90% decline over 10 years or 3 generations specieswide would trigger a Red List "critically endangered" listing (IUCN 2006)
Fish assemblage (multiple metrics)	20% change in overall biological condition from reference conditions considered evidence of degradation, corresponding with the 10th percentile of reference condition (Meador et al. 2008)
Fish assemblage (multiple metrics)	Index scores 25% less than the highest biological condition scores for reference conditions more than minimally disturbed, corresponding with ~25th percentile of reference conditions (Mebane et al. 2003)

for different endpoints. As previously noted, the approach of using two standard deviations from the mean of reference conditions to define the range of acceptable conditions may yield very large allowable effect sizes, ranging from about 30% to 100% declines. Thus, determinations of acceptable conditions and critical effects sizes must be situation-specific rather than generic.

Given these considerations, and the fact that laboratory toxicity testing may not accurately reflect the natural environment, field monitoring programs need to be carefully designed to have any reasonable chance of detecting population-level impacts if these are truly occurring. Critical steps in the design and interpretation of field monitoring programs for Se ecotoxicology include the following:

- Increase power to detect trends by monitoring a network of comparable reference and exposure sites rather than single sites.
- Ensure adequate frequency and duration of monitoring.
- Assess and quantify major sources of natural variation.
- Select appropriate error rates (Type I and Type II errors).
- Determine *a priori* the critical effect size that constitutes a population-level impact.
- Because chronic effects of Se primarily affect recruitment, focus on differences in the relative abundance of age-0 and age-1 fish both temporally and spatially; this will involve different methods than typically used for monitoring fish populations.

Note that, as documented above, error rates and critical effect sizes should be situation specific. As such they should be determined based on a consensus of those with relevant technical expertise. Further, even robust monitoring programs may not be able to convincingly detect declining abundance trends until several years of data have been collected. Thus, data from field monitoring programs should not be used in isolation, but rather in a weight-of-evidence determination along with exposure (Section 6.7.1) and reproductive effects data (Section 6.7.2). Moreover, as discussed above, alternative statistical strategies are appropriate such as comparing confidence intervals for nonequivalence or testing if linear trends are near zero (Parkhurst 2001; Dixon and Pechmann 2005; McGarvey 2007).

6.8 UNCERTAINTIES AND RECOMMENDATIONS FOR FURTHER RESEARCH

While the field of Se toxicity has been highly productive and prolific in recent years, numerous uncertainties, questions and hypotheses still remain. Table 6.9 outlines key uncertainties related to Se toxicity to aquatic organisms and provides associated recommendations for further research in these areas.

TABLE 6.9

Uncertainties and Opportunities for Future Research Pertaining to Se Toxicity

Aspect	Uncertainty	Recommendations for Further Research
Cellular mechanisms of Se toxicity	Studies investigating effects of Se on immunocompetence.	Laboratory and field studies investigating potential effects of Se on immune function.
Toxicokinetics and toxicodynamics	In egg-producing vertebrates, the relationships between reproductive strategy (e.g., oviparity vs. ovoviviparity, synchronous vs. asynchronous egg development) and deposition of Se into eggs (i.e., amount and timing of Se deposition). Underlying reasons for large differences in transfer efficiencies from body tissues (e.g., liver, muscle) to eggs among species.	Identify potentially susceptible species with different reproductive strategies and evaluate relative Se bioaccumulation in eggs. Evaluate how different variables affect Se deposition into the eggs, such as timing of dietary Se exposure relative to vitellogenesis and number of spawns/clutches per season.
Factors modifying Se toxicity	With the possible exception of interactions between organic Se and meHg, the mechanism(s) and extent of antagonistic reactions between Se and other factors (e.g., other elements, biotic and abiotic stressors) are unknown. Further, there are a few studies showing synergistic, not antagonistic, interactions that remain unexplained.	Determine the mechanism(s), extent and significance of antagonistic reactions between Se and other factors (chemical, biotic, and abiotic) for fish, waterbirds, and amphibians.
Nutritional factors	The mechanism(s) by which dietary factors can increase or reduce Se toxicity remain unknown and the extent and significance of these modifications of Se toxicity are uncertain.	Determine the mechanism(s), extent, and significance of dietary-based variations in Se toxicity for fish, water birds and amphibians.
Tolerance	Can fish, waterbirds, amphibians, and aquatic reptiles become tolerant (acclimation and/or adaptation) such that population-level impacts do not occur in highly Se-contaminated aquatic environments? If so, what are the underlying mechanisms and potential costs of such tolerance?	Determine whether oviparous vertebrates can become tolerant such that organic Se toxicity is reduced or eliminated, the types of tolerance possible (i.e., physiological or genetic), for what organisms, and the implications of such tolerance (including energetic or other costs) to populations exposed to increasing Se concentrations. Possible research includes side-by-side toxicity tests of suspected "tolerant" species and intolerant species, or ideally intraspecific comparisons among populations from "low" versus "high" Se environments.

Comparative sensitivity (protozoans)	Understanding of potential toxicity of Se to protozoans, which is currently based on a very small database.	Establishment of acute and chronic water concentration thresholds for protozoans, with potential standardized endpoints relating to behavior, growth, and survival, would be an important advance.
Comparative sensitivity (macroinvertebrates)	Although macroinvertebrate communities do not appear impacted by elevated concentrations of Se, sensitive species within those communities may be adversely affected. Dietary exposure and maternal transfer of Se to eggs in invertebrates has had little study.	Comparisons of macroinvertebrate taxa observed in areas with elevated Se concentrations to taxa expected in reference conditions may indicate potentially Se sensitive taxa; controlled exposures would be necessary to make any definitive conclusions of sensitivity.
Comparative sensitivity (fish)	The relative sensitivity of fish based on diet-only juvenile exposures and maternal transfer exposures.	Additional laboratory studies investigating effects of diet-only Se exposures by juvenile and adult animals relative to maternal transfer studies, and expanding testing to little-studied species or potential "sentinel" species.
Comparative sensitivity (amphibians and reptiles)	The relative sensitivity of understudied taxonomic groups.	More laboratory and field studies investigating species differences in sensitivity of amphibians and reptiles to Se.
Population implications of fish deformities	In the wild, how strongly is the development of subtle deformities (e.g., mild or moderate edema) in fish early-life stages related to their survival to older age classes through reproduction? When is the occurrence of Se-induced deformities a predictor of ultimate mortality versus a transient effect that fish may recover from without incurring lasting impairment? What environmental co-factors are important (e.g., velocity or flow regimes; thermal regimes; predator, prey, competitive, interactions), species combinations	In the wild, tracking the survival rates of early life stage (ELS) fish with or without subtle deformities to recruitment as reproductively fit adults is logistically highly challenging. Newly hatched ELS fish are very small (<15 to about 25 mm), and by the time they grow to sizes large enough to tag (>≈ 50 mm) fish afflicted by deformities may have already been lost from the cohort. If for greater experimental control, survival of ELS fish with and without deformities were tracked in quasi-natural experimental stream mesocosms, achieving realistic conditions and stresses for ELS fish would be challenging (e.g., prey capture, predator avoidance, currents). Alternatively, indirect correlative approaches such as monitoring emergent fry for deformity rates, proportions of young-of-year fish, and age-class strength may be more feasible to carry out but interpretation may be complicated by natural variability or factors such as density compensation (Section 6.6.3).

(continued)

TABLE 6.9 (CONTINUED)
Uncertainties and Opportunities for Future Research Pertaining to Se Toxicity

Aspect	Uncertainty	Recommendations for Further Research
Linkage between individual effects and impacts on populations	In different environments (e.g., freshwater: lotic versus lentic; estuarine; marine).	Population-level studies in different settings with elevated Se levels may provide useful information on population dynamics, compensation, and perhaps recovery. Extensive field monitoring has been conducted in areas with elevated Se concentrations, but much of this work languishes as poorly accessible grey literature. Review and publication of these studies in the primary literature could provide valuable information on patterns of ecosystems responses to elevated Se.

6.9 SUMMARY

Selenium is an essential nutrient that is incorporated into functional and structural proteins as selenocysteine. Several of these proteins are enzymes that provide cellular antioxidant protection. A key aspect of the toxicity of Se is the extremely narrow range between dietary essentiality and toxicity. Another important aspect of Se toxicity is that, although it is involved in antioxidant processes at normal dietary levels, it can become involved in the generation of reactive oxygen species at higher exposures, resulting in oxidative stress. Toxicity results from dietary exposure to organic Se compounds, predominantly selenomethionine, and the subsequent production of reactive oxygen species.

Oviparous (egg-laying) vertebrates such as fish and waterbirds are the most sensitive organisms to Se of those studied to date. Toxicity can result from maternal transfer of organic Se to eggs in oviparous vertebrates. Eggs are an important depuration pathway for fish but less so for birds. The most sensitive diagnostic indicators of Se toxicity in vertebrates occur when developing embryos metabolize organic Se present in egg albumen or yolk. Certain metabolites of organic Se can become involved in oxidation-reduction cycling, generating reactive oxygen species that can cause oxidative stress and cellular dysfunction. Toxicity endpoints include embryo mortality (which is the most sensitive endpoint in birds), and a characteristic suite of teratogenic deformities (such as skeletal, craniofacial, and fin deformities, and various forms of edema) that are the most useful indicators of Se toxicity in fish larvae.

Relative species sensitivities are not well understood but may be related to differences in reproductive physiology (e.g., the pattern of oogenesis or relative number of Se-containing amino acids in yolk), dynamics of Se transfer from diet or body tissues to eggs (i.e., dose), and/or differences in the capacity to metabolize Se to reactive forms (i.e., reactive oxygen species). Importantly, embryo mortality and severe malformations (developmental abnormalities) can result in impaired recruitment of individuals into populations and have caused population reductions of sensitive fish and bird species. These established linkages between the molecular/cellular mechanism of toxicity (oxidative stress), effects on individuals (early life stage mortality and deformities), and negative effects on populations and community structure provide one of the clearest examples in ecotoxicology of cause–effect relationships between exposure and altered population dynamics.

Similar to other toxicants, many factors can modify the toxicological responses of organisms to Se. Selenium interacts with many other inorganic and organic compounds, both in the aquatic environment and *in vivo*, in a predominantly antagonistic fashion. Nutritional factors such as dietary protein and carbohydrate content can modify Se toxicity. Abiotic factors such as temperature also appear to be important modifying factors of Se toxicity in both poikilotherms and homeotherms. Differences among freshwater, estuarine, and marine environments in the toxicological responses of organisms to Se are important considerations but have not been studied in great detail. The ecology of a species, particularly feeding niche, is a critical aspect related to its vulnerability to Se because of differential prey accumulation of organic Se and dietary exposure routes. Considerations of spatial and temporal variation in diet are important factors to consider when assessing potentially susceptible species; effects tend to be site specific.

Among taxa, there is a wide range of sensitivities to Se. Algae and plants are believed to be the least sensitive organisms. Very few studies have investigated the sensitivity of bacteria to Se, although they appear to be insensitive. Protozoans have also been understudied, and further work is needed investigating Se toxicity in this taxon. Most species of invertebrates, which are essential components of aquatic food webs and a key vector for transfer of organic Se to higher trophic levels, are also relatively insensitive to Se. As discussed above, oviparous vertebrates appear to be the most sensitive organisms. Although fish and waterbird sensitivities are well documented, there are reasons to suspect that amphibians and reptiles with oviparous modes of reproductive strategy are also sensitive. Compared to oviparous vertebrates, aquatic-dependent mammals do not appear to be sensitive to dietary organic Se exposure, further illustrating the importance of oviparity in Se toxicity. Although there have been suggestions of tolerance to Se (physiological acclimation or genetic adaptation) in certain biota, it is not known whether this is an actual phenomenon.

Selenium enrichment of reservoir environments (e.g., Belews Lake, Hyco Lake, Kesterson Reservoir) provide classic examples of adverse effects occurring through different levels of biological organization, comprising integrated whole-ecosystem examples of trophic transfer resulting in population-level reductions of resident species. Recovery from adverse effects on fish populations occurred once Se sources were eliminated. However, population-level effects from Se in natural ecosystems are difficult to detect. This difficulty reflects differences in species sensitivity as well as food web complexities and demographics where population-level effects are suspected. Few such widespread impacts on populations as documented at Belews, Hyco, and Kesterson reservoirs have been definitively documented in other ecosystems; however, population-level effects have been suspected at several other sites, including San Francisco Bay and Lake Macquarie, Australia.

Inability to observe population-level effects in the field can occur even when the species exposed in the field are the same or closely related to those for which adverse effects have been demonstrated in laboratory settings at lower Se tissue concentrations. In addition, several studies of aquatic ecosystems with naturally elevated Se concentrations have reported unaffected aquatic communities. Although statistical considerations and normal fish population monitoring design can preclude detection of low level (<10%) field population effects, these examples illustrate the critical importance of considering ecological and environmental factors when investigating potential Se toxicity in aquatic ecosystems.

REFERENCES

Adams WJ, Brix KV, Edwards M, Tear LM, DeForest DK, Fairbrother A. 2003. Analysis of field and laboratory data to derive selenium toxicity thresholds for birds. *Environ Toxicol Chem* 22:2020–2029.

Albers PH, Green DE, Sanderson CJ. 1996. Diagnostic criteria for selenium toxicosis in aquatic birds: dietary exposure, tissue concentrations, and macroscopic effects. *J Wildl Dis* 32:468–485.

Alfonso JM, Montero D, Robaina L, Astorga N, Izquierdo MS, Gines R. 2000. Association of a lordosis-scoliosis-kyphosis deformity in gilthead seabream (*Sparus aurata*) with family structure. *Fish Physiol Biochem* 22:159–163.

Al-Harbi AH. 2001. Skeletal deformities in cultured common carp *Cyprinus carpio* L. *Asian Fish Sci* 14:247–254.

Allan CB, Lacourciere GM, Stadtman TC. 1999. Responsiveness of selenoproteins to dietary selenium. *Annu Rev Nutr* 19:1–16.

Anastasia JR, Morgan SG, Fisher NS. 1998. Tagging crustacean larvae: assimilation and retention of trace elements. *Limnol Oceanogr* 43:362–368.

Andrades JA, Becerra J, Fernandez-Llebrez P. 1996. Skeletal deformities in larval, juvenile and adult stages of cultured gilthead sea beam (*Sparus aurata* L.). *Aquaculture* 141:1–11.

Arner ESJ, Holmgren A. 2000. Physiological functions of thioredoxin and thioredoxin reductase. *Eur J Biochem* 267:6102–6109.

Arteel GE, Sies H. 2001. The biochemistry of selenium and the glutathione system. *Environ Toxicol Pharmacol* 10:153–158.

Arukwe A, Goksøyr A. 2003. Eggshell and egg yolk proteins in fish: hepatic proteins for the next generation: oogenetic, population, and evolutionary implications of endocrine disruption. *Comp Hepatol* 2:4–21.

Atencio L, Moreno I, Jos A, Prieto AI, Moyano R, Blanco A, Camean AM. 2009. Effects of dietary selenium on the oxidative stress and pathological changes in tilapia (*Oreochromis niloticus*) exposed to a microcystin-producing cyanobacterial water bloom. *Toxicon* 53:269–282.

Bailey FC, Knight AW, Ogle RS, Klaine SJ. 1995. Effect of sulphate level on selenium uptake by *Ruppia maritima*. *Chemosphere* 30:579–591.

Baines SB, Fisher NS. 2001. Interspecific differences in the bioconcentration of selenite by phytoplankton and their ecological implications. *Mar Ecol Progr Ser* 213:1–12.

Baines SB, Fisher NS, Doblin MA, Cutter GA, Cutter LS, Cole B. 2004. Light dependence of selenium uptake by phytoplankton and implications for predicting selenium incorporation into food webs. *Limnol Oceanogr* 49:566–578.

Barnthouse LW, Christensen SW, Goodyear CP, Van Winkle W, Vaughan DS. 1984. Population biology in the courtroom: the Hudson River controversy. *BioScience* 34:14–19.

Barron MG, Carls MG, Heintz R, Rice SD. 2004. Evaluation of fish early life-stage toxicity models of chronic embryonic exposures to complex polycyclic aromatic hydrocarbon mixtures. *Toxicol Sci* 78:60–67.

Barwick DH, Harrell RD. 1997. Recovery of fish populations in Belews Lake following Se contamination. *Proc Ann Conf SE Assoc Fish Wildl Agencies* 51:209–216.

Bauder MB, Palace VP, Hodson PV. 2005. Is oxidative stress the mechanism of blue sac disease in retene-exposed trout larvae? *Environ Toxicol Chem* 24:694–702.

Baxter CV. 2002. Fish movement and assemblage dynamics in a Pacific Northwest riverscape. PhD Dissertation. Corvallis (OR, USA): Oregon State University.

Beckon WN, Parkins C, Maximovich A, Beckon AV. 2008. A general approach to modeling biphasic relationships. *Environ Sci Technol* 42:1308–1314.

Behne D, Kyriakopolis A. 2001. Mammalian selenium-containing proteins. *Annu Rev Nutr* 21:453–473.

Behne D, Pfeifer H, Rothlein D, Kyriakopoulos A. 2000. Cellular and subcellular distribution of selenium and selenium containing proteins in the rat. In: Roussel AM, Favier AE, Anderson RA, editors. Trace elements in man and animals 10. New York (NY, USA): Kluwer/Plenum, p 29–34.

Belzile N, Chen Y-W, Gunn JM, Tong J, Alarie Y, Delonchamp T, Land C-Y. 2006. The effect of selenium on mercury assimilation by freshwater organisms. *Can J Fish Aquat Sci* 63:1–10.

Bennett PM, Janz DM. 2007a. Bioenergetics and growth of young-of-the-year northern pike (*Esox lucius*) and burbot (*Lota lota*) exposed to metal mining effluent. *Ecotoxicol Environ Saf* 68:1–12.

Bennett PM, Janz DM. 2007b. Seasonal changes in morphometric and biochemical endpoints in northern pike (*Esox lucius*), burbot (*Lota lota*) and slimy sculpin (*Cottus cognatus*). *Freshwater Biol* 52:2056–2072.

Bennett WN, Brooks AS, Boraas ME. 1986. Selenium uptake and transfer in an aquatic food chain and its effects on fathead minnow (*Pimephales promelas*) larvae. *Arch Environ Contam Toxicol* 15:513–517.

Berkson J. 1942. Tests of significance considered as evidence. *J Amer Stat Assoc* 37:325–335 (reprinted in *Int J Epidemiol* 2003; 32:687–691).

Besser JM, Mebane CA, Mount DR, Ivey CD, Kunz JL, Greer EI, May TW, Ingersoll CG. 2007. Relative sensitivity of mottled sculpins (*Cottus bairdi*) and rainbow trout (*Oncorhynchus mykiss*) to toxicity of metals associated with mining activities. *Environ Toxicol Chem* 26:1657–1665.

Beyers DW, Sodergren C. 2001a. Evaluation of interspecific sensitivity to selenium exposure: larval razorback sucker versus flannelmouth sucker. Final Report to Recovery Implementation Program Project CAP-6 SE-NF. Dept. Fishery and Wildlife Biology, Colorado State Univ, Fort Collins, CO, USA.

Beyers DW, Sodergren C. 2001b. Assessment of exposure of larval razorback sucker to selenium in natural waters and evaluation of laboratory-based predictions. Final Report to Recovery Implementation Program Project CAP-6 SE. Dept. Fishery and Wildlife Biology, Colorado State Univ, Fort Collins, CO, USA.

Billiard SM, Querback K, Hodson PV. 1999. Toxicity of retene to early life stages of two freshwater fish species. *Environ Toxicol Chem* 18:2070–2077.

Bischoff K, Pichner J, Braselton WE, Counard C, Evers DC, Edwards WC. 2002. Mercury and selenium levels in livers and eggs of common loons (*Gavia immer*) from Minnesota. *Arch Environ Contam Toxicol* 42:71–76.

Blackburn DG. 1998. Structure, function and evolution of the oviducts of squamate reptiles, with special reference to viviparity and placentation. *J Exp Zool* 282:560–617.

Blackburn DG. 2000. Classification of the reproductive patterns of amniotes. *Herpetol Monogr* 14:371–377.

Bovee EC. 1978. Effects of heavy metals especially selenium, vanadium and zirconium on movement, growth and survival of certain animal aquatic life. PB-292563. Arlington (VA, USA): National Technical Information Service.

Bovee EC, O'Brien TL. 1982. Some effects of selenium, vanadium and zirconium on the swimming rate of *Tetrahymena pyriformis*: a bioassay study. *Univ Kansas Sci Bull* 52:39–44.

Bramblett RG, Fausch KD. 1991. Variable fish communities and the index of biotic integrity in a western Great Plains River. *Trans Am Fish Soc* 120:752–769.

Bramblett RG, Johnson TR, Zale AV, Heggem DG. 2005. Development and evaluation of a fish assemblage index of biotic integrity for northwestern Great Plains streams. *Trans Am Fish Soc* 134:624–640.

Brand GL, Horak RV, Le Bris N, Goffredi SK, Carney SL, Govenar B, Yancey PH. 2007. Hypotaurine and thiotaurine as indicators of sulfide exposure in bivalves and vestimentiferans from hydrothermal vents and cold seeps. *Mar Ecol—Evol Persp* 28:208–218.

Braune BA. 2007. Temporal trends of organochlorines and mercury in seabird eggs from the Canadian Arctic, 1975–2003. *Environ Pollut* 148:599–613.

Braune BM, Simon M. 2004. Trace elements and halogenated organic compounds in Canadian Arctic seabirds. *Mar Pollut Bull* 48:986–992.

Braune BM, Scheuhammer AM. 2008. Trace element and metallothionein concentrations in seabirds from the Canadian Arctic. *Environ Toxicol Chem* 27:645–651.

Braune BM, Donaldson GM, Hobson KA. 2002. Contaminant residues in seabird eggs from the Canadian arctic. II. Spatial trends and evidence from stable isotopes for intercolony differences. *Environ Pollut* 117: 133–145.

Brix KV, DeForest DK, Cardwell RD, Adams WJ. 2004. Derivation of a chronic site-specific water quality standard for selenium in the Great Salt Lake, Utah, USA. *Environ Toxicol Chem* 23:606–612.

Brown TA, Shrift A. 1982. Selenium: toxicity and tolerance in higher plants. *Biol Rev Cambridge Philosophical Soc* 57:59–84.

Bryan L, Hopkins WA, Baionno JA, Jackson BP. 2003. Maternal transfer of contaminants to eggs in common grackles (*Quiscalus quiscala*) nesting on coal fly ash basins. *Arch Environ Contam Toxicol* 45:273–277.

Bryson WT, Garrett WR, Mallin MA, MacPherson KA, Partin WE, Woock SE. 1984. Roxboro Steam Electric Plant 1982 Environmental Monitoring Studies, Volume II, Hyco Reservoir Bioassay Studies. Roxboro (NC, USA): Environmental Technology Section, Carolina Power & Light Company.

Bryson WT, Garrett WR, Mallin MA, MacPherson KA, Partin WE, Woock SE. 1985a. Roxboro Steam Electric Plant Hyco Reservoir 1983 Bioassay Report. Roxboro (NC, USA): Environmental Services Section, Carolina Power & Light Company.

Bryson WT, MacPherson KA, Mallin MA, Partin WE, Woock SE. 1985b. Roxboro Steam Electric Plant Hyco Reservoir 1984 Bioassay Report. Roxboro (NC, USA): Environmental Services Section, Carolina Power & Light Company.

Burger J, Trivedi CD, Gochfeld M. 2000. Metals in herring and great black-backed gulls from the New York bight: the role of salt gland in excretion. *Environ Monitor Assess* 64:569–581.

Burger J, Gochfeld M, Shukla T, Jeitner C, Burke S, Donio M, Shukla S, Snigaroff R, Snigaroff D, Stamm T, Volz C. 2007. Heavy metals in Pacific Cod (*Gadus macrocephalus*) from the Aleutians: location, age, size, and risk. *J Toxicol Environ Health A* 70:1897–1911.

Campbell LM, Norstrom RJ, Hobson KA, Muir DCG, Backus S, Fisk AT. 2005. Mercury and other trace elements in a pelagic Arctic marine food web (Northwater Polynya, Baffin Bay). *Sci Total Environ* 351:247–263.

Canton SP, Baker S. 2008. Part III. Field application of tissue thresholds: potential to predict fish population or community effects in the field. In: Selenium tissue thresholds: tissue selection criteria, threshold development endpoints, and potential to predict population or community effects in the field. Washington (DC, USA): North America Metals Council – Selenium Working Group; www.namc.org.

Carolina Power and Light. 1997. Largemouth bass selenium bioassay. Roxboro (NC, USA): Environmental Services Section.

CEC. 2004. Selenium bioaccumulation monitoring in Thompson Creek, Custer County, Idaho 2003 Report by Chadwick Ecological Consultants (Littleton, CO, USA) for Thompson Creek Mining (Challis, ID, USA).

Chan K-S, Stenseth NC, Lekve K, Gjøsæter J. 2003. Modeling pulse disturbance impact on cod population dynamics: the 1988 algal bloom of Skagerrak, Norway. *Ecol Monogr* 73:151–171.

Chapman DW. 1966. Food and space as regulators of salmonid populations in streams. *Amer Naturalist* 100:345–357.

Chapman GA. 1983. Do organisms in laboratory toxicity tests respond like organisms in nature? In: Bishop W, Cardwell R, Heidolph B, editors. Aquatic toxicology and hazard assessment: sixth symposium (STP 802). Volume STP 802. Philadelphia (PA, USA): American Society for Testing and Materials, p 315–327.

Chapman PM. 2007. Selenium thresholds for fish from cold freshwaters. *Human Ecol Risk Assess* 13:20–24.

Chapman PM. 2008. Environmental risks of inorganic metals and metalloids: a continuing, evolving scientific odyssey. *Human Ecol Risk Assess* 14:5–40.

Chatain B. 1994. Abnormal swimbladder development and lordosis in sea bass (*Dicentrarchus labrax*) and sea bream (*Sparus auratus*). *Aquaculture* 119:371–379.

Chen YW, Belzile N, Gunn JM. 2001. Antagonistic effect of selenium on mercury assimilation by fish populations near Sudbury metal smelters? *Limnol Oceanogr* 46:1814–1818.

Chiang H-C, Lo J-C, Yeh K-C. 2006. Genes associated with heavy metal tolerance and accumulation in Zn/Cd hyperaccumulator *Arabidopsis halleri*: a genomic survey with cDNA microarray. *Environ Sci Technol* 40: 6792-6798.

Clark DR Jr. 1987. Selenium accumulation in mammals exposed to contaminated California irrigation drainwater. *Sci Total Environ* 66:147–168.

Clark DR Jr, Ogasawara PA, Smith GJ, Ohlendorf HM. 1989. Selenium accumulation by raccoons exposed to irrigation drainwater at Kesterson National Wildlife Refuge, California, 1986. *Arch Environ Contam Toxicol* 18:787–794.

Cleveland L, Little EE, Buckler DR, Wiedmeyer RH. 1993. Toxicity and bioaccumulation of waterborne and dietary selenium in juvenile bluegill (*Lepomis macrochirus*). *Aquat Toxicol* 27:265–279.

Combs GF Jr, Combs SB. 1986. The role of selenium in nutrition. Orlando (FL, USA): Academic Pr.

Coughlan DJ, Velte JS. 1989. Dietary toxicity of selenium-contaminated red shiners to striped bass. *Trans Am Fish Soc* 118:400–408.

Coyle JJ, Buckler DR, Ingersoll CG, Fairchild JF, May TW. 1993. Effect of dietary selenium on the reproductive success of bluegills (*Lepomis macrochirus*). *Environ Toxicol Chem* 12:551–565.

Crutchfield JU Jr. 2000. Recovery of a power plant cooling reservoir ecosystem from selenium bioaccumulation. *Environ Sci Pol* 3:S145–S163.

Cumbie PM, Van Horn SL. 1978. Selenium accumulation associated with fish mortality and reproductive failure. *Proc Ann Conf SE Assoc Fish Wildl Agencies* 32:612–624.

Cunjak RA. 1996. Winter habitat of selected stream fishes and potential impacts from land-use activity. *Can J Fish Aquat Sci* 53:267–282.

Cunningham MK, Granberry WF Jr, Pope KL. 2000. Shrinkage of inland silverside larvae preserved in ethanol and formalin. *N Amer J Fish Manage* 20:816–818.

Curry RA, MacNeill WS. 2004. Population-level responses to sediment during early life in brook trout. *J N Am Benthol Soc* 23:140–150.

Cuvin-Aralar ML, Furnes RW. 1991. Mercury and selenium interaction: a review. *Ecotoxicol Environ Saf* 21:348–364.

Dabrowski K, Moerau R, El Saidy D. 1996. Ontogenetic sensitivity of channel catfish to ascorbic acid deficiency. *J Aquat Anim Health* 8:22–27.

Daoulas CH, Economou AN, Bantavas I. 1991. Osteological abnormalities in laboratory reared sea-bass (*Dicentrarchus labrax*) fingerlings. *Aquaculture* 97:169–180.

Dauwalter DC, Rahel FJ, Gerow KG. 2009. Temporal variation in trout populations: implications for monitoring and trend detection. *Trans Am Fish Soc* 138:38–51.

Davis RH, Fear J. 1996. Incorporation of selenium into egg proteins from dietary selenite. *British Poultry Sci* 37:197–211.

deBruyn AM, Chapman PM. 2007. Selenium toxicity to invertebrates: will proposed thresholds for toxicity to fish and birds also protect their prey? *Environ Sci Technol* 41: 1766–1770.

deBruyn AM, Hodaly A, Chapman PM. 2008. Tissue selection criteria: selection of tissue types for development of meaningful selenium tissue thresholds in fish. Part 1 of Selenium issue thresholds: tissue selection criteria, threshold development endpoints, and field application of tissue thresholds. Washington (DC, USA): North America Metals Council-Selenium Working Group; www.namc.org.

DeForest DK. 2008. Review of selenium tissue thresholds for fish: evaluation of the appropriate endpoint, life stage, and effect level and recommendation for a tissue-based criterion. Part 2 of Selenium tissue thresholds: tissue selection criteria, threshold development endpoints, and potential to predict population or community effects in the field. Washington (DC, USA): North America Metals Council-Selenium Working Group; www.namc.org.

DeForest DK, Brix KV, Adams WJ. 1999. Critical review of proposed residue-based selenium toxicity thresholds for freshwater fish. *Human Ecol Risk Assess* 5:1187–1228.

de Rosemond SC, Liber K, Rosaasen A. 2005. Relationship between embryo selenium concentration and early life stage development in white sucker (*Catostomus commersoni*) from a northern Canadian lake. *Bull Environ Contam Toxicol* 74:1134–1142.

Detwiler SJ. 2002. Toxicokinetics of selenium in the avian egg: comparisons between species differing in embryonic tolerance. PhD Dissertation. Davis (CA, USA): University of California.

DeVink J-MA, Clark RG, Slattery SM, Wayland M. 2008a. Is selenium affecting body condition and reproduction in boreal breeding scaup, scoters and ring-necked ducks? *Environ Pollut* 152:116–122.

DeVink J-MA, Clark RG, Slattery SM, Scheuhammer TM. 2008b. Effects of dietary selenium on reproduction and body mass of captive lesser scaup. *Environ Toxicol Chem* 27:471–477.

Dietz R, Riget F, Born EW. 2000. An assessment of selenium to mercury in Greenland marine animals. *Sci Tot Environ* 245:15–24.

Diplock AT, Hoekstra WG. 1976. Metabolic aspects of selenium action and toxicity. *CRC Crit Rev Toxicol* 5:271–329.

Divanach P, Papandroulakis N, Anastasiadis P, Koumoundouros G, Kentouri M. 1997. Effect of water currents during postlarval and nursery phase on the development of skeletal deformities in sea bass (*Dicentrarchus labrax* L.) with functional swimbladder. *Aquaculture* 156:145–155.

Dixon PM, Pechmann JHK. 2005. A statistical test to show negligible trend. *Ecology* 86:1751–1756.

Doroshov S, Van Eenennaam J, Alexander C, Hallen E, Bailey H, Kroll K, Restrepo C. 1992. Development of water quality criteria for resident aquatic species of the San Joaquin River; Part II, Bioaccumulation of dietary selenium and its effects on growth and reproduction in bluegill (*Lepomis macrochirus*). Sacramento (CA, USA): State Water Resources Control Board, State of California.

Driedger K, Weber LP, Rickwood CJ, Dubé MG, Janz DM. 2009. Overwinter alterations in energy stores and growth in juvenile fishes inhabiting areas receiving metal mining and municipal wastewater effluents. *Environ Toxicol Chem* 28:296–304.

Dubois KP, Moxon AL, Olson DE. 1940. Further studies on the effectiveness of arsenic in preventing selenium poisoning. *J Nutr* 19:477–482.

DuBowy PJ. 1989. Effects of diet on selenium bioaccumulation in marsh birds. *J Wildl Manage* 53:776–781.

Duellman WE, Trueb L. 1986. Biology of amphibians. Baltimore (MD, USA): The Johns Hopkins University Pr.

Eagles-Smith CA, Ackerman JT, Yee J, Adelsbach TL. 2009. Mercury demethylation in waterbird livers: dose-response thresholds and differences among species. *Environ Toxicol Chem* 28:568–577.

Egerer-Sieber C, Herl V, Muller-Uri F, Kreisb W, Muller YA. 2006. Crystallization and preliminary crystallographic analysis of selenomethionine-labelled progesterone 5b-reductase from *Digitalis lanata* Ehrh. *Acta Crystall Sec F* 62:186–188.

El-Begearmi MM, Combs JFJr. 1982. Dietary effects of selinite toxicity in the chick. *Poultry Sci* 61:770–776.

Eisler R. 2000. Handbook of chemical risk assessment. Boca Raton (FL, USA): Lewis.

Elliott, JM. 1989. Mechanisms responsible for population regulation in young migratory trout, *Salmo trutta*. I. The critical time for survival. *J Anim Ecol* 58:987–1001.

Environment Canada. 2002. Metal mining guidance document for aquatic environmental effects monitoring. Gatineau, QC, Canada. http://www.ec.gc.ca/esee-eem/.

Estay F, Díaz A, Pedrazza R, Colihueque N. 2003. Oogenesis and plasma levels of sex steroids in cultured females of brown trout (*Salmo trutta* Linnaeus, 1758) in Chile. *J Exp Zool* 298A:60–66.

EVS and PLA. 1998. Technical evaluation of fish methods in environmental monitoring for the mining industry in Canada. Unpublished report prepared for Aquatic Effect Technology Evaluation Program, Canmet, Natural Resources Canada. EVS Environment Consultants and Paine Ledge Associates, North Vancouver, BC, Canada.

Fairbrother A, Fowles J. 1990. Subchronic effects of sodium selenite and selenomethionine on several immune functions in mallards. *Arch Environ Contam Toxicol* 19:836–844.

Fairbrother A, Fix M, O'Hara T, Ribic C. 1994. Impairment of growth and immune function of avocet chicks from sites with elevated selenium, arsenic, and boron. *J Wildl Dis* 30:222–233.

Fairbrother A, Smits J, Grasman K. 2004. Avian immunotoxicology. *J Toxicol Environ Health B* 7:105–137.

Fan TWM, Teh SJ, Hinton DE, Higashi RM. 2002. Selenium biotransformations into proteinaceous forms by foodweb organisms of selenium-laden drainage waters in California. *Aquat Toxicol* 57:65–84.

Fernandez I, Hontoria F, Ortiz-Delgado JB, Kotzaminis Y, Estevez A, Zambonino-Infante JL, Gisbert E. 2008. Larval performance and skeletal deformities in farmed gilthead sea bream (*Sparus aurata*) fed with graded levels of Vitamin A enriched rotifers (*Brachionus plicatilis*). *Aquaculture* 283:102–115.

Fey DP. 1999. Effects of preservation technique on the length of larval fish: methods of correcting estimates and their implication for studying growth rates. *Arch Fish Marine Res* 47:17–29.

Field SA, O'Connor PJ, Tyre AJ, Possingham HP. 2007. Making monitoring meaningful. *Austral J Ecol* 32:485–491.

Fiess JC, Kunkel-Patterson A, Mathias L, Riley LG, Yancey PH, Hirano T, Grau FG. 2007. Effects of environmental salinity and temperature on osmoregulatory ability, organic osmolytes, and plasma hormone profiles in the Mozambique tilapia (*Oreochromis mossambicus*). *Comp Biochem Physiol A Mol Integr Physiol* 146:252–264.

Finley K, Garrett R. 2007. Recovery at Belews and Hyco Lakes: implications for fish tissue Se thresholds. *Integr Environ Assess Manag* 3:297–299.

Fisher NS, Hook SE. 2002. Toxicology tests with aquatic animals need to consider the trophic transfer of metal. *Toxicology* 181/182:531–536.

Fisher SJ, Anderson MR, Willis DW. 1998. Total length reduction in preserved yellow perch larvae. *N Amer J Fish Manage* 18:739–742.

Franson JC, Hollmén TE, Flint PL, Grand JB, Lanctot RB. 2004. Contaminants in molting long-tailed ducks and nesting common eiders in the Beaufort Sea. *Mar Pollut Bull* 48:504–513.

Franson JC, Wells-Berlin A, Perry MC, Shearn-Bochsler V, Finley DL, Flint PL, Hollmén T. 2007. Effects of dietary selenium on tissue concentrations, pathology, oxidative stress, and immune function in common eiders (*Somateria mollissima*). *J Toxicol Environ Health A* 70:861–874.

Freeman JL, Quinn CF, Marcus MA, Fakra S, Pilon-Smits EAH. 2006. Selenium tolerant diamondback moth disarms hyperaccumulator plant defense. *Current Biol* 16:2181–2192.

Gamberg M, Boila G, Stern G, Roach P. 2005. Cadmium, mercury and selenium concentrations in mink (*Mustela vison*) from Yukon, Canada. *Sci Total Environ* 351–352:523–529.

Ganser LR, Hopkins WA, O'Neil L, Hasse S, Roe JH, Sever DM. 2003. Liver histopathology of the southern watersnake, *Nerodia fasciata fasciata*, following chronic exposure to trace element-contaminated prey from a coal ash disposal site. *J Herpetol* 37:219–226.

Garifullina GF, Owen JD, Lindblom SD, Tufan H, Pilon M, Pilon-Smits EAH. 2003. Expression of a mouse selenocysteine lyase in *Brassica juncea* chloroplasts affects selenium tolerance and accumulation. *Physiol Plant* 118:538–544.

Geckler JR, Horning WB, Nieheisel TM, Pickering QH, Robinson EL, Stephan CE. 1976. Validity of laboratory tests for predicting copper toxicity in streams. EPA 600/3-76-116. Cincinnati (OH, USA): US EPA Ecological Research Service.

GEI Consultants. 2008a. Maternal transfer of selenium in fathead minnows, with modeling of ovary tissue-to-whole body concentrations. Unpublished report prepared on behalf of Conoco-Phillips. Littleton, CO, USA.

GEI Consultants. 2008b. Aquatic biological monitoring of Thompson Creek and Squaw Creek, Custer County, Idaho. Unpublished report prepared for Thompson Creek Mining. Littleton, CO, USA.

GEI Consultants. 2008c. Selenium bioaccumulation evaluation Thompson Creek, Custer County, Idaho, 2007. Unpublished report prepared for Thompson Creek Mining. Littleton, CO, USA.

Georgakopoulou E, Angelopoulou A, Kaspiris P, Divanach P, Koumoundouros G. 2007. Temperature effects on cranial deformities in European sea bass, *Dicentrarchus labrax* (L.). *J Appl Ichthyol* 23:99–103.

Gibbons WN, Munkittrick KR, Taylor WD. 1998. Monitoring aquatic environments receiving industrial effluent using small fish species. 2: Comparison between responses of trout-perch (*Percopsis omiscomaycus*) and white sucker (*Catostomus commersoni*) downstream of a pulp mill. *Environ Toxicol Chem* 17:2238–2245.

Gill CD, Fisk DM. 1966. Vertebral abnormalities in sockeye, pink, and chum salmon. *Trans Am Fish Soc* 129:754–770.

Gillespie RB, Baumann PC. 1986. Effects of high tissue concentrations of selenium on reproduction by bluegills. *Trans Am Fish Soc* 115:208–213.

Ginzburg LR, Ferson S, Akçakaya HR. 1990. Reconstructibility of density dependence and the conservative assessment of extinction risks. *Conserv Biol* 4:63–70.

Gortner RA Jr. 1940. Chronic selenium poisoning in rats as influenced by dietary protein. *J Nutr* 19:105–112.

Grand JB, Franson JC, Flint PL, Peterson MR. 2002. Concentrations of trace elements in eggs and blood of spectacled eiders on the Yukon-Kuskowim Delta, Alaska, USA. *Environ Toxicol Chem* 21:1673–1678.

Gray MA, Curry RA, Munkittrick KR. 2002. Non-lethal sampling methods for assessing environmental impacts using a small-bodied sentinel fish species. *Water Qual Res J Can* 37:195–211.

Grossman GD, Ratajczak RE Jr, Petty JT, Hunter MD, Peterson JT, Grenouillet G. 2006. Population dynamics of mottled sculpin (Pisces) in a variable environment: information theoretic approaches. *Ecol Monogr* 76:217–234.

Guy CS, Brown ML, editors. 2007. Analysis and interpretation of freshwater fisheries data. Bethesda (MD, USA): American Fisheries Society.

Ham KD, Pearsons TN. 2000. Can reduced salmonid population abundance be detected in time to limit management impacts? *Can J Fish Aquat Sci* 57:17–24.

Hamilton SJ. 2002. Rationale for a tissue-based selenium criterion for aquatic life. *Aquat Toxicol* 57:85–100.

Hamilton SJ. 2003. Review of residue-based selenium toxicity thresholds for freshwater fish. *Ecotoxicol Environ Saf* 56:201–210.

Hamilton SJ, Buhl KJ, Faerber NL, Wiedmeyer RH, Bullard FA. 1990. Toxicity of organic selenium in the diet to chinook salmon. *Environ Toxicol Chem* 9:347–358.

Hamilton SJ, Holley KM, Buhl KJ, Bullard FA, Weston LK, McDonald SF. 2005a. Selenium impacts on razorback sucker, Colorado River, Colorado. I. Adults. *Ecotoxicol Environ Saf* 61:7–31.

Hamilton SJ, Holley KM, Buhl KJ, Bullard FA. 2005b. Selenium impacts on razorback sucker, Colorado River, Colorado. II. Eggs. *Ecotoxicol Environ Saf* 61:32–43.

Hamilton SJ, Holley KM, Buhl KJ, Bullard FA. 2005c. Selenium impacts on razorback sucker, Colorado River, Colorado. III. Larvae. *Ecotoxicol Environ Saf* 61:168–169.

Hansen LD, Maier KJ, Knight AW. 1993. The effect of sulphate on the bioconcentration of selenate by *Chironomus decorus* and *Daphnia magna*. *Arch Environ Contam Toxicol* 25:72–78.

Harding LE. 2008. Non-linear uptake and hormesis effects of selenium in red-winged black-birds (*Agelaius phoeniceus*). *Sci Total Environ* 389:350–366.

Harding LE, Graham M, Paton D. 2005. Accumulation of selenium and lack of severe effects on productivity of American dippers (*Cinclus mexicanus*) and spotted sandpipers (*Actitis macularia*). *Arch Environ Contam Toxicol* 48:414–423.

Hardy RW, Oram L, Möller G. 2005. Effects of dietary selenium on cutthroat trout (*Oncorhynchus clarki bouvieri*) growth and reproductive performance over a life cycle. *Arch Environ Contam Toxicol* DOI:10.1007/S00244-009-9392-X.

Hartikainen H, Xue T. 1999. The promotive effect of selenium on plant growth as triggered by ultraviolet irradiation. *J Environ Qual* 28:1372–1375.

He PP, Lu XZ, Wang GY. 2004. Effects of Se and Zn supplementation on the antagonism against Pb and Cd in vegetables. *Environ Int* 30:167–172.

Heinz GH. 1996. Selenium in birds. In: Beyer WN, Heinz GH, Redmon-Norwood AW, editors. Environmental contaminants in wildlife: interpreting tissue concentrations. Boca Raton (FL, USA): CRC Lewis. p 447–458.

Heinz GH, Fitzgerald MA. 1993. Overwinter survival of mallards fed selenium. *Arch Environ Contam Toxicol.* 25:90–94.

Heinz GH, Hoffman DJ. 1996. Comparison of the effects of seleno-l-methionine, seleno-dl-methionine and selenized yeast on reproduction in mallards. *Environ Pollut* 91:169–175.

Heinz GH, Hoffman DJ. 1998. Methylmercury chloride and selenomethionine interactions on health and reproduction in mallards. *Environ Toxicol Chem* 17:139–145.

Heinz GH, Hoffman DJ, Krynitsky AJ, Weller DMG. 1987. Reproduction in mallards fed selenium. *Environ Toxicol Chem* 6:423–433.

Heinz GH, Hoffman DJ, Gold LG. 1989. Impaired reproduction of mallards fed an organic form of selenium. *J Wildl Manage* 53:418–428.

Heinz GH, Pendleton GW, Krynitsky AJ, Gold LG. 1990. Selenium accumulation and elimination in mallards. *Arch Environ Contam Toxicol* 19:374–379.

Henny CJ, Rudis DD, Roffe TJ, Robinson-Wilson E. 1995. Contaminants and sea ducks in Alaska and the circumpolar region. *Environ Health Perspect* 103:41–49.

Henny CJ, Grove RA, Bentley VR. 2001. Effects of selenium, mercury, and boron on waterbird egg hatchability at Stillwater, Malheur, Seedskadee, Ouray, and Benton Lake National Wildlife Refuges and surrounding vicinities. Information Report No. 5. Denver (CO, USA): National Irrigation Water Quality Program, United States Bureau of Reclamation.

Hermanutz RO. 1992. Malformation of the fathead minnow (*Pimephales promelas*) in an ecosystem with elevated selenium concentrations. *Bull Environ Contam Toxicol* 49: 290–294.

Hermanutz RO, Allen KN, Roush TH, Hedtke SF. 1992. Effects of elevated selenium concentrations on bluegills (*Lepomis macrochirus*) in outdoor experimental streams. *Environ Toxicol Chem* 11:217–224.

Hermanutz RO, Allen KN, Detenbeck NE, Stephan CE. 1996. Exposure of bluegill (*Lepomis macrochirus*) to selenium in outdoor experimental streams. Duluth (MN, USA): Mid-Continent Ecology Division, US Environmental Protection Agency.

Hesketh J. 2008. Nutrigenomics and selenium: gene expression patterns, physiological targets, and genetics. *Annu Rev Nutr* 28:157–177.

Hicks BD, Hilton JW, Ferguson HE. 1984. Influence of dietary selenium on the occurrence of nephrocalcinosis in the rainbow trout, *Salmo gairdneri* Richardson. *J Fish Dis* 7: 379–389.

Hill CH. 1975. Interrelationships of selenium with other trace elements. *Fed Proc* 34:2096–2100.

Hill CH. 1979. The effect of dietary protein levels on mineral toxicity in chicks. *J Nutr* 109:501–507.

Hill J, Grossman GD. 1987. Home range estimates for three North American stream fishes. *Copeia* 1987:376–380.

Hilton JW, Hodson PV. 1983. Effect of increased dietary carbohydrate on selenium metabolism and toxicity in rainbow trout (*Salmo gairdneri*). *J Nutr* 113:1241–1248.

Hilton JW, Hodson PV, Slinger SJ. 1980. The requirement and toxicity of selenium in rainbow trout (*Salmo gairdneri*). *J Nutr* 110:2527–2535.

Hodson PV, Hilton JW. 1983. The nutritional requirements and toxicity to fish of dietary and water-borne selenium. *Ecol Bull* 35:335–340.

Hoffman DJ. 1990. Embryotoxicity and teratogenicity of environmental contaminants to bird eggs. *Rev Environ Contam Toxicol* 115:40–89.

Hoffman DJ. 2002. Role of selenium toxicity and oxidative stress in aquatic birds. *Aquat Toxicol* 57:11–26.

Hoffman DJ, Heinz GH. 1988. Embryonic and teratogenic effects of selenium in the diet of mallards. *J Toxicol Environ Health* 24:477–490.

Hoffman DJ, Heinz GH. 1998. Effects of mercury and selenium on glutathione metabolism and oxidative stress in mallard ducks. *Environ Toxicol Chem* 17:161–166.

Hoffman DJ, Heinz GH, Krynitsky AJ. 1989. Hepatic glutathione metabolism and lipid peroxidation in response to excess dietary selenomethionine and selenite in mallard ducklings. *J Toxicol Environ Health* 27:263–271.

Hoffman DJ, Heinz GH, LeCaptain LJ, Bunck CM, Green DE. 1991a. Subchronic hepatotoxicity of selenomethionine in mallard ducks. *J Toxicol Environ Health* 32:449–464.

Hoffman DJ, Sanderson CJ, LeCaptain LJ, Cromartie E, Pendleton GW. 1991b. Interactive effects of boron, selenium, and dietary protein on survival, growth, and physiology in mallard ducklings. *Arch Environ Contam Toxicol* 20:288–294.

Hoffman DJ, Sanderson CJ, LeCaptain LJ, Cromartie E, Pendleton GW. 1992a. Interactive effects of arsenate, selenium, and dietary protein on survival, growth, and physiology in mallard ducklings. *Arch Environ Contam Toxicol* 22:55–62.

Hoffman DJ, Sanderson CJ, LeCaptain LJ, Cromartie E, Pendleton GW. 1992b. Interactive effects of selenium, methionine, and dietary protein on survival, growth, and physiology in mallard ducklings. *Arch Environ Contam Toxicol* 23:163–171.

Hoffman DJ, Heinz GH, LeCaptain LJ, Eisemann JD, Pendleton GW. 1996. Toxicity and oxidative stress of different forms of organic selenium and dietary protein in mallard ducklings. *Arch Environ Contam Toxicol* 31:120–127.

Hoffman DJ, Marn CM, Marois KC, Sproul E, Dunne M, and Skorupa JP. 2002. Sublethal effects in avocet and stilt hatchlings from selenium-contaminated sites. *Environ Toxicol Chem* 21:561–566.

Holm J. 2002. Sublethal effects of selenium on rainbow trout (*Oncorhynchus mykiss*) and brook trout (*Salvelinus fontinalis*). MSc Thesis. Winnipeg (MN, Canada): University of Manitoba.

Holm J, Palace VP, Wautier K, Evans RE, Baron CL, Podemski C, Siwik P, Sterling G. 2003. An assessment of the development and survival of wild rainbow trout (*Oncorhynchus mykiss*) and brook trout (*Salvelinus fontinalis*) exposed to elevated selenium in an area of active coal mining. In: The Big Fish Bang. Proceedings of the 26th Annual Larval Fish Conference, July 22–26, Bergen (Norway). p 257–273.

Holm J, Palace VP, Siwik P, Sterling G, Evans R, Baron C, Werner J, Wautier K. 2005. Developmental effects of bioaccumulated selenium in eggs and larvae of two salmonid species. *Environ Toxicol Chem* 24:2373–2381.

Hopkins WA. 2000. Reptile toxicology: challenges and opportunities on the last frontier of vertebrate ecotoxicology. *Environ Toxicol Chem* 19:2391–2393.

Hopkins WA. 2006. Use of tissue residues in reptile ecotoxicology: a call for integration and experimentalism. In: Gardner S, Oberdorster E, editors, New perspectives: toxicology and the environment. Volume 3, Reptile toxicology. London (UK): Taylor and Francis. p 35–62.

Hopkins WA. 2007. Amphibians as models for studying environmental change. *ILAR J* 48:270–277.

Hopkins WA, Mendonça MT, Congdon JD. 1997. Increased circulating levels of testosterone and corticosterone in southern toads, *Bufo terrestris*, exposed to coal combustion waste. *Gen Comp Endocrinol* 108:237–246.

Hopkins WA, Mendonça MT, Rowe CL, Congdon JD. 1998. Elevated trace element concentrations in southern toads, *Bufo terrestris*, exposed to coal combustion wastes. *Arch Environ Contam Toxicol* 35:325–329.

Hopkins WA, Mendonça MT, Congdon JD. 1999a. Responsiveness of the hypothalamo-pituitary-interrenal axis in an amphibian (*Bufo terrestris*) exposed to coal combustion wastes. *Comp Biochem Physiol C-Toxicol Pharmacol* 122:191–196.

Hopkins WA, Rowe CL, Congdon JD. 1999b. Elevated trace element concentrations and standard metabolic rate in banded water snakes (*Nerodia fasciata*) exposed to coal combustion wastes. *Environ Toxicol Chem* 18:1258–1263.

Hopkins WA, Congdon JD, Ray JK. 2000. Incidence and impact of axial malformations in larval bullfrogs (*Rana catesbeiana*) developing in sites polluted by a coal burning power plant. *Environ Toxicol Chem* 19:862–868.

Hopkins WA, Roe JH, Snodgrass JW, Jackson BP, Kling DE, Rowe CL, Congdon JD. 2001. Nondestructive indices of trace element exposure in squamate reptiles. *Environ Pollut* 115:1–7.

Hopkins WA, Roe JH, Snodgrass JW, Staub BP, Jackson BP, Congdon JD. 2002. Trace element accumulation and effects of chronic dietary exposure on banded water snakes (*Nerodia fasciata*). *Environ Toxicol Chem* 21:906–913.

Hopkins WA, Staub BP, Baionno JA, Jackson BP, Roe JH, Ford NB. 2004. Trophic and maternal transfer of selenium in brown house snakes (*Lamprophis fuliginosus*). *Ecotox Environ Saf* 58:285–293.

Hopkins WA, Staub BP, Baionno JA, Jackson BP, Talent LG. 2005a. Transfer of selenium from prey to predators in a simulated terrestrial food chain. *Environ Pollut* 134:447–456.

Hopkins WA, Snodgrass JW, Baionno JA, Roe JH, Staub BP, Jackson BP. 2005b. Functional relationships among selenium concentrations in the diet, target tissues, and nondestructive tissue samples of two species of snakes. *Environ Toxicol Chem* 24:344–351.

Hopkins WA, DuRant SE, Staub BP, Rowe CL, Jackson BP. 2006. Reproduction, embryonic development, and maternal transfer of contaminants in an amphibian *Gastrophryne carolinensis*. *Environ Health Perspect* 114:661–666.

House R. 1995. Temporal variation in abundance of an isolated population of cutthroat trout in western Oregon, 1981–1991. *N Am J Fish Manage* 15:33–41.

Howell GO, Hill CH. 1978. Biological interaction of selenium with other trace elements in chicks. *Environ Health Perspect* 25:147–150.

Hurst TP. 2007. Causes and consequences of winter mortality in fishes. *J Fish Biol* 71:315–345.

IUCN. 2006. Standards and Petitions Working Group of the IUCN SSC Biodiversity Assessments Sub-Committee in December 2006, World Conservation Union (IUCN). 2006. Guidelines for using the IUCN Red List Categories and Criteria. Version 6.2, Gland, Switzerland. 60 p. www.iucn.org.

Jacobs K, Shen L, Benemariya H, Deelstra H. 1993. Selenium distribution in egg white proteins. *Z Lebensm Unters Forsch* 196:236–238.

Janz DM. 2008. A critical evaluation of winter stress syndrome. In: Selenium tissue thresholds: Tissue selection criteria, threshold development endpoints, and potential to predict population or community effects in the field. Washington (DC, USA): North America Metals Council—Selenium Working Group; www.namc.org.

Janz DM, Muscatello JR. 2008. Standard operating procedure for evaluating selenium-induced deformities in early life stages of freshwater fish. In: Selenium tissue thresholds: tissue selection criteria, threshold development endpoints, and potential to predict population or community effects in the field. Washington (DC, USA): North America Metals Council—Selenium Working Group; www.namc.org.

Jihen EH, Imed M, Fatima H, Abdelhamid K. 2008. Protective effects of selenium (Se) and zinc (Zn) on cadmium (Cd) toxicity in the liver of the rat: histology and Cd accumulation. *Ecotoxicol Environ Saf* 46: 3522–3527.

Jihen EH, Imed M, Fatima H, Abdelhamid K. 2009. Protective effects of selenium (Se) and zinc (Zn) on cadmium (Cd) toxicity in the liver of the rat: effects on oxidative stress. *Ecotoxicol Environ Saf* 72:1559–1564.

Johannessen SC, Miller WL, Cullen JJ. 2003. Calculation of UV attenuation and colored dissolved organic matter absorption spectra from measurements of ocean color. *J Geophys Res-Oceans* 108:3301–3313.

Johnson DH. 1999. The insignificance of statistical significance testing. *J Wildl Manage* 63:763–772.

Johnson DH, Shrier BM, O'Neal J, Knutzen JA, Xanthippe A, O'Neill TA, Pearsons TN, editors. 2007. Salmonid field protocols handbook: techniques for assessing status and trends in salmon and trout populations. Bethesda (MD, USA): American Fisheries Society.

Kelly JM, Janz DM. 2008. Altered energetics and parasitism in juvenile northern pike (*Esox lucius*) inhabiting metal-mining contaminated lakes. *Ecotoxicol Environ Saf* 70:357–369.

Kelly JM, Janz DM. 2009. Assessment of oxidative stress and histopathology in juvenile northern pike (*Esox lucius*) inhabiting lakes downstream of a uranium mill. *Aquat Toxicol* 92:240–249.

Kelly BC, Ikonomou MG, Higgs DA, Oakes J, Dubetz C. 2008. Mercury and other trace elements in farmed and wild salmon from British Columbia, Canada. *Environ Toxicol Chem* 27:1361–1370.

Kennedy CJ, McDonald LE, Loveridge R, Strosher MM. 2000. The effects of bioaccumulated selenium on mortalities and deformities in the eggs, larvae, and fry of a wild population of cutthroat trout (*Oncorhynchus clarki lewisi*). *Arch Environ Contam Toxicol* 39:46–52.

Khan MAK, Wang F. 2009. Mercury-selenium compounds and their toxicological significance: Toward a molecular understanding of the mercury-selenium antagonism. *Environ Toxicol Chem* 28:1567–1577.

Kilgour BW, Munkittrick KR, Portt CB, Hedley K, Culp JM, Dixit S, Pastershank G. 2005. Biological criteria for municipal wastewater effluent monitoring programs. *Water Qual Res J Can* 37:374–387.

Kilgour BW, Dubé MG, Hedley K, Portt CB, Munkittrick KR. 2007. Aquatic environmental effects monitoring guidance for environmental assessment practitioners. *Environ Mon Assess* 130:423–436.

Kime DE. 1998. Endocrine disruption in fish. Norwell (MA, USA): Kluwer Academic Publishers.

Kingsford MJ, Suthers IM, Gray CA. 1996. Exposure to sewage plumes and the incidence of deformities in larval fishes. *Mar Pollut Bull* 33:201–212.

Koeman JH, Peeters WHM, Koudstaal-Hol DHM, Tjioe PS, deGoeij JJM. 1973. Mercury-selenium correlations in marine mammals. *Nature* 245:385–386.

Koller LD, Exon JH, Talcot PT, Osborne CH, Henningson GM. 1986. Immune responses in rats supplemented with selenium. *Clinical Exp Immunol* 63:570–576.

Koumoundouros G, Maingot E, Divanach P, Kentouri M. 2002. Kyphosis in reared sea bass (*Dicentrarchus labrax* L.): ontogeny and effects on mortality. *Aquaculture* 209:49–58.

Kovacic P, Somanathan R. 2006. Mechanism of teratogenesis: electron transfer, reactive oxygen species, and antioxidants. *Birth Defects Res C: Embryo Today: Rev* 78: 308–325.

Kraus RJ, Ganther HE. 1989. Synergistic toxicity between arsenic and methylated selenium compounds. *Biol Trace Elem Res* 20:105–113.

Kroll KJ, Doroshov SI. 1991. Vitellogenin: potential vehicle for selenium in oocytes of the white sturgeon (*Acipenser transmontanus*). In: Williot P, editor, Acipenser. Montpellier (France): Cemagref Publishers, p 99–106.

Lam JCW, Tanabe S, Lam MHW, Lam PKS. 2005. Risk to breeding success of waterbirds by contaminants in Hong Kong: evidence from trace elements in eggs. *Environ Pollut* 135:481–490.

Latshaw JD, Osman M. 1975. Distribution of selenium into egg proteins after feeding natural and synthetic selenium compounds. *Poultry Sci* 54:1244–1252.

Latshaw JD, Biggert MD. 1981. Incorporation of selenium into egg proteins after feeding selenomethionine or sodium selenite. *Poultry Sci* 60:1309–1313.

LeBoeuf RA, Zentr KL, Hoekstra WG. 1985. Effects of dietary selenium concentration and duration of selenium feeding on hepatic glutathione concentrations in rats. *Proc Soc Exp Biol Med* 180:348–352.

Lemly AD. 1985. Toxicology of selenium in a freshwater reservoir: Implications for environmental hazard evaluation and safety. *Ecotoxicol Environ Saf* 10:314–338.

Lemly AD. 1993a. Teratogenic effects of selenium in natural populations of freshwater fish. *Ecotoxicol Environ Saf* 26:181–204.

Lemly AD. 1993b. Guidelines for evaluating selenium data from aquatic monitoring and assessment studies. *Environ Mon Assess* 28:83–100.

Lemly AD. 1993c. Metabolic stress during winter increases the toxicity of selenium to fish. *Aquat Toxicol* 27:133–158.

Lemly AD. 1996a. Selenium in aquatic organisms. In: Beyer WN, Heinz GH, Redmon-Norwood AW, editors, Environmental contaminants in wildlife—interpreting tissue concentrations. New York (NY, USA): Lewis. p 427–445.

Lemly AD. 1996b. Winter stress syndrome: An important consideration for hazard assessment of aquatic pollutants. *Ecotoxicol Environ Saf* 34:223–227.

Lemly AD. 1997a. A teratogenic deformity index for evaluating impacts of selenium on fish populations. *Ecotoxicol Environ Saf* 37:259–266.

Lemly AD. 1997b. Ecosystem recovery following Se contamination in a freshwater reservoir. *Ecotoxicol Environ Saf* 36:275–281.

Lemly AD. 2002. Symptoms and implications of selenium toxicity in fish: the Belews Lake case example. *Aquat Toxicol* 57:39–49.

Lemly AD, Skorupa JP. 2007. Technical issues affecting the implementation of US Environmental Protection Agency's proposed fish tissue–based aquatic criterion for selenium. *Integr Environ Assess Manag* 3:552–558.

Levander OA, Morris VG. 1970. Interactions of methionine, vitamin E, and antioxidants on selenium toxicity in the rat. *J Nutr* 100:1111–1118.

Levesque HM, Moon TW, Campbell PGC, Hontela A. 2002. Seasonal variation in carbohydrate and lipid metabolism of yellow perch (*Perca flavescens*) chronically exposed to metals in the field. *Aquat Toxicol* 60:257–267.

Li H, Zhang J, Wang T, Luo W, Zhou Q, Jiang G. 2008. Elemental selenium particles at nano-size (Nano-Se) are more toxic to Medaka (*Oryzias latipes*) as a consequence of hyper-accumulation of selenium: a comparison with sodium selenite. *Aquat Toxicol* 89:251–256.

Liermann M, Hilborn R. 2001. Depensation: evidence, models and implications. *Fish and Fisheries* 2:33–58.

Lin Y, Spallholz JE. 1993. Generation of reactive oxygen species from the reaction of selenium compounds with thiols and mammary tumor cells. *Biochem Pharmacol* 45:429–437.

Lindesjoo E, Thulin J. 1992. A skeletal deformity of Northern pike (*Esox lucius*) related to pulp mill effluents. *Can J Fish Aquat Sci* 49:166–172.

Lindesjoo E, Thulin J, Bengtsson B, Tjarnlund U. 1994. Abnormalities of a gill cover bone, the operculum, in perch *Perca fluviatilis* from a pulp mill effluent area. *Aquat Toxicol* 28:189–207.

Lobón-Cerviá J. 2009. Why, when and how do fish populations decline, collapse and recover? The example of brown trout (*Salmo trutta*) in Rio Chaballos (northwestern Spain). *Freshw Biol* 54:1149–1162.

Luoma SN, Clements WH, DeWitt T, Gerritsen J, Hatch A, Jepson P, Reynoldson TB, Thom RM. 2001. Role of environmental variability in evaluating stressor effects. In: Baird DJ, Burton GA Jr, editors, Ecological variability: separating natural from anthropogenic causes of ecosystem impairment. Pensacola (FL, USA): SETAC Pr. p. 141–178.

Lynn DH, Gilron GL. 1993. A brief review of approaches using ciliated protists to assess aquatic ecosystem health. *J Aquat Ecosystem Health* 4:263–270.

Maier KJ, Knight AW. 1994. Ecotoxicology of selenium in freshwater systems. *Rev Environ Contam Toxicol* 134:31–48.

Marn CM. 2003. Post-hatching survival and productivity of American avocets at drainwater evaporation ponds in the Tulare Basin, California. PhD Dissertation. Corvallis (OR, USA): Oregon State University.

Mapstone BD. 1995. Scalable decision rules for environmental impact studies: effect size, type I, and type II errors. *Ecol Appl* 5:401–410.

Matthews WJ. 1987. Physicochemical tolerance and selectivity of stream fishes as related to their geographic ranges and local distributions. In: Matthews WJ, Heins DC, editors, Community and evolutionary ecology of North American stream fishes. Norman (OK, USA): University of Oklahoma Pr. p 111–120.

Maxwell D, Jennings S. 2005. Power of monitoring programmes to detect decline and recovery of rare and vulnerable fish. *J Appl Ecol* 42:25–37.

Mayland H. 1994. Selenium in plant and animal nutrition. In: Frankenberger WT Jr, Benson SJr, editors, Selenium in the environment. New York (NY, USA). p 29–45.

McCloskey DN. 1995. The insignificance of statistical significance. *Sci Amer* 272:32–33.

McDonald BG, Chapman PM. 2007. Selenium effects: a weight of evidence approach. *Integr Environ Assess Manage* 3:129–136.

McDonald, BG, Chapman PM. 2009. The need for adequate QA/QC measures for Se larval deformity assessments: implications for tissue residue guidelines. *Integr Environ Assess Manage* 5: 470–475.

McFadden JT, Alexander GR, Shetter DS. 1967. Numerical changes and population regulation in brook trout *Salvelinus fontinalis*. *J Fish Res Board Can* 24:1425–1459.

McGarvey DJ. 2007. Merging precaution with sound science under the Endangered Species Act. *BioScience* 57:65–70.

McIntyre DO, Pacheco MA, Garton MW, Wallschläger D, Delos CG. 2008. Effect of selenium on juvenile bluegill sunfish at reduced temperatures. EPA-822-R-08-020. Washington (DC, USA): Office of Water, US Environmental Protection Agency.

McMeans BC, Borga K, Bechtol WR, Higginbotham D, Fisk AT. 2007. Essential and nonessential element concentrations in two sleeper shark species collected in arctic waters. *Environ Pollut* 148:281–290.

Meador MR, Whittier TR, Goldstein JN, Hughes RM, Peck DV. 2008. Evaluation of an index of biotic integrity approach used to assess biological condition in western U.S. streams and rivers at varying spatial scales. *Trans Am Fish Soc* 137:13–22.

Mebane CA. 2000. Evaluation of proposed new point source discharges to a special resource water and mixing zone determinations: Thompson Creek Mine, upper Salmon River subbasin, Idaho. Boise (ID, USA): Idaho Department of Environmental Quality. http://deq.idaho.gov/water/data_reports/surface_water/monitoring/mixing_zones.cfm.

Mebane CA, Essig DA. 2003. Concepts and recommendations for using the "natural conditions" provisions of the Idaho Water Quality Standards. Boise (ID, USA): Idaho Department of Environmental Quality. http://www.deq.state.id.us/water/data_reports/surface_water/monitoring/natural_background_paper.pdf.

Mebane CA, Maret TR, Hughes RM. 2003. An index of biological integrity (IBI) for Pacific Northwest rivers. *Trans Am Fish Soc* 132:239–261.

Mechaly A, Teplitsky A, Belakhov V, Baasov T, Shoham G, Shoham Y. 2000. Overproduction and characterization of seleno-methionine xylanase T-6. *J Biotech* 78:83–86.

Mehrle PM, Haines TA, Hamilton S, Ludke JL, Maye TL, Ribick MA. 1982. Relation between body contaminants and bone development in east-coast striped bass. *Trans Am Fish Soc* 111:231–241.

Miki K, Mingxu X, Gupta A, Ba Y, Tan Y, Al-Refaie W, Bouvet M, Makuuchi M, Moosa AR, Hoffman RM. 2001. Methioninase cancer gene therapy with selenomethionine as suicide prodrug substrate. *Cancer Res* 61:6805–6810.

Miller LL, Wang F, Palace VP, Hontela A. 2007. Effects of acute and subchronic exposures to waterborne selenite on the physiological stress response and oxidative stress indicators in juvenile rainbow trout. *Aquat Toxicol* 83:263–271.

Moksnes K, Norheim G. 1982. Selenium concentrations in tissues and eggs of growing and laying chickens fed sodium selenite at different levels. *Acta Vet Scand* 23:368–379.

Mommsen TP, Walsh PJ. 1988. Vitellogenesis and oocyte assembly. In: Hoar WS, Randall DJ, editors, Fish physiology Vol XIA. New York (NY, USA): Academic Pr. p 347–406.

Moxon AL. 1938. The effects of arsenic on the toxicity of seleniferous grains. *Science* 88:91.

Muir D, Braune B, DeMarch B, Norstrom R, Wagemann R, Lockhart L, Hargrave B, Bright D, Addison R, Payne J, Reimer K. 1999. Spatial and temporal trends and effects of contaminants in the Canadian Arctic marine ecosystem: a review. *Sci Tot Environ* 230:83–144.

Munkittrick KR, Dixon DG. 1989. A holistic approach to ecosystem health using fish population characteristics. *Hydrobiologia* 188/189:123–135.

Munkittrick KR, McMaster ME, Van der Kraak G, Portt C, Gibbons WN, Farwell A, Gray MA. 2001. Development of methods for effects-driven cumulative effects assessment using fish populations: Moose River project. Pensacola (FL, USA): SETAC Pr.

Munkittrick KR, Arens CJ, Lowell RB, Kaminski GP. 2009. A review of potential methods for determining critical effect size for designing environmental monitoring programs. *Environ Toxicol Chem* 28:1361–1371.

Munther, GL. 1970. Movement and distribution of smallmouth bass in the middle Snake River. *Trans Am Fish Soc* 99:44–53.

Muscatello JR. 2009. Selenium accumulation and effects in aquatic organisms downstream of uranium mining and milling operations in northern Saskatchewan. PhD Dissertation. Saskatoon (SK, Canada): University of Saskatchewan.

Muscatello JR, Janz DM. 2009. Assessment of larval deformities and selenium accumulation in northern pike (*Esox lucius*) and white sucker (*Catostomus commersoni*) exposed to metal mining effluent. *Environ Toxicol Chem* 28:609–618.

Muscatello J, Bennett PM, Himbeault KT, Belknap AM, Janz DM. 2006. Larval deformities associated with selenium accumulation in northern pike (*Esox lucius*) exposed to metal mining effluent. *Environ Sci Technol* 40:6506–6512.

Myers RA, Hutchings JA, Barrowman NJ. 1997. Why do fish stocks collapse? The example of cod in Atlantic Canada. *Ecol Appl* 7:91–106.

Nagelkerke LAJ, van Densen WLT. 2007. Serial correlation and inter-annual variability in relation to the statistical power of monitoring schemes to detect trends in fish populations. *Environ Monit Assess* 125:247–256.

Nagle RD, Rowe CL, Congdon JD. 2001. Accumulation and selective maternal transfer of contaminants in the turtle *Trachemys scripta* associated with coal ash deposition. *Arch Environ Contam Toxicol* 40:531–536.

Neumann PM, De Souza MP, Pickering IJ, Terry N. 2003. Rapid microalgal metabolism of selenate to volatile dimethylselenide. *Plant Cell Environ* 26:897–905.

NewFields. 2009. Brown trout laboratory reproduction studies conducted in support of development of a site-specific selenium criterion. Unpublished draft report prepared for J.R. Simplot Company, Smoky Canyon Mine, Pocatello, ID, USA.

Newman, MC. 2008. "What exactly are you inferring?" A closer look at hypothesis testing. *Environ Toxicol Chem* 27:1013–1019.

Nigro M, Leonzio C. 1996. Intracellular storage of mercury and selenium in different marine vertebrates. *Marine Ecol Progr Ser* 135:137–143.

NRC. 2006. Nutrient requirements of small ruminants: sheep, goats, cervids, and New World camelids. Washington (DC, USA): National Research Council.

Nuutinen S, Kukkonen JV. 1998. The effect of selenium and organic material in lake sediments on the bioaccumulation of methylmercury by *Lumbriculus variegatus* (Oligochaeta). *Biogeochemistry* 40:267–278.

O'Toole D, Raisbeck MF. 1997. Experimentally-induced selenosis of adult mallard ducks: clinical signs, lesions, and toxicology. *Vet Pathol* 34:330–340.

O'Toole D, Raisbeck MF. 1998. Magic numbers, elusive lesions: Comparative pathology and toxicology of selenosis in waterfowl and mammalian species. In: Frankenberger WT Jr, Engberg RA, editors, Environmental chemistry of selenium. New York (NY, USA): Marcel Dekker. p 355–395.

Ogle RS, Knight AW. 1996. Selenium bioaccumulation in aquatic ecosystems: 1. Effects of sulphate on the uptake and toxicity of selenate in *Daphnia magna*. *Arch Environ Contam Toxicol* 30: 274–279.

Ohlendorf HM. 1989. Bioaccumulation and effects of selenium in wildlife. In Jacobs LW, editor, Selenium in agriculture and the environment. Special publication 23. Madison (WI, USA): American Society of Agronomy and Soil Science Society of America. p 133–177.

Ohlendorf HM. 2003. Ecotoxicology of selenium. In Hoffman DJ, Rattner BA, Burton GA Jr, Cairns J Jr, editors, Handbook of ecotoxicology. 2nd edition. Boca Raton (FL, USA): CRC. p 465–500.

Ohlendorf HM. 2007. Threshold values for selenium in Great Salt Lake: selections by the science panel. Unpublished report prepared for the Great Salt Lake Science Panel. Sacramento, CA, USA. http://www.deq.utah.gov/Issues/GSL_WQSC/selenium.htm.

Ohlendorf HM, Harrison CS. 1986. Mercury, selenium, cadmium and organochlorines in eggs of three Hawaiian seabird species. *Environ Pollut B* 11:169–191.

Ohlendorf HM, Hothem RL. 1995. Agricultural drainwater effects on wildlife in central California. In Hoffman DJ, Rattner BA, Burton GAJr, Cairns J Jr, editors. Handbook of ecotoxicology. Boca Raton (FL USA): Lewis Publishers. p 577–595.

Ohlendorf HM, Heinz GH. In press. Selenium in birds. In: Beyer WN, Meador JP, editors, Environmental contaminants in biota: interpreting tissue concentrations. 2nd edition. Boca Raton (FL, USA): Taylor and Francis.

Ohlendorf HM, Hoffman DJ, Saiki MK, Aldrich TW. 1986. Embryonic mortality and abnormalities of aquatic birds: apparent impacts of selenium from irrigation drainwater. *Sci Total Environ* 52:49–63.

Ohlendorf HM, Kilness AW, Simmons JL, Stroud RK, Hoffman DJ, Moore JF. 1988. Selenium toxicosis in wild aquatic birds. *J Toxicol Environ Health* 24:67–92.

Ohlendorf HM, Covington S, Byron E, Arenal C. 2008. Approach for conducting site-specific assessments of selenium bioaccumulation in aquatic systems. Washington (DC, USA): North America Metals Council – Selenium Working Group; www.namc.org.

Oremland RS, Herbel MJ, Blum JS, Langley S, Beveridge TJ, Ajayan PM, Sutto T, Ellis AV, Curran S. 2004. Structural and spectral features of selenium nanospheres produced by Se-respiring bacteria. *Appl Environ Microbiol* 70:52–60.

Orr PL, Guiguer KR, Russell CK. 2006. Food chain transfer of selenium in lentic and lotic habitats of a western Canadian watershed. *Ecotoxicol Environ Saf* 63:175–188.

Osmundson BC, May T, Skorupa J, Krueger R. 2007. Selenium in fish tissue: prediction equations for conversion between whole body, muscle, and eggs. Poster presentation at the 2007 Annual Meeting of the Society of Environmental Toxicology and Chemistry, Milwaukee, WI, USA.

Palace VP, Spallholz JE, Holm J, Wautier K, Evans RE, Baron CL. 2004. Metabolism of sele-nomethionine by rainbow trout (*Oncorhynchus mykiss*) embryos can generate oxidative stress. *Ecotoxicol Environ Saf* 58:17–21.

Palace VP, Halden NM, Yang P, Evans RE, Sterling GL. 2007. Determining residence patterns of rainbow trout using laser ablation inductively coupled plasma mass spectrometry (LA-ICP-MS) analysis of selenium in otoliths. *Environ Sci Technol* 41:3679–3683.

Palmer IS, Thiex N, Olson OE. 1983. Dietary selenium and arsenic effects in rats. *Nutr Rep Int* 27:249–258.

Paradis Y, Brodeur P, Mingelbier M, Magnan P. 2007. Length and weight reduction in larval and juvenile yellow perch preserved with dry ice, formalin, and ethanol. *N Amer J Fish Manage* 27:1004–1009.

Parkhurst, DF. 2001. Statistical significance tests: equivalence and reverse tests should reduce misinterpretation. *BioScience* 51:1051–1057.

Patching SG, Gardiner PH. 1999. Recent developments in selenium metabolism and chemical speciation: a review. *J Trace Elem Med Biol* 13:193–214.

Paulsson K, Lindbergh K. 1989. The selenium method for treatment of lakes for elevated lev-els of mercury in fish. *Sci Total Environ* 87–88:495–507.

Paveglio FL, Bunck CM, Heinz GH. 1992. Selenium and boron in aquatic birds from central California. *J Wildl Manage* 56:31–42.

Peakall DB, Fox GA. 1987. Toxicological investigations of pollutant-related effects in Great Lakes gulls. *Environ Health Persp* 71:187–193.

Peterson SA, Ralston NVC, Peck DV, Van Sickle J, Robertson JD, Spate VL, Morris JS. 2009. How might selenium moderate the toxic effects of mercury in stream fish of the western US? *Environ Sci Technol* 43: 3919–3925.

Pilon-Smits, E. 2005. Phytoremediation. *Ann Rev Plant Biol* 56: 15–39.

Poole GC, Dunham JB, Keenan DM, Sauter ST, McCullough DA, Mebane CA, Lockwood JC, Essig DA, Hicks MP, Sturdevant DJ, Materna EJ, Spalding SA, Risley JC, Deppman M. 2004. The case for regime-based water quality standards. *BioScience* 54:155–161.

Post JR, Parkinson EA. 2001. Energy allocation strategy in young fish: Allometry and sur-vival. *Ecology* 82:1040–1051.

Pratt JR, NJ Bowers. 1990. Effect of selenium on microbial communities in laboratory micro-cosms and outdoor streams. *Tox Assess* 5:293–308.

Presser TS, Luoma SN. 2006. Forecasting selenium discharges to the San Francisco Bay-Delta Estuary: ecological effects of a proposed San Luis Drain Extension. Professional Paper 1646. Reston (VA, USA): US Geological Survey.

Puls R. 1988. Mineral levels in animal health: Diagnostic data. Clearbrook (BC, Canada): Sherpa International.

Punshon T, Jackson BP, Lanzirotti A, Hopkins WA, Bertsch PM, Burger J. 2005. Application of synchrotron x-ray microbeam spectroscopy to the determination of metal distribution and speciation in biological tissues. *Spectr Lett* 38:343–363.

Pyron M, Beitinger TL. 1989. Effect of selenium on reproductive behaviour and fry of fathead minnows. *Bull Environ Contam Toxicol* 42:609–613.

Quinn CF, Galeas ML, Freeman JL, Pilon-Smith EAH. 2007. Selenium: deterrence, toxicity and adaptation. *Integr Environ Assess Manage* 3:460–462.

Raimondo SM, Rowe CL, Congdon JD. 1998. Exposure to coal ash impacts swimming per-formance and predator avoidance in larval bullfrogs (*Rana catesbeiana*). *J Herpetol* 32:289–292.

Raisbeck MF, Schaumber RA, Belden EL. 1998. Immunotoxic effects of selenium in mammals. In: Garland T, Barr AC, editors. Toxic plants and other natural toxicants. Wallingford (UK): CABI. p 260–266.

Ralston CR, Blackwell JLIII, Ralston NVC. 2006. Effects of dietary selenium and mercury on house crickets (*Acheta domesticus* L.): implication of environmental co-exposures. *Environ Bioind* 1:98–109.

Ralston NVC, Ralston CR, Blackwell JLIII, Raymond LJ. 2008. Dietary and tissue selenium in relation to methylmercury toxicity. *NeuroToxicology* 29:802–811.

Reddy CC, Massaro EJ. 1983. Biochemistry of selenium: an overview. *Fundam Appl Toxicol* 3:431–436.

Reed JM, Blaustein AR. 1997. Biologically significant population declines and statistical power. *Conserv Biol* 11:281–282.

Reilly C. 2006. Selenium in food and health. New York (NY, USA): Springer.

Rhian M, Moxon AL. 1943. Chronic selenium poisoning in dogs and its prevention by arsenic. *J Pharmacol Exp Ther* 78:249.

Riedel GF, Sanders JG. 1996. The influence of pH and media composition on the uptake of inorganic selenium by *Chlamydomonas reinhardtii*. *Environ Toxicol Chem* 15:1577–1583.

Rinchard J, Kestemont P. 2005. Comparative study of reproductive biology in single- and multiple-spawner cyprinid fish. I. Morphological and histological features. *J Fish Biol* 49:883–894.

Roe JH, Hopkins WA, Baionno JA, Staub BP, Rowe CL, Jackson BP. 2004. Maternal transfer of selenium in *Alligator mississippiensis* nesting downstream from a coal burning power plant. *Environ Toxicol Chem* 23:1969–1972.

Roe JH, Hopkins WA, Jackson BP. 2005. Species- and stage-specific differences in trace element tissue concentrations in amphibians: implications for the disposal of coal-combustion wastes. *Environ Pollut* 136:353–363.

Roe JH, Hopkins WA, DuRant SE, Unrine JM. 2006. Effects of competition and coal combustion wastes on recruitment and life history characteristics of salamanders in temporary wetlands. *Aquat Toxicol* 79:176–184.

Romano M, Rosnova P, Anteo C, Limatola E. 2004. Vertebrate yolk proteins: a review. *Mol Repro Dev* 69:109–116.

Rose KA, Cowan JH Jr, Winemiller KO, Myers RA, Hilborn R. 2001. Compensatory density dependence in fish populations: importance, controversy, understanding and prognosis. *Fish and Fisheries* 2:293–327.

Rowe CL, Hopkins WA, Coffman VR. 2001. Failed recruitment of southern toads (*Bufo terrestris*) in a trace element-contaminated breeding habitat: direct and indirect effects that may lead to a local population sink. *Arch Environ Contam Toxicol* 40:399–405.

Rowe CL, Kinney OM, Fiori AP, Congdon JD. 1996. Oral deformities in tadpoles (*Rana catesbeiana*) associated with coal ash deposition: effects on grazing ability and growth. *Freshw Biol* 36:723–730.

Rowe CL, Kinney OM, Nagle RD, Congdon JD. 1998. Elevated maintenance costs in an anuran (*Rana catesbeiana*) exposed to a mixture of trace elements during the embryonic and early larval periods. *Physiol Zool* 71:27–35.

Rudolph B-L, Andreller I, Kennedy CJ. 2008. Reproductive success, early life stage development, and survival of westslope cutthroat trout (*Oncorhynchus clarki lewisi*) exposed to elevated selenium in an area of active coal mining. *Environ Sci Technol* 42: 3109–3114.

Rush SA, Borgå K, Dietz R, Born EW, Sonne C, Evans T, Muir DCG, Letcher RJ, Norstrom RJ, Fisk AT. 2008. Geographic distribution of selected elements in the livers of polar bears from Greenland, Canada and the United States. *Environ Pollut* 153:618–626.

Saiki MK, Jennings MR, May TW. 1992. Selenium and other elements in fresh-water fishes from the irrigated San Joaquin Valley, California. *Sci Total Environ* 126:109–137.

Saiki MK, Martin BA, May TM. 2004. Reproductive status of western mosquitofish inhabiting selenium-contaminated waters in the grassland water district, Merced County, California. *Arch Environ Contam Toxicol* 47:363–369.

Sanders RW, Gilmour, CC. 1994. Accumulation of selenium in a model freshwater microbial food web. *Appl Environ Microbiol* 60:2677–2683.

Santolo GM, Yamamoto JT, Pisenti JM, Wilson BW. 1999. Selenium accumulation and effects on reproduction in captive American kestrels fed selenomethionine. *J Wildl Manage* 63:502–511.

Sarret G, Avoscan L, Carriere M, Collins R, Geoffroy N, Carrot F, Coves J, Gouget B. 2005. Chemical forms of selenium in the metal-resistant bacterium *Ralstonia metallidurans* CH34 exposed to selenite and selenate. *Appl Environ Microbiol* 71:2331–2337.

Scheuhammer AM, Perreault JA, Bond DE. 2001. Mercury, methylmercury, and selenium concentrations in eggs of common loons (*Gavia immer*) from Canada. *Environ Monitor Assess* 72:79–94.

Scheuhammer AM, Wong AHK, Bond D. 1998. Mercury and selenium accumulation in common loons (*Gavia immer*) and common mergansers (*Mergus merganser*) from eastern Canada. *Environ Toxicol Chem* 17:197–201.

Schmetterling DA, Adams SB. 2004. Summer movements within the fish community of a small montane stream. *N Am J Fish Manage* 24:1163–1172.

Schrauzer GN. 2000. Selenomethionine: a review of its nutritional significance, metabolism and toxicity. *J Nutr* 130:1653–1656.

Schultz R, Hermanutz R. 1990. Transfer of toxic concentrations of selenium from parent to progeny in the fathead minnow (*Pimephales promelas*). *Bull Environ Contam Toxicol* 45: 568–573.

Sedell JR, Reeves GH, Hauer FR, Stanford JA, Hawkins CP. 1990. Role of refugia in recovery from disturbances: modern fragmented and disconnected river systems. *Environ Manage* 14:711–724.

Seiler RL, Skorupa JP, Naftz DL, Nolan BT. 2003. Irrigation-induced contamination of water, sediment, and biota in the western United States—synthesis of data from the National Irrigation Water Quality Program. Professional Paper 1655. Denver (CO, USA): National Irrigation Water Quality Program, US Geological Survey.

Sfakianakis DG, Georgakopoulou E, Papadakis I, Divanach P, Kentouri M, Koumoundouros G. 2006. Environmental determinants of hemal lordosis in European sea bass, *Dicentrarchus labrax* (Linnaeus, 1758). *Aquaculture* 254:54–64.

Shine R. 1985. The evolution of viviparity in reptiles: an ecological analysis. In: Gans G, Billet F, editors, Biology of the reptilia, Vol. 15 (Development B). New York (NY, USA): Academic Pr. p 605–694.

Skorupa, JP. 1998a. Selenium poisoning of fish and wildlife in nature: lessons from twelve real-world experiences. In: Frankenberger WT Jr, Engberg RA, editors, Environmental chemistry of selenium. New York (NY, USA): Marcel Dekker. p 315–354.

Skorupa JP. 1998b. Risk assessment for the biota database of the National Irrigation Water Quality Program. Washington (DC, USA): National Irrigation Water Quality Program, US Department of the Interior.

Skorupa JP. 1999. Beware of missing data and undernourished statistical models: comment on Fairbrother et al.'s critical evaluation. *Human Ecol Risk Assess* 5:1255–1262.

Skorupa JP, Ohlendorf HM. 1991. Contaminants in drainage water and avian risk thresholds. In: Dinar A, Ziberman D, editors, The economics and management of water and drainage in agriculture. Boston (MA, USA): Kluwer Academic Publishers, p. 345–368.

Smith GJ, Heinz GH, Hoffman DJ, Spann JW, Krynitsky AJ. 1988. Reproduction in black-crowned night herons fed selenium. *Lake Reserv Manage* 4:175–180.

Snodgrass JW, Hopkins WA, Roe JH. 2003. Relationships among developmental stage, metamorphic timing, and concentrations of elements in bullfrogs (*Rana catesbeiana*). *Environ Toxicol Chem* 22:1597–1604.

Snodgrass JW, Hopkins WA, Broughton J, Gwinn D, Baionno JA, Burger J. 2004. Species-specific responses of developing anurans to coal combustion wastes. *Aquat Toxicol* 66:171–182.

Snodgrass JW, Hopkins WA, Jackson BP, Baionno JA, Broughton J. 2005. Influence of larval period on responses of overwintering green frog (*Rana clamitans*) larvae exposed to contaminated sediments. *Environ Toxicol Chem* 24:1508–1514.

Sorensen, EMB. 1991. Metal Poisoning in Fish. Boca Raton (FL, USA): CRC Pr.

Southworth GR, Peterson MJ, Turner RR. 1994. Changes in concentrations of selenium and mercury in largemouth bass following elimination of fly ash discharge to a quarry. *Chemosphere* 29:71–79.

Southworth GR, Peterson MJ, Ryon MG. 2000. Long-term increased bioaccumulation of mercury in largemouth bass follows reduction of waterborne selenium. *Chemosphere* 41:1101–1105.

Spallholz JE, Hoffman DJ. 2002. Selenium toxicity: cause and effects in aquatic birds. *Aquat Toxicol* 57:27–37.

Spallholz JE, Shriver BJ, Reid TW. 2001. Dimethyldiselenide and methylseleninic acid generate superoxide in an in vitro chemiluminescence assay in the presence of glutathione: implications for the anticarcinogenic activity of L-selenomethionine and L-Se methylselenocysteine. *Nutr Cancer* 40:34–41.

Spallholz JE, Palace VP, Reid TW. 2004. Methioninase and selenomethionine but not Se-methylselenocysteine generate methylselenol and superoxide in an in vitro chemiluminescence assay: experiments and review. *Biochem Pharamacol* 67:547–554.

Srivastava M, Ma LQ, Rathinasabapathi B, Srivastava P. 2009. Effects of selenium on arsenic uptake in arsenic hyperaccumulator *Pteris vittata* L. *Bioresource Technol* 100:1115–1121.

Stadtman TC. 1996. Selenocysteine. *Ann Rev Biochem* 65:83–100.

Stanley TR Jr, Spann JW, Smith GJ, Rosscoe R. 1994. Main and interactive effects of arsenic and selenium on mallard reproduction and duckling growth and survival. *Arch Environ Contam Toxicol* 26:444–451.

Stanley TR Jr, Smith GJ, Hoffman DJ, Heinz H, Rosscoe R. 1996. Effects of boron and selenium on mallard reproduction and duckling growth and survival. *Environ Toxicol Chem* 15:1124–1132.

Staub BP, Hopkins WA, Novak J, Congdon JD. 2004. Respiratory and reproductive characteristics of eastern mosquitofish (*Gambusia holbrooki*) inhabiting a coal ash settling basin. *Arch Environ Contam Toxicol* 46:96–101.

Stewart AR, Luoma SN, Schlekat CE, Doblin MA, Hieb KA. 2004. Food web pathway determines how selenium affects aquatic ecosystems: a San Francisco Bay case study. *Environ Sci Technol* 38:4519–4526.

Stoddard JL, Larsen DP, Hawkins CP, Johnson RK, Norris RH. 2006. Setting expectations for the ecological condition of streams: the concept of reference condition. *Ecol Appl* 16:1267–1276.

Stout JH, Trust KA, Cochrane JF, Suydam RS, Quakenbush LT. 2002. Environmental contaminants in four eider species from Alaska and arctic Russia. *Environ Pollut* 119:215–226.

Stryer L. 1995. Biochemistry. New York (NY, USA): WH Freeman.

Stuart SN, Chanson JS, Cox NA, Young BE, Rodrigues ASL, Fischman DL, Waller RW. 2004. Status and trends of amphibian declines and extinctions worldwide. *Science* 306:1783–1786.

Sunde RA. 1984. The biochemistry of selenoproteins. *J Am Org Chem* 61:1891–1900.

Sunde RA. 1997. Selenium. In: O'Dell BO, Sunde RA, editors, Handbook of nutritionally essential minerals. Marcel Dekker (New York, NY, USA). p 493–557.

Suter GW II, Barnthouse LW, Breck JE, Gardner RH, O'Neill RV. 1985. Extrapolating from the laboratory to the field: how uncertain are you? In: Cardwell RD, Purdy R, Bahner RC, editors, Aquatic toxicology and hazard assessment: seventh symposium, ASTM STP 854. Philadelphia (PA, USA): American Society for Testing and Materials. p 400–413.

Suter GW II, Barnthouse LW, Efroymson RA, Jager H. 1999. Ecological risk assessment in a large river-reservoir: 2. Fish community. *Environ Toxicol Chem* 18:589–598.

Tashjian DH, Teh SJ, Sogomoyan A, Hung SSO. 2006. Bioaccumulation and chronic toxicity of dietary L-selenomethionine in juvenile white sturgeon (*Acipenser transmontanus*). *Aquat Toxicol* 79:401–409.

Teh SJ, Deng X, Deng D-F, Teh F-C, Hung SSO, Fan TW, Liu J, Higasi RM. 2004. Chronic effects of dietary selenium on juvenile Sacramento splittail (*Pogonichthys macrolepidotus*). *Environ Sci Technol* 38:6085–6593.

Terborgh J. 1989. Where have all the birds gone? Princeton (NJ, USA): Princeton University Pr.

Thapar NT, Guenther E, Carlson CW, Olson OE. 1969. Dietary selenium and arsenic additions to diets for chickens over a life cycle. *Poultry Sci* 48:1988–1993.

Thompson MB, Stewart JR, Speake BK. 2000. Comparison of nutrient transport across the placenta of lizards differing in placental complexity. *Comp Biochem Physiol* 127A:469–479.

Tinkle DW, Gibbons JW. 1977. The distribution and evolution of viviparity in reptiles. Misc Publication # 154. Ann Arbor (MI, USA): Museum of Zoology, University of Michigan.

Tillitt DE, Papoulias DM. 2002. 2,3,7,8-Tetrachlorodibenzo-p-dioxin toxicity in the zebrafish embryo: local circulation failure in the dorsal midbrain is associated with increased apoptosis. *Toxicol Sci* 69:1–2.

Trust KA, Rummel KT, Scheuhammer AM, Brisbin Jr IL, Hooper MJ. 2000. Contaminant exposure and biomarker responses in spectacled eiders (*Somateria fischeri*) from St. Lawrence Island, Alaska. *Arch Environ Contam Toxicol* 38:107–113.

Tully WC, Franke KW. 1935. A new toxicant occurring naturally in certain samples of plant foodstuffs. VI. A study of the effect of affected grains on growing chicks. *Poultry Sci* 14:280–284.

Unrine JM, Jackson BP, Hopkins WA, Romanek C. 2006. Isolation and partial characterization of proteins involved in maternal transfer of selenium in the western fence lizard (*Sceloporus occidentalis*). *Environ Toxicol Chem* 25:1864–1867.

Unrine JM, Jackson BP, Hopkins WA. 2007a. Selenomethionine biotransformation and incorporation into proteins along a simulated terrestrial food chain. *Environ Sci Technol* 41:3601–3606.

Unrine JM, Hopkins WA, Romanek CS, Jackson BP. 2007b. Bioaccumulation of trace elements in omnivorous amphibian larvae: Implications for amphibian health and contaminant transport. *Environ Pollut* 149:182–192.

Ursini F, Maiorino M, Gregolin C. 1985. The selenoenzyme phospholipid hydroperoxide glutathione peroxidase. *Biochim Biophys Acta* 839:62–70.

USDOI. 1998. Guidelines for interpretation of the biological effects of selected constituents in biota, water, and sediment. Bureau of Reclamation, Fish and Wildlife Service, Geological Survey, Bureau of Indian Affairs. National Irrigation Water Quality Program Information Report No. 3. Denver (CO, USA): US Department of the Interior.

USEPA (United States Environmental Protection Agency). 1998. Report of the peer consultation workshop on selenium aquatic toxicity and bioaccumulation. EPA-822-R-98-007. Office of Water, Washington, DC, USA.

USEPA. 2004. Draft aquatic life water quality criteria for selenium—2004. Office of Water, Office of Science and Technology, Washington, DC.

USEPA. 2007. Ecological soil screening levels for selenium. Interim Final. OSWER Directive 9285.7-72. http://www.epa.gov/ecotox/ecossl/.

Van Horn S. 1978. Piedmont fisheries investigation. Federal Aid in Fish Restoration Project F-23, Final Report, Study I. Development of the sport fishing potential of an industrial cooling lake. Raleigh (NC, USA): North Carolina Wildlife Resources Commission.

Van Kirk RW, Hill SL. 2007. Demographic model predicts trout population response to selenium based on individual-level toxicity. *Ecol Model* 206:407–420.

Vidal D, Bay SM, Schlenk D. 2005. Effects of dietary selenomethionine on larval rainbow trout (*Oncorhynchus mykiss*). *Arch Environ Contam Toxicol* 49:71–75.

Villeneuve DL, Curtis LR, Jenkins JJ, Warner KE, Tilton F, Kent ML, Watral VG, Cunningham ME, Markle DF, Sethajintanin D, Krissanakriangkrai O, Johnson ER, Grove R, Anderson KA. 2005. Environmental stresses and skeletal deformities in fish from the Willamette River, Oregon. *Environ Sci Technol* 39:3495–3506.

Wake DB, Vredenburg VT. 2008. Are we in the midst of the sixth mass extinction? A view from the world of amphibians. *Proc Nat Acad Sci* 105:11466–11473.

Wang Z, Jiang C, Lu J. 2002. Induction of caspase-mediated apoptosis and cell-cycle G1 arrest by selenium metabolite methylselenol. *Mol Carcinog* 34:113–120.

Waters TF. 1999. Long-term trout production dynamics in Valley Creek, Minnesota. *Trans Am Fish Soc* 128:1151–1162.

Watson RR. 1996. Trace elements in laboratory rodents. Boca Raton (FL, USA): CRC Pr.

Wayland M, Gilchrist HG, Marchant T, Keating, Smits JE. 2002. Immune function, stress response and body condition in arctic-breeding common eiders in relation to cadmium, mercury and selenium concentrations. *Environ Res* 90:47–60.

Weber LP, Dubé MG, Rickwood CJ, Driedger KL, Portt C, Brereton CI, Janz DM. 2008. Effects of multiple effluents on resident fish from Junction Creek, Sudbury, Ontario. *Ecotoxicol Environ Saf* 70:433–445.

Welker TL, Congleton JL. 2009. Effect of dietary α-tocopherol + ascorbic acid, selenium, and iron on oxidative stress in sub-yearling Chinook salmon (*Oncorhynchus tshawytscha* Walbaum). *J Anim Physiol Anim Nutrit* 93:15–25.

Wells PG, Kim PM, Laposa RR, Nicol CJ, Parmana T, Winn LM. 1997. Oxidative damage in chemical teratogenesis. *Mutation Res* 396:65–78.

Wells PG, McCallum GP, Chen CS, Henderson JT, Lee CJJ, Perstin J. 2009. Oxidative stress in developmental origins of disease: teratogenesis, neurodevelopmental deficits, and cancer. *Tox Sci* 108:4–18.

Werner J, Wautier K, Mills K, Chalanchuk S, Kidd K, Palace VP. 2005. Reproductive fitness of lake trout (*Salvelinus namaycush*) exposed to environmentally relevant concentrations of the potent estrogen ethynylestradiol (EE2) in a whole lake exposure experiment. *Scientia Marina* 70S2:59–66.

Wiemeyer SN, Hoffman DJ. 1996. Reproduction in eastern screech-owls fed selenium. *J Wildl Manage.* 60:332–341.

Wilber CG. 1980. Toxicology of selenium—a review. *Clinic Toxicol* 17:171–230.

Williams ML, Hothem RL, Ohlendorf HM. 1989. Recruitment failure in American avocets and black-necked stilts nesting at Kesterson Reservoir, California, 1984–1985. *Condor* 91:797–802.

Woock SE, Garrett WR, Partin WE, Bryson WT. 1987. Decreased survival and teratogenesis during laboratory selenium exposures to bluegill, *Lepomis macrochirus*. *Bull Environ Contam Toxicol* 39:998–1005.

Wren CD. 1984. Distribution of metals in tissues of beaver, raccoon and otter from Ontario, Canada. *Sci Total Environ* 34:177–184.

Wren CD, Stokes PM, Fischer KL. 1986. Mercury levels in Ontario mink and otter relative to food levels and environmental acidification. *Can J Zool* 64:2854–2859.

Yang D-Y, Chen Y-W, Gunn JM, Belzile N. 2008. Selenium and mercury in organisms: interactions and mechanisms. *Environ Rev* 16:71–92.

Yuan T, Weljie AM, Vogel HJ. 1998. Tryptophan fluorescence quenching by methionine and selenomethionine residues of calmodulin: orientation of peptide and protein binding. *Biochemistry* 37:3187–3195.

7 Selenium Risk Characterization

Peter V. Hodson, Robin J. Reash, Steven P. Canton, Patrick V. Campbell, Charles G. Delos, Anne Fairbrother, Nathaniel P. Hitt, Lana L. Miller, Harry M. Ohlendorf

CONTENTS

7.1 INTRODUCTION

This SETAC Workshop on Ecological Assessment of Se in the Aquatic Environment was structured around the framework of an ecological risk assessment (ERA). The purpose was to focus thinking on those aspects of the science of selenium (Se) that are critical to an understanding and management of the adverse effects of Se on aquatic ecosystems. This chapter provides an overview of the risk characterization stage of risk assessment, placing emphasis on the environmental characteristics of Se that require special consideration in characterizing risk. Most important, it recommends an approach to risk characterization that compares Se concentrations in the tissues of fish and aquatic birds, the most sensitive receptors, to known benchmarks of tissue Se associated with embryotoxicity. This approach provides the highest degree of confidence in the assessed risk and the least uncertainty for the protection of these critical receptors. It also creates a starting point for estimating the concentrations of Se in water, sediments, and prey associated with critical tissue residues, using a model of Se accumulation in aquatic food webs (Chapter 5). These predicted Se concentrations in other media create the necessary link between tissue residues and environmental management of Se-contaminated ecosystems.

7.1.1 What Is the Function of Risk Characterization?

Risk Characterization is the phase of an ERA that integrates analyses of potential exposures of aquatic organisms to Se with what is known about its effects (see previous chapters). The value of a Se Risk Characterization depends heavily on a clear definition during Problem Formulation (Chapter 3). Previous paradigms provide general guidance for conducting ERAs (Stahl et al. 2001; USEPA 1997, 1998a), and USEPA (2007) provides specific guidance for metals, but they all lack specific considerations important for Se.

Selenium risk assessment is particularly challenging because of the complexity of Se chemistry, exposure, and differences in the concentrations or dosages associated with effects (even among closely related biological species). Exposure levels are not easy to quantify, because they depend upon the organisms of concern, biological and hydrologic connectivity, geographic location, physical and chemical parameters of the system, and accuracy in analytical approaches.

Risk Characterization for Se in the aquatic environment can serve a wide range of functions, such as prospective and retrospective risk assessments (Section 7.2.3). Identification of assessment and measurement endpoints and consideration of appropriate conceptual models are especially important for the Risk Characterization to be meaningful. This chapter describes the particular ways in which Risk Characterization could be conducted for Se to guide management decisions concerning risks associated with aquatic ecosystems.

7.1.2 What Is Unique about Se?

Selenium is a natural non-metal present in varying concentrations among soil and geological strata (Chapter 3). It is also an essential micronutrient with a relatively narrow margin between nutritionally optimal and potentially toxic concentrations (Chapter 6). The complexity of Se biogeochemistry and the well-known influence of site-specific factors complicate Se risk assessments.

There are many well-documented cases describing the severity of toxic effects that Se can cause in aquatic ecosystems (see case studies in Appendix A). Many of these toxic episodes occurred in high-exposure settings where bioaccumulation via trophic transfer was efficient and rapid due to long hydraulic residence times and/or site-specific trophic transfer mechanisms (e.g., Belews Lake, Hyco Lake, and Kesterson Reservoir). Potential Se risk appears to differ among habitat types (e.g., lotic versus lentic exposure settings) and receptors. Therefore, considerations of habitat types, Se vulnerability of receptors, and ancillary, modifying physicochemical parameters (e.g., hydraulic flushing rate, sediment particle size, organic carbon characteristics, ratio of particulate to dissolved Se) are important when estimating the risk of potential adverse effects. Accurate estimates of exposure pathways and trophic transfers are crucial for an accurate Se ERA.

Bioavailability of selenite and selenate depends on physical-chemical processes in the ecosystem and uptake and assimilation efficiencies of organisms. The transfer of Se through the food web is dominated by bioaccumulative enrichment in the primary producers at the base of the food web. Conversion to organic forms of Se at this trophic level further enhances bioavailability to higher trophic levels (Chapters 4 and 5). In practice, risk assessors will often need to address total Se in the various media, although the predominant chemical form at the site can influence the rate at which Se enters the food web.

The primary route of exposure is dietary, although water can contribute up to 15% to 20% of the amount accumulated by invertebrates (Chapters 4 and 5). Dietary exposure is more critical from an effects perspective because about two-thirds (66%) of the Se in prey will be bioavailable (Dubois and Hare 2009). This requires risk assessors to understand food web structure to estimate the risk of waterborne Se. Deriving protective benchmarks or threshold values based on a prey concentration (or consumer tissue concentration, such as that in whole-body fish or bird eggs) will require research to further develop those benchmarks and apply them in risk assessments.

There are similarities but also important differences between Se and mercury (Hg). For both, dietary uptake can adversely affect higher trophic levels. While both Se and Hg bioaccumulate, Hg is a nonessential trace metal, and its propensity to biomagnify as methyl mercury (MeHg) is much greater. This does not imply that bioaccumulation of Se in aquatic food webs has a lower absolute risk relative to Hg; rather, the patterns and modes of bioaccumulation are different. The bioavailable (and most toxic) form of Hg is primarily one organometallic compound (MeHg), whereas it is believed that several organo-Se compounds have the potential to cause toxic effects in vivo (Chapter 6).

7.2 PRINCIPAL TECHNICAL ISSUES

7.2.1 PROBLEM FORMULATION

7.2.1.1 Conceptual Model

The conceptual model for Se risk assessment is superficially similar to parts of the model developed for the United States Environmental Protection Agency's (USEPA's) Framework for Metals Risk Assessment (USEPA 2007) as summarized in Figure 7.1. As with trace metals, it is necessary to consider Se loadings to the aquatic system, transport and fate within the system, water and sediment exposures, bioaccumulation, dietary exposures, tissue residues, and possible effects on individuals, populations, or communities of aquatic species. However, the relative importance and complexity of some components of the conceptual model are greater for Se than for other inorganic contaminants, as described in previous chapters.

Among the workshop conclusions were that 1) embryotoxicity to birds and oviparous fish represents the most important adverse effect of Se; and 2) toxicity follows accumulation of organic forms of Se, almost exclusively from the diet. There is a link between the concentrations of Se in water or sediment and accumulation to toxic concentrations in fish and bird tissues. However, the transfer of Se from water or sediment to the lower trophic members of aquatic food webs occurs at rates determined by specific enrichment functions (EFs) that are large and can vary more than 10-fold among ecosystems. Similarly, the transfer of Se among species within a food web varies according to trophic transfer functions (TTFs) that can vary by up to 5-fold. As a result, the same concentration of waterborne Se in a river and an estuary may cause more than a 10-fold difference in Se concentrations in fish tissues among habitats, as discussed in Chapter 5.

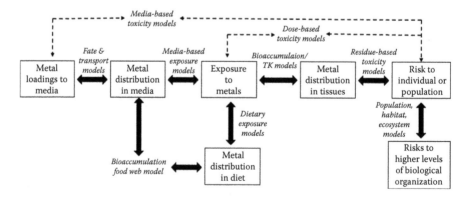

FIGURE 7.1 A conceptual model for metals risk assessment. (Adapted from USEPA 2007.)

For many risk assessments, risk characterization might compare the measured concentrations of a chemical in water or sediments to benchmarks of chronic toxicity, such as the water or sediment concentration causing specified effects. For Se, such estimates of risk would carry a high degree of uncertainty. For instance, the amount of Se transferred to fish or birds at any given waterborne concentration cannot be estimated accurately without a detailed knowledge of the EF relating waterborne to algal Se and the TTFs that describe transfer from algae through the food web to fish or aquatic birds.

To improve the accuracy and efficacy of Se risk assessments, and to first answer the question of whether Se poses a problem in an aquatic ecosystem, risk assessments should ideally start with analyses of fish and/or bird tissues. If concentrations are close to, or in excess of, benchmark values for embryotoxicity (see Chapter 6), further investigation would be initiated to characterize the rates of Se transfer through the food web to the receptor. This approach does not mean that Se data for water or sediments are not useful, but rather that the uncertainty of risk estimates based on these data is much greater than for estimates based on tissue Se concentrations.

This approach of basing risk assessment, and ultimately environmental management, primarily on the concentrations of Se accumulated in fish or aquatic bird tissues is an important departure from the risk assessment approach applied to most other elements. It reflects planning and development of tissue-based environmental quality guidelines for Se by the USEPA, and incorporates elements of recent recommendations for tissue-based guidelines by McDonald and Chapman (2007) and Lemly and Skorupa (2007). A conceptual model of this approach is summarized in Figure 7.2.

Figure 7.2 depicts the relative certainty of Se concentrations measured in different environmental compartments related to a prediction or characterization of risk. The critical medium (i.e., the compartment that yields the greatest amount of certainty and insight into potential adverse effects) is biological tissue in the most vulnerable aquatic organisms.

The risk assessor must carefully consider the bioavailability of Se from water or sediment to biota and the transfer from one trophic level to the next (i.e., EFs and TTFs) to estimate dietary exposure of aquatic animals to Se. The wide variation in foraging ranges, feeding modes, and digestive physiology across species limits the ability to make generalizations for Se exposure. While an analysis of Se at any level can be used to measure Se concentrations and to estimate the risk of embryotoxicity, uncertainty is least with measures of tissue Se in the receptors. Consequently, Se concentrations in tissues (especially eggs of fish or birds) are more useful for assessing risk from Se than tissue data for many other chemicals.

7.2.1.2 Assessment Endpoints: Level of Ecological Organization

Selenium risk can be assessed at any level of biological or ecological organization, although the information needs may differ significantly depending upon the assessment endpoints that are chosen. A good Problem Formulation for Se will set up the Risk Characterization by stipulating what parts of the ecosystem are of concern (e.g., fish, birds, and other oviparous vertebrates), the time frame for which risk should be considered, and the level of biological organization upon which decisions will be

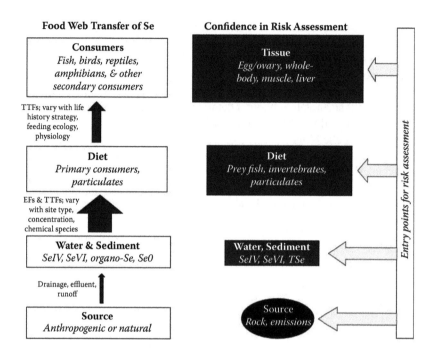

FIGURE 7.2   Conceptual diagram of Se transfer from water to species that experience embryotoxicity and associated benchmarks. The size of the arrows in the left column indicates the relative rates of transfer, and the size of the compartment in the right column indicates the relative confidence for deriving estimated risks.

made. The answers to these and similar questions may differ depending upon the management objectives for a particular site and the legal or regulatory framework under which the assessment is being conducted. Associated measurement endpoints are discussed in the following text and in Chapter 3. As indicated in Chapter 6, there are equivalent concerns for reproduction of other oviparous vertebrate species, such as amphibians and reptiles. However, the limited toxicity data base for amphibians and reptiles does not currently support their use in ERA to elucidate the probable effects of Se based on concentrations in their tissues.

When considering the ecological risk of a Se-induced change in reproductive fitness, risk assessors should consider that there are emergent properties of populations and communities that make them more than simply the sum of effects at the organismal level. For example, increasing mortality of a group of fish as a function of contaminant exposure in a toxicity test (or observing dead fish in the field) cannot be assumed to cause a population decline (Fairbrother 2001). Contaminant-induced mortality may not be additive to other mortality in the population, and there are other compensatory mechanisms that may occur (e.g., increased fecundity by adults). Furthermore, other non-contaminant-related factors may limit the population, so "harvest" by contamination could have no measurable effect. On the other hand, there may be populations that are very sensitive to changes in reproductive rates, thus responding very quickly to contaminant-induced reductions. Without knowing the life history strategies and ecological context of the species of concern, inferences made from one level of organization to the next will necessarily have great uncertainty. Thus, if the management goal is to protect populations so that the number and density of individuals remain relatively constant over time, making risk management decisions on the basis of population characteristics will provide greater accuracy than decisions based on organismal attributes (e.g., mortality or reproduction). However, if the risk management goal is to protect individuals and increase the overall population size, a greater emphasis on organism-level risk considerations may be warranted.

Measurement endpoints should be as closely associated with the assessment endpoints as possible, which for Se is reproduction. Thus, measurement endpoints should include Se concentrations and toxicity benchmarks in eggs of fish or birds (or other oviparous species) or in whole-body fish. The species and endpoints that are quantified during the risk analysis should be directly related to the species and level of biological organization of concern. For instance, if the goal of the assessment is to reduce the risk to fish from Se exposure, it makes sense to collect information about the species of fish at the site that are most vulnerable, either because their exposure pathways suggest the potential for high uptake or because they are a sensitive species. This is particularly important for species that migrate into and out of an area contaminated by Se because their dietary exposure depends on the spatial heterogeneity of their prey as well as their own movements. Migrations for feeding, breeding, and refugia may also help to explain metapopulation dynamics (i.e., recolonization). For example, documenting that some fish populations are sustained by continual immigration from noncontaminated areas can avoid arriving at the erroneous conclusion that Se poses no risk because fish are present.

If a species of particular concern is not present, surrogate species may be used as representatives. This adds uncertainty to the risk assessment, as relative sensitivities between the surrogate and desired species may not be known and there may be differences in dietary composition that would affect total Se exposure. Such uncertainties must be acknowledged in the final risk characterization and will contribute to the degree of conservatism in management decisions.

7.2.2 Additional Considerations in Se Risk Assessment

7.2.2.1 Scale or Source Type

In large and complex systems with multiple or diffuse sources of Se (e.g., an estuary), risk characterization of Se can proceed along the same tiered approach as for smaller systems with identifiable single-point sources. However, it will likely be necessary to divide the larger system into smaller units within which physical or chemical characteristics are similar. Some species of concern such as fish or birds also may differ in their use of various parts of a large ecosystem, thus changing exposure and subsequent risk factors depending upon where they are within the system. Exchanges between different parts of the system need to be addressed at some point in the risk characterization, but even that will be simplified by addressing it as input/output boundary conditions for a smaller unit. In this manner, the risk characterization explicitly acknowledges the spatial differences in potential for risk in different parts of a large system.

7.2.2.2 What to Measure

Various abiotic and biotic compartments can be assessed to determine whether there is a potential for unacceptable adverse effects to occur. Elevated waterborne Se concentrations may be an important indicator of a potential problem, but the primary exposure route is the diet. Measuring the Se content of invertebrates will help determine the Se exposure of both fish and birds feeding in the contaminated water body. Selenium concentrations in fish and bird eggs, and under some circumstances in muscle or liver, are robust measures for use in assessing potential adverse effects.

While water, diet, and tissue Se concentrations are key in determining the potential for adverse effects, the most important step in the risk pathway is the initial transfer from water to primary producers; that is, the site-specific EF (Chapter 5). TTFs are orders of magnitude less than EFs and are, therefore, of less influence in determining the potential for bioaccumulation and resultant risk.

7.2.2.3 Evidence of Long-Term Previous Exposure

Some researchers have suggested that the long-term persistence of Se-sensitive species in relatively contaminated freshwater settings indicates possible compensatory mechanisms (individual acclimation, long-term selection of tolerant genotypes, or shifts in life history attributes) in response to chronic Se exposure. In most cases, the hypotheses of adaptation, acclimation, or tolerance reflect the observation that Se-sensitive fish populations are abundant and temporally persistent and that individuals have tissue concentrations of Se that exceed a benchmark or threshold (Lohner et al. 2001a, 2001b; Reash et al. 2006; see also Appendix A). Certain sunfish species (bluegill, green sunfish) may persist in Se-contaminated settings for several years, as documented by their presence in routine monitoring studies (Reash et al. 1988).

While it is possible that some species have acquired genetic or physiological tolerance to long-term exposure, there is little evidence for or against this possibility. The observed presence of a particular species may be a consequence of open-system immigration (ecological connectivity) discussed earlier. Although it could be applicable to prospective studies (Section 7.2.3), considerations of long-term acquired tolerance are typically absent during a risk characterization, and additional research is needed to determine if the basis for any apparent tolerance is ecosystem-specific, affecting exposure rates, exposure routes, and forms of Se, or if it is biological, related to physiological or genetic adaptation.

7.2.3 PROSPECTIVE AND RETROSPECTIVE APPROACHES FOR SE RISK ASSESSMENT

In determining when or if a Se risk assessment is warranted, it is necessary to evaluate the potential for adverse effects. Such an evaluation can be prospective, predicting the risk of adverse effects from Se exposure before initiation of an action; or retrospective, documenting the actual risk of Se exposure based on empirical measures of abiotic and biotic components due to a current or past action. Both approaches are used frequently in assessing Se. An "action" in the context of Se risk assessment could be any activity that will result in the release or increased mobilization of Se, or that has done so in the past.

7.2.3.1 Prospective Se Risk Assessment

Some of the key issues to address in a prospective Se risk characterization include the following:

> *Ecosystem Vulnerability* — Past studies have shown that certain ecosystems are more vulnerable to Se risk than others (Simmons and Wallschläger 2005; EPRI 2006; Orr et al. 2006; Canton et al. 2008; Chapters 3 and 5). There are important differences between lentic and lotic environments as to how biogeochemical processes affect the chemical form and propensity for Se to bioaccumulate. Simmons and Wallschläger (2005) recognized the scarcity of data available for lotic systems as compared to lentic systems and suggested that the ecology, hydrology, and biochemistry of these systems are different and affect bioaccumulation and food web transfer (Chapter 4).
>
> Lentic ecosystems (lakes, reservoirs, ponds, wetlands) are considered to be at increased risk of Se-caused adverse effects. Conditions in these systems maximize the mobility of Se into the food web, increasing the chance of elevated exposures, and thereby increasing the potential risk. In contrast, there is a lower probability that Se in lotic ecosystems (especially those with higher gradients, coarse sediments, and low sediment organic carbon) would be mobilized into food webs. Hydrological connectivity is an important consideration because Se inputs from upstream areas may increase risk in downstream settings or habitats, or increased flows may decrease concentrations.
>
> *Species Vulnerability* — Species native to Se-contaminated environments must be considered, and once the potential ecological receptors have been identified, the vulnerability of each species to Se can be assessed. Chapters 5 and 6 outlines such an evaluation and points out that vulnerability to Se toxicity is not confined simply to "sensitivity" (e.g., tissue EC_x) but also includes life history traits that affect its potential for bioaccumulative exposure, such as trophic level, habitat preferences, migration potential, and diet. Appropriate consideration of these factors allows an evaluation of the species' vulnerability in terms of both exposure and effects measures.
>
> *Form of Se* — The potential for risk is related, in part, to the form of Se released to the environment. The general consensus is that the potential risk of Se forms is (in decreasing order of toxicological significance): organo-Se > selenite > selenate > selenide complexes. Of course, while the form of Se may affect exposure, it cannot be evaluated apart from the environmental conditions and species present, as noted earlier. Despite the original form released, Se taken up in food chains can be metabolized to a bioavailable organo-Se form, with the uptake rate (= EF) and transfer rate (= TTF) dependent upon the initial Se form and other ecosystem attributes (water residence time, planktonic, and bacterial,

as well as invertebrate species present, etc.), as described in Chapter 5. Thus, the risk assessment may not be concerned with the form of Se released, but rather with the form to which the receptor, such as fish or birds, are exposed.

In summary, a proposed action resulting in new or increased releases of Se to a vulnerable system (based on the factors defined earlier) would require a prospective quantitative Se risk assessment. The results of such a risk assessment, along with other considerations (such as balancing risks against potential habitat benefits), may be used to establish appropriately protective exposure levels for future conditions. Predictive risk assessment may also be useful for evaluating future risks in areas with increasing waterborne concentrations of Se over time, or at a site for creation of wetland habitat (e.g., Lemly and Ohlendorf 2002).

7.2.3.2 Retrospective Se Risk Assessment

Historically, risk often has been assessed after Se release, when there was a suspicion of adverse effects. In other cases, adverse effects may have been documented, but the causes were not specifically known. In these situations, a retrospective risk assessment would be used to evaluate site-specific sources and exposure pathways. As with any risk characterization, the assessor must understand the full suite of potential stressors to assess risk from Se exposure.

Examples of retrospective risk assessments include evaluations of the nature and magnitude of effects at contaminated sites, or the potential effects of Se on the status of a declining species. Another application of Se risk assessment is to compare the risks of Se concentrations at a site to other beneficial values of the setting. Such an assessment might not be required by regulations, but the risk assessment process would help to determine current or potential future liabilities. Note that ecosystem vulnerability, species vulnerability, and the form of Se are as important in a retrospective risk assessment as in prospective assessments.

7.3   RISK CHARACTERIZATION

7.3.1   GENERAL APPROACHES

Once the Problem Formulation has identified the sources of contamination and assessment endpoints, several options are available for beginning a Se risk assessment. One can begin with whatever data are available, whether for water, sediment, invertebrates (representing potential diet for fish or birds), fish (whole-body or tissue), or birds (eggs or tissues), by comparing the measured concentrations to available benchmarks (Figure 7.2). The results of such a comparison can indicate whether 1) available data are adequate to conclude that risks to fish, amphibians, reptiles, and birds are unlikely, and no further assessment is warranted; or 2) available data are inadequate to reach conclusions and a more comprehensive risk assessment is needed.

Two general approaches for site-specific assessment of Se have recently been described (Figure 7.3). One approach compares available Se concentrations to benchmarks in a stepwise fashion (e.g., Lemly and Skorupa 2007; Canton et al. 2008).

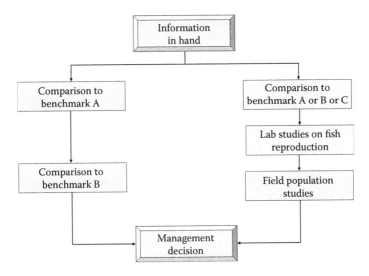

FIGURE 7.3 Conceptual approaches to risk assessment and management.

A variant of the first approach, a composite index (i.e., multimetric), has also been proposed (e.g., Lemly 1995, 2007). The second general approach requires field studies of fish populations and reproductive success studies before reaching risk conclusions (e.g., McDonald and Chapman 2007).

These approaches share several attributes. They are data-driven and rely on comparison to benchmarks at various levels of the decision pathway. Both implement a dichotomous decision framework, whereby decisions at one level constrain subsequent alternative actions and potential decisions. They emphasize the collection of data at increasing resolution and detail (i.e., both move from general information to specific information) in a tiered approach.

The two general types of decision frameworks described above are 1) exposure-based frameworks that compare measured Se concentrations in a medium (e.g., water, sediment, or tissue) with benchmark concentrations selected for that medium; or 2) effects-based frameworks that measure actual effects or biological quality.

Exposure-based decision frameworks require agreement on the benchmark concentration. Setting the benchmark too high will fail to address problems in some sites where they exist (Type II errors), whereas setting it too low will cause so many sites to fail that the screen may be ineffective in focusing attention on the actual problem sites (Type I errors). The more TTF steps between the medium and the target species, the less reliable the benchmark may become. Because waterborne concentrations are often poorly correlated to Se dietary intake and effects (Chapters 5 and 6), it may be difficult to select a single water benchmark concentration that will be applicable across a broad geographic area. Not only do exposure-based decision frameworks depend on agreement regarding concentration, they must also agree on the media.

For fish, effects-based decision criteria for Se exposure fall in 2 categories: 1) those measuring reproductive anomalies, particularly deformities and edema, in the offspring of females resident at a site, and 2) those measuring community structure and function. Neither approach is likely to trigger a Se assessment until concentration benchmark decision criteria have been exceeded. The measurement of reproductive effects in fish is being standardized, and laboratory studies of the developmental success of field-collected eggs are increasingly common in North America for a variety of species (Chapter 6). Survival of offspring and incidence and severity of deformities and edema are the endpoints of interest and may be assessed using reference versus exposed or concentration-gradient study designs. Typically, effects are portrayed as a function of measured egg (and possibly, in the case of fish, adult female whole-body) concentration, not as a comparison between reference and contaminated sites.

Use of reproductive effect-based decision criteria assumes that reproductive effects occur at lower Se exposures than do other adverse effects. The primary uncertainties arise from extrapolation from tested to untested species because of the species-specific nature of these assessments. Additionally, impaired or highly variable reproductive success can be induced by a variety of factors other than Se.

Fish population assessments may be included further along the decision framework to directly determine if general biological goals are being attained, but they are not diagnostic of cause. The choice of reference sites is a critical study design decision, but Se cause-and-effect links may not be possible as there is considerable potential for either Type I or II error. The concern about false positives is not about fisheries impairment, but whether Se is the cause. However, if used in concert with the weight of evidence from other Se decision criteria (as mentioned earlier), there is a low probability of Type I error. Conversely, the concern about false negatives (inadequate sensitivity) might be a major factor affecting the acceptance of results. Risk assessors should strive to reduce the probability of false negatives in the absence of measurable reproductive effects.

7.3.2 WIDELY APPLICABLE CONCENTRATION BENCHMARKS

Appropriate approaches for deriving concentration benchmarks or goals may differ depending on the intended use of the benchmark. If a benchmark is used as a screening value to initiate an assessment of risk at a site, its intent may be termed protective, and false negatives would be unlikely. If a benchmark is directly applied to determining pollution control requirements, its intent may be termed predictive. The regulatory context requires benchmarks that are not overly protective but sufficient to achieve biological quality goals.

Because all risk assessment approaches rely at some point on comparisons of exposure estimates to benchmarks, it is imperative that benchmark values be meaningful, accurate, and relevant. In 1998, an expert workshop evaluated water, sediment, and fish tissue as potential media for application of Se benchmarks and concluded that tissue benchmarks were the most relevant to assessing risks to fish and birds (USEPA 1998b).

Although interest in tissue concentrations arises because they are viewed as being more uniformly predictive of effects, some deviations appear. For example, fish whole-body and egg-ovary effects thresholds appear to be significantly higher when derived from organisms collected from systems such as Belews Lake, previously contaminated with Se but no longer receiving Se inputs. A mechanistic explanation or a quantitative normalization for this phenomenon is not possible at this time, and it represents an important uncertainty in predicting the extent of recovery (i.e., reduced risk) associated with remedial actions.

Comparisons of data for various co-occurring species suggest that the differences in accumulation of Se among freshwater species are not great (Chapter 5), thus somewhat limiting the potential errors of a tissue-based benchmark. An example is egg-ovary tissue concentrations, which provide better prediction of reproductive effects than whole-body or muscle concentrations, with values generally consistent among fish species (Chapter 6). Translation of whole-body or muscle measurements into egg-ovary estimates can be made using species-specific relationships but are available for only a limited number of species. Such relationships may introduce potential errors due to a lack of consistency of results within a species (Holm et al. 2005). Use of these relationships for other species would incorporate additional uncertainties.

7.4 OPTIONS FOR RISK MANAGEMENT

7.4.1 SELENIUM RISK MANAGEMENT/RESOURCE MANAGEMENT TOOLS

Approaches to regulating pollution vary across the world, ranging from establishing nationally recommended numeric criteria for multiple *priority pollutants* in the United States, to little or no regulation in some developing countries. Canada, Australia, and the European Union have distinct ways of establishing limits that set in motion differing levels of study, remediation, and/or effluent limitations. In countries with active pollution regulation, differing terminologies such as thresholds, guidelines, criteria, trigger values, and benchmarks are often used. While the terminologies and risk management approaches are different, a common feature is the use of science to predict effects on aquatic life. These predictions may trigger actions such as limiting loads, reducing concentrations, mitigating risk, and/or conducting additional monitoring or studies.

When Se may adversely affect aquatic life, basic pollution-prevention principles should not be overlooked. First, all stakeholders (e.g., regulators, resource managers, dischargers, landowners) should explore ways to prevent and then to minimize Se discharge. Best management practices and technologies that prevent entry of Se into aquatic systems should be considered preferred strategies.

7.4.1.1 United States Management Process

In the United States, this workshop came at a time when the national recommended aquatic life criteria for Se were under review. Information on the current and proposed USEPA *National Recommended Water Quality Criteria* for Se can be found at www.epa.gov/waterscience/criteria. The federal Clean Water Act requires the USEPA to develop criteria for "priority pollutants," among which Se is included. States and tribes must either adopt the national recommended criteria or develop their own. After criteria are adopted, point source dischargers with reasonable potential to cause exceedance of criteria are given appropriate effluent limits in their

National Pollutant Discharge Elimination System (NPDES) permits. Beyond that, waters listed as "impaired" (not attaining goals) are subject to total maximum daily load (TMDL) restrictions on both point and nonpoint sources.

7.4.1.2   Canadian Management Process

In Canada, aquatic ecosystems are protected in part by the *Fisheries Act*, which prohibits the deposit of substances that are deleterious to fish into waters frequented by fish (Fisheries Act 1985). While this is a federal regulation, it is delegated to some provinces, which can authorize the deposits of deleterious substances through various effluent permitting processes. Permits are negotiated by the stakeholder and the government and may contain specific limits on chemical concentrations and toxicity of the effluent, and directions on monitoring and/or compliance requirements. With respect to Se, permits may require stakeholders to monitor levels in water or the biota or possibly comply with the *Canadian Water Quality Guideline* of 1 µg/L total Se in surface waters (CCME 1999). If dischargers are in noncompliance with permits or a deleterious substance has been deposited into fish habitat, fines of up to 1 million dollars or jail sentences of up to 3 years may be applied (Fisheries Act 1985).

7.4.1.3   Australian and New Zealand Management Processes

Australia and New Zealand regulate water quality through the *Australian and New Zealand Guidelines for Fresh and Marine Water Quality*. This approach uses a trigger value to activate further study or management action (ANZECC 2000). Trigger values are derived using a risk assessment approach and standard laboratory toxicity tests. For Se, freshwater trigger values are 5, 11, 18, and 34 µg/L set to protect 99%, 95%, 90%, or 80% of species exposed, respectively (ANZECC 2000). Currently, there are no marine trigger values. Once a trigger value has been exceeded, further action is required. This may include monitoring, a site-specific risk assessment, biological effects assessment, or management and remedial actions. Site-specific investigations of risk and biological effect may also be used to set a site-specific Se guideline for further use.

7.4.1.4   European Union Management Process

In the European Union, water quality is managed through the Water Directive Framework, which protects surface, ground, and marine waters across international borders and ensures they attain good ecological and chemical status by 2015 (Water Framework Directive 2000). Currently, Se is regulated by the *Water Pollution by Discharges of Certain Dangerous Substances Directive*, but this will be integrated into the current water directive framework (European Commission 1976). Selenium is listed as a substance that has a deleterious effect on the environment and whose characteristics may depend on the specific environment. Member states are required to reduce the release of Se to meet water quality standards set by the member states.

Regardless of the country or management scheme, risk management considerations are inherent in setting any value from which management action may ensue. Each jurisdiction will have to determine, via their respective political and resource management processes, whether they are protecting a population, the individuals within a species, the water's designated use, or some other appropriate goal.

7.4.2 HOW SHOULD FOLLOW-UP MONITORING BE DESIGNED?

From a multi-stakeholder perspective, it is desirable to evaluate the environmental benefits of management actions implemented to reduce or eliminate the source of Se. Pertinent management actions could include, but are not limited to 1) remediation of contaminated sites required under certain statutory regulations (e.g., *Superfund* regulations and the *Toxic Substances Control Act* in the United States); 2) reduction of wastewater levels via permitting programs; and 3) actions imposed due to demonstrated adverse effects on specially protected fauna. No matter what jurisdiction takes regulatory action, the end goal is the same: decreased exposure of vulnerable environmental receptors to Se. The follow-up question then becomes "How do we measure the ecological benefits of investments or requirements that reduced Se input?"

The most direct answer is to measure the receptor that was assumed to be affected, or was vulnerable to a high probability of effects. For Se, direct measurement of the applicable media used to establish the benchmark or threshold is most appropriate. While measurement of ecological receptors immediately following the management action is logical, there are some practical factors to consider when designing a "postimpact" assessment for Se:

- Spatial scale of the contamination problem: Se concentrations in abiotic and biotic compartments vary spatially. Thus, it is important to sample at the same spatial scale as the distribution of Se and to consistently document the degree of spatial variability to enable the description of temporal trends of Se distribution, based on past or future studies.
- Food web considerations: As summarized in Chapters 3, 4, and 5, trophic transfer of Se is the typical means through which species are exposed to potentially harmful tissue concentrations. Thus, sampling is needed to determine whether tissue concentrations decrease following remediation.
- Temporal factors: Based on the Belews and Hyco Lake examples (Finley and Garrett 2007), ecosystem responses to reduced Se loads can be rapid. Therefore, monitoring should be at frequent intervals to detect this trend.
- Considerations of other contaminants: Se is typically discharged by industries in complex mixtures. In some cases, remediation or elimination of Se inputs has the cobenefit of removing other contaminants. Thus, monitoring of cocontaminants is recommended.

7.5 AREAS OF CONSENSUS AND UNCERTAINTIES

Throughout the workshop, a series of ideas emerged that described unique aspects of the transport, accumulation, and effects of Se that must be taken into account when characterizing risk and managing contaminated environments. There was a remarkable degree of consensus on most of these ideas; only a few ideas defied consensus. Nevertheless, even though there was general support for these ideas, significant uncertainties were often identified about aspects that might modify the assessment of exposures and impacts, the estimation of risk, or the options available for environmental managers. To ensure that risk characterization recognizes and accounts for these important ideas, the primary areas of consensus (and no consensus) have been summarized in Tables 7.1 to 7.5, grouped according to their relevance to the risk assessment process. The tables also include a summary of some of the uncertainties associated with each idea that would naturally require further research. Since these research needs are listed in other chapters, they are not repeated here.

TABLE 7.1
Consensus on Se Issues Related to Problem Formulation

Technical Issue	Consensus	Uncertainties	Importance in Risk Assessment
Conceptual models	Must recognize differences in chemical species due to sources (Chapter 4).	Limited availability of information about environmental biogeochemistry of organic chemical species in different ecosystems.	Moderate
Conceptual models	Must recognize that tissue benchmarks provide the most certainty for assessing risk of effects on embryotoxicity of fish and birds.	More data on previously untested fish and bird species and other vertebrates (amphibians, reptiles) would decrease uncertainty.	High
Assessment endpoints	Fish and bird reproductive fitness is a critical assessment endpoint because of embryotoxicity due to maternally transferred Se.	Mechanism of toxicity to amphibians and reptiles is not well defined.	High
Assessment endpoints	It is essential to have clear definition of issues and concerns to be addressed so the risk characterization will provide useful information.	Communication between risk assessors and managers may be lacking.	High
Nutrition versus toxicity	Given the multiple dimensions to ecological niches, organisms live in suboptimal conditions because not all requirements can be optimized simultaneously. For Se, the trade-off is between deficiency and toxicity. These trade-offs include other trace elements.	Environmental variability of those characteristics makes it difficult to incorporate them in risk assessment.	Low

TABLE 7.2

Consensus on Se Issues Related to Environmental Cycling

Technical Issue	Consensus	Uncertainties	Importance in Risk Assessment
Sulfur interactions	Sulfate is a major modifier of cycling and exposure in marine ecosystems. The sulfur-to-Se ratio influences Se accumulation.	The mechanism and receptor where interactions with Se uptake and molecular transformations occur.	Moderate
Particulate versus dissolved Se	Uptake of dissolved Se into microflora is the entry point for Se from water to the food web.	The nature of the particulates, including algae, bacteria, and detritus, is not well known, nor are the rates of transfer to higher-level organisms.	High
Se transformations	Transformation of Se to organic forms is critical for food web transfer and effects at higher levels of organization.	The organisms, molecular mechanisms, location of transformation, and nature of transformation products are not well known. In particular, the role of the microbial and algal community associated with biofilms is poorly understood.	Moderate

7.6 SUMMARY AND CONCLUSIONS

Traditional frameworks for ERA encompass the activities needed for risk assessment and characterization of Se. However, 1) the focus of a Se risk assessment and characterization is primarily on tissue concentrations rather than on water or sediment, and 2) characterizing risks from Se in the aquatic environment requires site-specific risk assessments to a much greater extent than many other contaminants.

Selenium risk assessment is particularly challenging because of the complexity of Se chemistry and differences in dosages associated with effects, even among closely related species. Historical perspectives on Se-induced adverse effects should be an important consideration for the risk assessor. The magnitude and severity of adverse effects observed in unique high-exposure settings (e.g., Kesterson Reservoir and Belews and Hyco Lakes [USA]; see previous chapters) were unprecedented because of scientific unknowns concerning how Se could bioaccumulate and elicit toxic effects. Considerable advancements in understanding the environmental chemistry and biogeochemical cycling, dose–response relationships, elucidation of vulnerable species and habitats, and toxic endpoints or benchmarks of Se have been made in the past 25 years. This chapter identifies the principal procedural steps and scientific knowledge required for a defensible risk characterization in settings where Se is known to occur, or has the potential to occur, above normal "background" levels. The primary goal is to ensure that historical unintended adverse effects are not repeated.

TABLE 7.3

Consensus on Se Issues Related to Exposure Assessment

Technical Issue	Consensus	Uncertainties	Importance in Risk Assessment
Bioaccumulation	The enrichment function for Se uptake by particulates is a critical process that influences the level of contamination of the food web.	The relative rate of uptake from water to particulates varies widely within and among ecosystems, and the causes of this variation are uncertain.	High
Exposure route	Dietary exposure is far more important than water uptake for higher trophic level animals.	The relative contribution of waterborne Se to Se exposure of invertebrates is about 15% (Chapter 5). Data are needed for other genera.	High
Exposure estimation	EFs, TTFs, and food web transfer models provide a useful approach for translating between tissue and water concentrations.	Propagation of errors over multiple trophic levels causes an unknown final error in the estimates; limited availability of EFs and TTFs for extrapolations; EF variation within and among ecosystems.	High
$TTF = Se_{pred}/Se_{prey}$	TTFs are similar across species, with the exception of bivalves, where TTFs are 5- to 6-fold higher.	Expansion of the data base to encompass more ecosystems and species would reduce uncertainties.	High
Site specificity	Each ecosystem has a unique combination of EFs and TTFs.	Enrichment functions are highly variable within ecosystems. The limited data base of EFs and TTFs limits the capacity to extrapolate.	Very high

7.6.1 IMPORTANCE OF PROBLEM FORMULATION

The value of the risk characterization depends heavily on information developed in earlier phases of the risk assessment. The problem formulation is particularly important, because it must clearly define the issues or concerns to be addressed and lead to appropriate analyses of exposures and effects of Se on aquatic organisms and aquatic-dependent wildlife. Identification of assessment and measurement endpoints, as well as a conceptual model that considers sources, speciation, transport and environmental partitioning, and ecological exposures are critical. Having this information available, the risk assessor can then consider the unique attributes and challenges associated with Se.

TABLE 7.4
Consensus on Se Issues Related to Effects Assessment

Technical Issue	Consensus	Uncertainties	Importance in Risk Assessment
Tissue benchmarks	Benchmark Se concentrations in fish and bird tissues that are associated with embryo toxicity vary among species.	The number of species for which tissue benchmarks have been established is limited; test conditions may affect the outcome of estimated benchmarks; extrapolation to other species increases uncertainty. Species sensitivity distributions, based on tissue benchmarks, are not well described.	Very high
Site specificity	Site specificity of Se effects is due to the feeding ecology, trophic status, and sensitivity to toxicity of receptor species in each ecosystem.	Feeding ecology (diet, movements) of resident oviparous vertebrates is often not considered in field assessments.	High
Marine versus freshwater species sensitivity	Marine birds may be less sensitive than freshwater birds.	Differences between marine and freshwater fish species in relative sensitivity are not well known (see benchmarks given earlier).	Moderate
Acute toxicity	Events of acute toxicity are rare.	Negligible.	Low
Ecosystem recovery after site remediation	Fish biodiversity can recover to precontamination levels within several years after Se sources are controlled.	Residues remain high in recovered fish populations; the recovery rate of bird reproduction in field settings is unknown.	High
Mechanism of reproductive toxicity	Embryo toxicity follows maternal transfer of Se associated with egg proteins and is likely due to oxidative stress.	Identity and nature of Se transport proteins among fish and bird species.	Moderate
Mixture toxicity	Se toxicity is influenced by other elements and vitamins.	Mechanisms of toxic interactions are not well understood.	Moderate

TABLE 7.5

Consensus on Se Issues Related to Risk Characterization

Technical Issue	Consensus	Uncertainties	Importance in Risk Assessment
Risk characterization approach	The most effective and efficient approach is a tiered risk assessment with multiple lines of evidence and a weight of evidence integration.	Availability of benchmark values for large numbers of species is limited; population studies can have high variance and may be difficult to interpret due to fish migration, etc.	Moderate (well established)
Species vulnerability	The most sensitive species are not always the most at risk (and vice versa) — life history characteristics have important effects on the extent of exposure.	Limited data exist that link vulnerability to Se to life history attributes of fish and birds. Little or no information exists for amphibians and reptiles.	High
Ecosystem type	Lentic ecosystems are more vulnerable to Se than lotic due to a greater exposure of receptors.	There are few lotic studies, and a high degree of connectivity among lentic and lotic systems creates uncertainty about exposure of mobile species; there is inadequate information on recycling of Se from sediments.	High

7.6.2 Risk Characterization: Unique Challenges Concerning Se

For the risk assessor, Se presents unique problems and considerations, including the following:

1) Se is a natural element (essential for metabolic function in vertebrates) and bioaccumulates in freshwater and marine environments.
2) The range between nutritional requirements and the onset of adverse effects is comparatively narrow, leaving a small margin of error for risk characterization.
3) Biogeochemical cycling and bioaccumulation dynamics can be complex and are highly site specific.
4) The forms of Se released to the environment vary widely, and the rates of transformation to organo-Se, the form that is most toxic, depend on site-specific factors; however, the risk assessment may not be concerned with the form of Se released, but with the forms to which the receptors (e.g., fish, waterbirds) are exposed.
5) In contrast to cationic trace metals, the primary route of exposure that leads to adverse effects to aquatic organisms is dietary.

Unique features of Se biogeochemistry and ecotoxicology in the environment require refocusing of the conceptual risk characterization typically implemented for trace metals. Figure 7.2 shows the recommended conceptual model for conducting risk characterizations for Se. This figure depicts the relative certainty with which Se concentrations in compartments can be assessed concerning prediction or characterization of risk. The critical medium (i.e., the compartment that yields the greatest amount of certainty and insight into potential adverse effects) is biological tissue in the most vulnerable aquatic organisms. There is consensus that fish and bird reproduction are the critical assessment endpoints, and that larval or embryonic survival and Se concentrations in eggs are the appropriate measurement endpoints in terms of assessing or predicting a problem at a given location because measured levels in these tissues are often strongly linked to adverse effects. Limited information concerning Se effects on amphibians and reptiles suggests that concentrations in their eggs might likewise be reasonable predictors of toxicity. Because Se is bioaccumulative, the risk assessment should focus on longer-term average exposures via diet.

Other indicators of exposure, such as Se concentrations in diet, particulate phases, and water or sediment, and their associated benchmarks, can be the starting point for an initial risk characterization. There is little confidence, however, in predicting risk based on information for waterborne Se concentrations alone (see Figure 7.2).

The vulnerability of a species depends on its propensity to bioaccumulate Se, the transfer rate into eggs, and the species' sensitivity to each unit of concentration in eggs. Risk characterization may start with Se concentrations in any environmental compartment, but uncertainty about potential adverse effects is least when the concentrations in reproductive tissue are known.

Selenium risk characterization may also involve direct measures of reproductive effects. Results from such studies, when well conducted, may be the definitive indicator of the occurrence of effects in a target species at the study site. For conducting such studies with fish, standard protocols are essential.

7.6.3 RISK MANAGEMENT

Risk management approaches for Se vary among countries, with differing terminologies, trigger values, guidelines or criteria, and resource management goals. Regardless, all risk management decisions should be based on the best available science. For Se, this means putting risk management activities into an ecosystem context through integration of biological, ecological, and chemical data, with emphasis on the biological and ecological data as they relate to Se-specific assessment and measurement endpoints.

Best management practices and technologies that prevent entry of Se into aquatic systems should be considered as preferred management approaches. A good Se risk assessment that identifies the frequency and magnitude of potential effects, and the particular locations, receptors, and endpoints that would be affected, will provide the risk manager with a scientific foundation for selecting appropriate pollution prevention or remedial options.

All regulatory approaches for managing risks of Se in water bodies would benefit from incorporation of the key findings discussed throughout this workshop. Numerical benchmarks are typically used by all jurisdictions to trigger either further assessment of potential for risk or required pollution reduction actions. The efficacy of management decisions depends upon knowledge of how Se risks occur and what changes may be expected following source reduction. Thus, the scientific foundation laid down in this workshop is directly relevant to risk reduction and risk management under a wide range of regulatory jurisdictions, regardless of their location.

7.6.4 UNCERTAINTIES

Because Se is a naturally occurring substance and is essential for animal nutrition, assessing risks resulting from new or additional Se inputs is highly complex. There is a robust scientific literature on Se behavior in selected freshwater and marine systems, yet there remain significant uncertainties about its transport and environmental partitioning in different types of (relatively unstudied) ecosystems.

Knowledge about toxicity mechanisms is limited, and the relative sensitivity of organisms to Se exposures is limited to only a few animal groups. Whole taxonomic classes, such as amphibians and aquatic reptiles, are not well represented in the effects database, so the entire range of sensitivities remains elusive. Although it is now recognized that the most important step in Se becoming bioavailable is the initial uptake of waterborne Se into small organisms at the base of the food web, our ability to precisely predict when, where, and how much bioaccumulation will occur is limited.

A better understanding of the recycling of Se via sediments is needed, as this may influence recovery rates. Similarly, we know little about how adaptation or acclimation to Se can occur in chronic, high-exposure settings. Selenium enrichment often occurs in systems that are contaminated with other chemicals, but little is known about how these substances interact to either potentiate or ameliorate Se effects. Different types of ecosystems are known to be more or less vulnerable to the potential risks associated with Se, with slow-moving water having greater vulnerability than large, faster-moving streams or rivers. However, hydrological and ecological connectivity among systems, and differences in habitats within each type of system, make it difficult to generalize across systems with a high degree of accuracy. Because of these and other uncertainties, an important conclusion from the workshop deliberations is that Se requires site-specific risk assessments to a much greater extent than most other contaminants.

7.7 REFERENCES

[ANZECC] Australian and New Zealand Environment and Conservation Council. 2000. Australian and New Zealand guidelines for fresh and marine water quality. Volume 1: The guidelines. Artarmon (NSW, AUS): Australian Water Association.

[CCME] Canadian Council of Ministers of the Environment. 1999. Canadian water quality guidelines for the protection of aquatic life: summary table. In: Canadian environmental quality guidelines. Winnipeg (SK, Canada): CCME.

Canton SP, Fairbrother A, Lemly AD, Ohlendorf HM. 2008. Experts workshop on the evaluation and management of selenium in the Elk Valley, British Columbia: Workshop summary report. Submitted to Environmental Protection Division, Kootenay and Okanagan Regions, B.C. Ministry of the Environment, Nelson, BC, Canada; www.env.gov.bc.ca/eirs/epd.

Dubois M, Hare L. 2009. Selenium assimilation and loss by an insect predator and its relationship to selenium subcellular partitioning in prey tissue. *Environ Pollut* 157:772–777.

[EPRI] Electric Power Research Institute. 2006. Fate and effects of selenium in lentic and lotic systems. EPRI Technical Report 1005315. Prepared by Great Lakes Environmental Center. Palo Alto, CA, USA: EPRI.

European Commission. 1976. Council Water Framework Directive of 4 May 1976 on pollution caused by certain dangerous substances discharged into the aquatic environment of the community. *Offic J European Communities* 129.

Fairbrother A. 2001. Putting the impacts of environmental contamination in perspective. In Shore RE, Rattner BA, editors, Ecotoxicology of wild mammals. Ecological and Environmental Toxicology Series. Chichester, UK: John Wiley. p 671–689.

Finley K, Garrett R. 2007. Recovery at Belews and Hyco Lakes: implications for fish tissue selenium thresholds. *Integr Environ Assess Manage* 3:297–299.

Fisheries Act (R.S., 1985, c.F-14). 1985. Government of Canada Act current to February 4, 2009.

Holm J, Palace V, Siwik P, Sterling G, Evans R, Baron C, Werner J, Wautier K. 2005. Developmental effects of bioaccumulated selenium in eggs and larvae of two salmonid species. *Environ Toxicol Chem* 24:2373–2381.

Lemly AD. 1995. A protocol for aquatic hazard assessment of selenium. *Ecotox Environ Safety* 32:280–288.

Lemly AD. 2007. A procedure for NEPA assessment of selenium hazards associated with mining. *Environ Monit Assess* 125:361–375.

Lemly AD, Ohlendorf HM. 2002. Regulatory implications of using constructed wetlands for treatment of selenium-laden wastewater. *Ecotox Environ Safety* 52:46–56.

Lemly AD, Skorupa JP. 2007. Technical issues associated with implementation of U.S. Environmental Protection Agency's proposed fish tissue-based aquatic criterion for selenium. *Integr Environ Assess Manage* 3:552–558.

Lohner, TW, Reash RJ, Willet VE, Rose LA. 2001a. Assessment of tolerant sunfish populations (*Lepomis* sp.) inhabiting selenium-laden coal ash effluents 1. Hematological and population level assessment. *Ecotox Environ Safety* 50:203–216.

Lohner TW, Reash, RJ, Willet VE, Rose LA. 2001b. Assessment of tolerant sunfish populations (*Lepomis* sp.) inhabiting selenium-laden coal ash effluents 2. Tissue biochemistry evaluation. *Ecotox Environ Safety* 50:217–224.

McDonald BG, Chapman PM. 2007. Selenium effects: a weight-of-evidence approach. *Integr Environ Assess Manage* 3:129–136.

Orr PL, Guiguer KR, Russell CK. 2006. Food-chain transfer of selenium in lentic and lotic habitats of a western Canadian watershed. *Ecotox Environ Safety* 63:175–188.

Reash RJ, Van Hassel JH, Wood KV. 1988. Ecology of a southern Ohio stream receiving fly ash pond discharge: changes from acid mine drainage conditions. *Arch Environ Contam Toxicol* 17:543–554.

Reash RJ, Lohner TW, KV Wood. 2006. Selenium and other trace metals in fish inhabiting a fly ash stream: implications for regulatory tissue thresholds. *Environ Pollut* 142:397–408.

Simmons DBD, Wallschläger D. 2005. A critical review of the biogeochemistry and ecotoxicology of selenium in lotic and lentic environments. *Environ Toxicol Chem* 24:1331–1343.

Stahl RG Jr, Backman RA, Barton AL, Clark JR, deFur PL, Ells SJ, Pittinger CA, Slimak MW, Wentsel RS. 2001. Risk management: ecological risk-based decision making. Pensacola (FL, USA): SETAC Pr.

[USEPA] U.S. Environmental Protection Agency. 1997. Ecological risk assessment guidance for Superfund: Process for designing and conducting ecological risk assessments. EPA 540-R-97-006. Washington (DC, USA): Office of Solid Waste and Emergency Response.

[USEPA] U.S. Environmental Protection Agency. 1998a. Final guidelines for ecological risk assessment. EPA/630/R-95/002F. Washington (DC, USA): Risk Assessment Forum.

[USEPA] U.S. Environmental Protection Agency. 1998b. Report of the peer consultation workshop on selenium aquatic toxicity and bioaccumulation. EPA-822-R-98-007. Washington (DC, USA): Office of Water.

[USEPA] U.S. Environmental Protection Agency. 2007. Framework for metals risk assessment. EPA 120/R-07/001. Washington (DC, USA): Risk Assessment Forum.

Water Framework Directive. 2000. Directive 2000/60/EC of the European Parliament and of the Council of 23 Oct 2000 establishing a framework for Community action in the field of water policy. *Offic J European Communities* 327.

Appendix A: Selected Case Studies of Ecosystem Contamination by Se

Terry F. Young, Keith Finley, William J. Adams, John Besser, William A. Hopkins, Dianne Jolley, Eugenia McNaughton, Theresa S. Presser, D. Patrick Shaw, Jason Unrine

CONTENTS

A1.0   BELEWS LAKE, NORTH CAROLINA

A1.1   SELENIUM SOURCE

In the 1970s, Se was introduced into the newly impounded Belews Lake, a 1560-ha coal-burning power plant cooling reservoir, via water from an ash settling basin. Coal ash collected from power plant flue gases was transported to the ash basin, where the coarse (bottom ash) and fine (fly ash) particulates settled from the water column. Clarified ash sluice water was returned to Belews Lake, an impoundment having limited natural surface water inflows. As an unforeseen result of this design, highly concentrated Se-laden wastewater (100 to 200 µg/L) was introduced into the lake over several years, leading to increased Se loading, elevated Se concentrations in the water column, and ultimately elevated Se concentrations in the sediment and biota. In 1985, the primary Se source was terminated with the installation of a dry fly ash collection system at the power plant, and the rerouting of the NPDES-permitted ash basin effluent (which continued to receive minor Se contributions from power plant bottom ash) to the nearby Dan River.

A1.2   FATE AND TRANSFORMATION

Selenium from coal combustion wastes occurred primarily as selenite (Cutter 1991). Lakewide water-column Se concentrations averaging 10 µg/L were observed in the late 1970s, with rapidly declining concentrations by the early 1980s as various control measures were implemented (Figure A.1). Sediment Se concentrations ranged mostly from about 4 to near 30 mg/kg dw, with a few samples having concentrations considerably greater (up to 100 mg/kg; Duke Energy 2006). During the contamination episode, the entire lake was affected with the Se-laden effluents except for a small remote headwater region of the reservoir.

Monitoring conducted a year before the impoundment showed that Belews Lake supported a diverse warm-water fishery typical of the ecoregion (Figure A.2). The lake's fish community included several trophic levels, including typical warm-water reservoir fish such as shad (*Dorosoma* spp.; Clupeidae) and blueback herring (*Alosa aestivalis;* Clupeidae), common carp (*Cyprinus carpio;* Cyprinidae) and minnows (Cyprinidae), largemouth bass (*Micropterus salmoides;* Centrarchidae), crappie (*Pomoxis* spp.; Centrarchidae), and a variety of sunfish (Centrarchidae) as well as several catfish (Ictaluridae). When maximal Se inputs occurred, Se concentrations ranged from 41 to 97 mg/kg dw in zooplankton and phytoplankton; 15 to 57 mg/kg dw in benthic macroinvertebrates; and 40 to 159 mg/kg dw in fish egg/ovary tissue (Lemly 1997a).

A1.3   EFFECTS

Adverse impacts were initially observed in 1976, with a widespread reproductive failure of the fishery, except at the remote uplake headwater area (Cumbie and

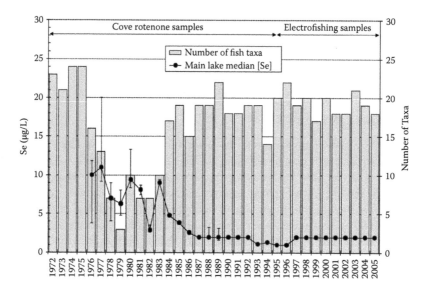

FIGURE A.1   Belews Lake monitoring data characterizing the decline and recovery of the warm-water fish community, quantified by the number of fish taxa collected annually in rotenone (1972–1994) and electrofishing samples (post-1994). In the years prior to 1977, a total of 29 fish taxa were collected. Water-column Se concentrations after 1986 primarily reflect variability in the analytical laboratory reporting limit. (Data from Van Horn 1978; Barwick and Harrell 1997; Duke Energy 2006.)

Van Horn 1978). Out of as many as 29 resident species documented prior to contamination, by 1977 only common carp, catfish, and fathead minnows (*Pimephales promelas*) were found in the remaining downlake fish community. Piscivore and insectivore species were virtually absent from the downlake area.

In addition to reproductive failure, adult mortality was hypothesized to have affected downlake fish populations at the height of the Se exposure, although no notable adult fish kills were reported. As the lakewide Se contamination diminished over the next 2 decades, populations of the more sensitive fish species migrating from the upper watershed gradually became reestablished in Belews Lake (Barwick and Harrell 1997). As the fish community was reestablished, green sunfish (*Lepomis cyanellus*) were notable among centrarchids in their apparent relative tolerance to higher environmental Se exposures, and in their slightly lower propensity to bioaccumulate Se. By the mid-1990s, the Belews Lake fishery once again represented a diverse warm water community, including sensitive fish species, although marginally elevated rates of deformities in larval fish were reported as late as 1996 (Barwick and Harrell 1997; Lemly 1997a).

A1.4   Lessons Learned

Waterborne Se concentrations in the lake never reached the then-applicable USEPA chronic Se criterion for the protection of aquatic life (35 µg/L Se as selenite, dissolved

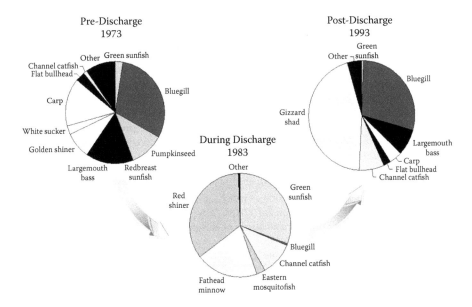

FIGURE A.2   Changes in Belews Lake warm-water fish community composition observed in samples collected at selected 10-year intervals, including sampling prior to, during, and after termination of a coal ash pond discharge to the lake. (Adapted from Barwick and Harrell 1997.)

exposure only), yet the fish community was decimated as a result of food web Se bioaccumulation. USEPA revised the chronic Se criterion for the protection of aquatic life to 5 µg/L using data from Belews Lake. Use of field data implicitly incorporated protection based on dietary exposure. Both Belews Lake and Hyco Lake (see next case study) demonstrated that discharging effluent from coal ash settling ponds into lentic systems can produce undesirable outcomes, due in part to prolonged retention of Se within the system.

Embryolarval stages of fish were clearly more sensitive to Se when compared to adults, as evidenced by the lack of adult fish kills and population structures missing young-of-year classes (Lemly 2002). Monitoring annual recruitment and fish population structure helped to establish the date of the onset and progression of this particular contamination episode. Toxic effects on benthic invertebrates were not documented. Se-related deformities were observed in field-collected specimens, and the development of a standard index of malformations was explored as an ecosystem monitoring tool (Lemly 1997b).

A 1996 retrospective comparison of lake conditions, ten years after waste disposal had ceased, showed a largely recovered but changed community structure (Figure A.2) and higher-than-normal incidence of deformed fry in some species due to the recycling of legacy Se from sediments into food webs (Lemly 1997a). Even as fish populations have recovered, Se concentrations in fish tissues, sediments, and the benthic food web remain substantially elevated with respect to reference sites.

A1.5   References

Barwick DH, Harrell RD. 1997. Recovery of fish populations in Belews Lake following sele-
    nium contamination. *Proc Ann Conf SE Assoc Fish Wildl Agencies* 51:209–216.
Cumbie PM, Van Horn SL. 1978. Selenium accumulation associated with fish mortality and
    reproductive failure. *Proc Ann Conf SE Assoc Fish Wildl Agencies* 32:612–624.
Cutter GA. 1991. Selenium biogeochemistry in reservoirs. Volume 1: Time series and mass balance
    results. Research Project 2020-1. Charlotte (NC, USA): Electric Power Research Institute.
Duke Energy. 2006. Assessment of balanced and indigenous populations in Belews Lake for
    Belews Creek Steam Station: NPDES No. NC0024406. Huntersville (NC, USA): Duke
    Energy Corporate EHS Services.
Lemly AD. 1997a. Ecosystem recovery following selenium contamination in a freshwater
    reservoir. *Ecotoxicol Environ Saf* 36:275–281.
Lemly AD. 1997b. A teratogenic deformity index for evaluating impacts of selenium on fish
    populations. *Ecotoxicol Environ Saf* 37:259–266.
Lemly AD. 2002. Symptoms and implications of selenium toxicity in fish: the Belews Lake
    example. *Aquat Toxicol* 57:29–49.
Van Horn SL. 1978. Development of the sport fishing potential of an industrial cooling lake.
    Federal Aid in Fish Restoration Project F-23. Raleigh (NC, USA): North Carolina
    Wildlife Resources Commission.

A2.0   HYCO LAKE, NORTH CAROLINA

A2.1   Selenium Source

Hyco Lake was impounded in 1964 to provide a cooling water supply for a coal-fired electric generating station. Similar to nearby Belews Lake, Hyco Lake received Se-laden effluent from an ash settling pond associated with disposal of fly ash extracted from power plant flue gas, as well as boiler bottom ash. As increased generation capacity was added to the power plant in the 1970s, Hyco Lake began receiving significantly increased loading of Se via ash pond effluent. Fly ash–associated Se inputs to the reservoir were eventually eliminated with the 1990 installation of a dry fly ash collection system at the power plant.

A2.2   Fate and Transformation

Selenium was predominantly in the soluble selenite form as it entered Hyco Lake (Cutter 1991). The degree of Se contamination varied within the lake, with headwater areas and portions of the lake most distant from the effluent outfall being less impacted, as compared to areas closer to the ash pond discharge. Water-column Se concentrations in areas near the effluent source were as high as 7 to 14 µg/L, to near 1 µg/L or less elsewhere. Sediment concentrations ranged over an order of magnitude from 3.6 to 35 mg/kg dw, depending on proximity to the effluent source.

Selenium concentrations in phytoplankton during the height of the contamination ranged from about 5 to slightly above 60 mg/kg dw. Selenium concentrations in benthic invertebrates ranged from 30 to a maximum of 88 mg/kg dw. Fish tissue concentrations were monitored in bluegill (*Lepomis macrochirus*) and largemouth bass

(*Micropterus salmoides*) during the maximal Se inputs into the reservoir. Bluegill muscle concentrations at the peak of the contamination episode ranged between 22 and 68 mg/kg dw, and liver concentrations ranged between 27 and 239 mg/kg dw. Largemouth bass tissues, sampled after dry fly ash collection methods were adopted at the power plant and effluent Se loading to the reservoir was curtailed, were lower, in part reflecting lower lakewide environmental concentrations of Se. Largemouth bass muscle Se ranged from 7 to 24 mg/kg dw, with liver concentrations ranging between approximately 10 and 35 mg/kg dw.

A2.3   EFFECTS

Fish community surveys documented declines in the Hyco Lake fishery in the late 1970s and early 1980s (Crutchfield 2000). Following start-up of an additional large power plant generating unit in September 1980, a month-long fish kill occurred in the reservoir (Carolina Power and Light Corporation 1981). Whereas Hyco Lake had previously supported a diverse warm-water fishery, comprising sunfish (*Lepomis* spp.), largemouth bass, crappie (*Pomoxis* spp.), yellow perch (*Perca flavescens*), and sucker species (Catostomidae), fish recruitment failure and the massive fish kill had by 1980 led to a decimated fishery. As a result of Se bioaccumulation via food web exposures, an assemblage of more Se-tolerant species, including green sunfish (*L. cyanellus*), satinfin shiner (*Cyprinella analostana*), gizzard shad (*Dorosoma cepedianum*), mosquito fish (*Gambusia affinis*), and introduced redbelly tilapia (*Tilapia zillii*) became dominant in the reservoir. Selenium concentrations in the reservoir were reduced beginning in the late 1980s and 1990s with the implementation of a dry fly ash collection system at the power plant. Gradual reductions in Se exposure via the food web led to the reestablishment of a diverse Hyco Lake fish community similar to the period prior to Se impact.

A2.4   LESSONS LEARNED

As in the closely parallel Belews Lake case, the primary exposure of fish to Se in Hyco Lake occured via the food web, and reduced fish population recruitment for sensitive species, due to early life stage susceptibility, was the first observed impact. Laboratory studies subsequently confirmed the importance of maternal transfer via ovary/egg to larvae in fish reproductive failure (Baumann and Gillespie 1986; Woock et al. 1987). Based on fish monitoring data from areas of Hyco Lake having successful recruitment, a warm-water fish muscle Se threshold of 8 mg/kg dw and fish Se liver threshold of 12 mg/kg dw were proposed for protection of reproductive impairment (Crutchfield 2000).

A2.5   REFERENCES

Baumann PC, Gillespie RB. 1986. Selenium bioaccumulation in gonads of largemouth bass and bluegill from three power plant cooling reservoirs. *Environ Toxicol Chem* 5:695–701.
Carolina Power and Light Corporation 1981. Hyco Reservoir environmental report 1979–1980. Volume III: Chemical and biological studies. New Hill, NC, USA.

Crutchfield JU Jr. 2000. Recovery of a power plant cooling reservoir ecosystem from selenium bioaccumulation. *Environ Sci Policy* 3:S145–S163.

Cutter GA. 1991. Selenium biogeochemistry in reservoirs. Volume 1: Time series and mass balance results. Research Project 2020-1. Charlotte (NC, USA): Electric Power Research Institute.

Woock SE, Garrett WR, Partin WE, Bryson WT. 1987. Decreased survival and teratogenesis during laboratory selenium exposures to bluegill, *Lepomis macrochirus*. *Bull Environ Contam Toxicol* 39:998–1005.

A3.0   MARTIN CREEK RESERVOIR, TEXAS

A3.1   SELENIUM SOURCE

Martin Creek Reservoir is a large (2000 ha) reservoir used by a coal-fired power plant for cooling water. In 1978, the reservoir received surface water discharge for 8 months from 2 fly ash settling ponds associated with the power plant. Whereas most fly ash–contaminated sites receive contaminant input for years or even decades, Martin Creek Reservoir provided the unique opportunity to examine the impacts of a relatively brief input of Se into an aquatic system. Fish in the system were monitored during the year prior to effluent discharge, and for several years after discharge ceased.

A3.2   FATE AND TRANSFORMATION

As soon as effluent discharge started, dissolved Se concentrations rose rapidly in the fly ash ponds, and discharges from the ponds into the reservoir reached concentrations as high as 2700 µg/L (Garrett and Inman 1984). As a result, concentrations in redear (*Lepomis microlophus*) liver rose to greater than 10 mg/kg ww (40 mg/kg dw if 75% moisture is assumed) (Sorenson et al. 1982a,b). Despite the fact that the discharge only lasted 8 months, Se continued to cycle within aquatic food webs for years thereafter. For example, Se concentrations in fish muscle exceeded 7–9 mg/kg ww (28–36 mg/kg dw if 75% moisture is assumed) following one year of recovery, declining to approximately 3–4 mg/kg ww (12–16 mg/kg dw if 75% moisture is assumed) after 3 years (Garrett and Inman 1984). However, in some cases (e.g., redear sunfish), tissue concentrations still exceeded 7 mg/kg ww (28 mg/kg dw if 75% moisture is assumed) 7 years after discharge ceased.

A3.3   EFFECTS

Numerous studies in Martin Creek Reservoir have documented histopathologic abnormalities in fish tissues following the discharge of power plant effluent into the system. In all cases, authors concluded that abnormalities were attributable to Se because it was the only coal-derived trace element that was accumulated by fish in the reservoir (Sorenson et al. 1983a; Garrett and Inman 1984). Liver, kidney, heart, and gonadal abnormalities were the most commonly documented aberrations (Sorensen et al. 1982a,b, 1983a,b, 1984; Sorenson 1988). For some species (e.g., redear sunfish) overall health was compromised (reduced condition factor) for more than 8 years after discharge ceased (Sorenson 1988). More important, population decline was documented in several fish species, drastically altering the aquatic community in Martin Creek Reservoir. Planktivorous and carnivorous species were most affected, with gizzard shad populations experiencing some of the most drastic declines, from 890 individuals/ha in 1977 to 182 individuals/ha in 1979. Recovery of this species was slow, reaching only 264 individuals/ha in 1981 (Garrett and Inman 1984). For several species, small size classes were reduced for years thereafter, providing indirect evidence of reproductive and/or recruitment problems.

A3.4 LESSONS LEARNED

What we learned from the Martin Creek Reservoir study is that, under certain circumstances, ecological damage can occur quickly once Se is discharged into an aquatic system, but recovery tends to be slow after discharge has ceased. Similar observations were made in Belews Lake. Numerous histopathologic changes were identified that, when accompanied with tissue residue data, build a compelling case that Se is the causative agent contributing to population declines.

A3.5 REFERENCES

Garrett GP, Inman CR. 1984. Selenium-induced changes in fish populations in a heated reservoir. *Proc Ann Conf SE Assoc Fish Wildl Agencies* 38:291–301.

Sorensen EMB. 1988. Selenium accumulation, reproductive status, and histopathological changes in environmentally exposed redear sunfish. *Arch Toxicol* 61:324–329.

Sorensen EMB, Harlan CW, Bell JS. 1982a. Renal changes in selenium-exposed fish. *Amer J Forensic Med Pathol* 3:123–129.

Sorensen EMB, Bauer TL, Bell JS, Harlan CW. 1982b. Selenium accumulation and cytotoxicity in teleosts following chronic, environmental exposure. *Bull Environ Contam Toxicol* 29:688–696.

Sorensen EMB, Bauer TL, Harlan CW, Pradzynski AH, Bell JS. 1983a. Hepatocyte changes following selenium accumulation in a freshwater teleost. *Amer J Forensic Med Pathol* 4:25–32.

Sorensen EMB, Bell JS, Harlan CW. 1983b. Histopathological changes in selenium exposed fish. *Amer J Forensic Med Pathol* 4:111–123.

Sorensen EMB, Cumbie PM, Bauer TL, Bell JS, Harlan CW. 1984. Histopathological, hematological, condition-factor, and organ weight changes associated with selenium accumulation in fish from Belews Lake, NC. *Arch Environ Contam Toxicol* 13:152–162.

A4.0 D-AREA POWER PLANT, SAVANNAH RIVER, SOUTH CAROLINA

A4.1 SELENIUM SOURCE

The D-Area power plant has been supplying electricity to the Savannah River Site for more than 50 years. The 70 MW power plant utilizes a wet waste handling system, where slurried fly ash is deposited in a series of large settling basins before surface waters are expelled into Beaver Dam Creek, a tributary of the Savannah River. The configuration of the basins has changed over the decades, with several basins being filled and capped. At one point in the 1950s, the ash slurry was pumped directly into the Savannah River Floodplain, where it settled across a 40-ha natural depression up to 2.7 m deep (Roe et al. 2005). These historical practices resulted in significant surface contamination that persists today. Because the site (both the active basins and the legacy floodplain contamination) is located in an area with rich invertebrate and vertebrate biodiversity, a wide variety of organisms are exposed to the waste materials.

A4.2   FATE AND TRANSFORMATION

Selenium concentrations in the ash have fluctuated over the years, but dissolved concentrations as high as 100–110 μg/L have been documented in portions of the drainage system from 1973 to 1979 (Cherry et al. 1976, 1979a,b; Guthrie and Cherry 1979; Cherry and Guthrie 1977; Rowe et al. 2002). Typical aqueous concentrations at the outfall of the settling basin system are presently much lower, in the range of 2 μg/L. Concentrations are rapidly diluted in Beaver Dam Creek before reaching the Savannah River, but total loading of Se may be an important consideration given the duration of continuous discharge. Selenium in the coal ash is predominately Se (IV) (Jackson and Miller 1998). Se speciation in the aqueous phase has not been investigated at this site.

From an ecological perspective, the D-area site is among the best-studied coal ash disposal sites in the world. Since the 1970s, numerous studies have documented the accumulation of Se and other coal-derived elements in flora and fauna. At the base of the aquatic food chain, algal and plant resources contain concentrations ranging from 1 mg/kg in leaves of trees and emerging macrophytes to 6 mg/kg (all concentrations dry weight) in filamentous algae and 12 mg/kg in submerged macrophytes (Unrine unpublished data). The periphyton, an important food resource for grazing species in the system, contains Se concentrations reaching 7 mg/kg (Unrine unpublished data). Concentrations in aquatic invertebrates have been documented as high as 20 mg/kg ranging from 0.7 mg/kg in chironomids to 20 mg/kg in bivalves (Unrine et al. 2007; Nagle et al. 2001; Guthrie and Cherry 1979). At higher trophic levels, Se is clearly bioavailable to most species of fish, bird, amphibian, and reptile studied to date. For example, studies have documented whole-body concentrations of Se from 15 mg/kg in fish and 25 mg/kg in amphibian larvae at the site (Unrine et al. 2007; Rowe et al. 2002). The highest concentrations of Se have been found in individual organs from higher trophic level predators such as water snakes (exceeding 100 mg/kg in liver) that feed upon fish and amphibians in the basins (Hopkins et al. 2002). Some of these prey species have whole-body concentrations of Se that also exceed 100 mg/kg (Hopkins et al. 2006). When Se concentrations in whole bodies of other animals were compared across trophic levels using stable isotopes, there was no evidence that Se had biomagnified in this system (Unrine et al. 2007). Concentrations of other trace elements such as As, Cd, Sr, and Cu are also present at elevated concentrations in food webs at the site, but the interactions of these elements with Se have not been rigorously studied in this system.

Many of the species studied to date (amphibians, fish [*Gambusia*], turtles, alligators, and birds [green herons and grackles]) maternally transfer Se to their eggs (Nagle et al. 2001; Bryan et al. 2003; Roe et al. 2004; Staub et al. 2004; Hopkins et al. 2006). Of particular interest are species such as birds and amphibians that move into the site seasonally to breed. For example, narrow-mouth toads transfer

53% of their total body burden of Se to their eggs, resulting in concentrations as high as 100 mg/kg dw in eggs (Hopkins et al. 2006).

Selenium in animal tissues from the site has primarily been shown to consist of Se (-II) in proteinaceous forms or as selenomethionine, supporting the notion that uptake, reductive metabolism, and assimilation of Se at the base of food webs is an important factor in determining bioavailability. High concentrations of Se (-II) were present in deformed mouth parts in amphibian larvae (Punshon et al. 2005). Another study indicated that Se in amphibian larvae primarily consisted of selenomethionine and proteinaceous forms of Se with a relatively small amount of selenite present (Jackson et al. 2005).

A4.3   Effects

Accumulation of Se and other coal-derived elements has adverse effects on a variety of fauna in D-area. Sublethal effects include changes in endocrinology, metabolism, performance, predator–prey interactions, behavior, reproductive success, and histological aberrations (Rowe et al. 2002; Bryan et al. 2003; Ganser et al. 2003; Hopkins et al. 2003, 2004, 2005, 2006; Roe et al. 2004, 2006; Snodgrass et al. 2003, 2004, 2005; Staub et al. 2004). In several cases, direct lethality in amphibian larvae (*Bufo terrestris*) and juvenile lake chubsucker (*Erimyzon sucetta*) has been observed. Indirect effects appear to be extremely common in the system, particularly for higher trophic-level species that suffer from reduced food resources in the site. In one of the more interesting studies, microbial communities in downstream areas were also affected by the effluent (Stepanauskas et al. 2005). Specifically, coevolution of metal resistance and antibiotic resistance was documented downstream from D-area as well as downstream from other coal-fired power plants.

A4.4   Lessons Learned

Studies in D-area suggest that benthic feeding species (e.g., certain amphibian larvae and benthic feeding fish) and fossorial species (e.g., adult burrowing amphibians) may be particularly at risk, accumulating as much or more Se in their tissues than higher trophic level species. The D-area studies also highlight connections between aquatic and terrestrial Se exposure pathways from disposal of coal ash in settling basins. Some limited information on Se transformation from inorganic forms in fly ash to amino acid and proteinaceous forms in vertebrates also exists for D-area (Punshon et al. 2005; Jackson et al. 2005), which is lacking for most sites.

Like many other sites, the presence of other contaminants in the effluent has complicated the process of ascribing adverse effects to Se alone. The indirect effects of the waste stream, including cascading food web effects, further complicate the risk assessment process.

A4.5   References

Bryan L, Hopkins WA, Baionno JA, Jackson BP. 2003. Maternal transfer of contaminants to eggs in common grackles (*Quiscalus quiscala*) nesting on coal fly ash basins. *Arch Environ Contam Toxicol* 45:273–277.
Cherry DS, Guthrie RK. 1977. Toxic metals in surface waters from coal ash. *Water Resour Bull* 13:1227–1236.

Cherry DS, Guthrie RK, Rogers JH Jr, Cairns J Jr, Dixon KL. 1976. Responses of mosquitofish (*Gambusia affinis*) to ash effluent and thermal stress. *Trans Amer Fish Soc* 105:686–694.

Cherry DS, Larrick SR, Guthrie RK, Davis EM, Sherberger FF. 1979a. Recovery of invertebrate and vertebrate populations in a coal ash-stressed drainage system. *J Fish Res Board Can* 36:1089–1096.

Cherry DS, Guthrie RK, Sherberger FF, Larrick SR. 1979b. The influence of coal ash and thermal discharges upon the distribution and bioaccumulation of aquatic invertebrates. *Hydrobiologia* 62:257–267.

Ganser LR, Hopkins WA, O'Neil L, Hasse S, Roe JH, Sever DM. 2003. Liver histopathology of the southern watersnake, *Nerodia fasciata fasciata*, following chronic exposure to trace element-contaminated prey from a coal ash disposal site. *J Herpetol* 37:219–226.

Guthrie RK, Cherry DS. 1979. Trophic level accumulation of heavy metals in a coal ash basin drainage system. *Water Resour Bull* 15:244–248.

Hopkins WA, Roe JH, Snodgrass JW, Staub BP, Jackson BP, Congdon JD. 2002. Trace element accumulation and effects of chronic dietary exposure on banded water snakes (*Nerodia fasciata*). *Environ Toxicol Chem* 21:906–913.

Hopkins WA, Snodgrass JW, Staub BP, Jackson BP, Congdon JD. 2003. Altered swimming performance of benthic fish (*Erimyzon sucetta*) exposed to contaminated sediments. *Arch Environ Contam Toxicol* 44:383–389.

Hopkins WA, Staub BP, Snodgrass JW, Taylor BE, DeBiase AE, Roe JH, Jackson BP, Congdon JD. 2004. Responses of benthic fish exposed to contaminants in outdoor microcosm—examining the ecological relevance of previous laboratory toxicity test. *Aquat Toxicol* 68:1–12.

Hopkins WA, Snodgrass JW, Baionno JA, Roe JH, Staub BP, Jackson BP. 2005. Functional relationships among selenium concentrations in the diet, target tissues, and nondestructive tissue samples of two species of snakes. *Environ Toxicol Chem* 24: 344–351.

Hopkins WA, DuRant SE, Staub BP, Rowe CL, Jackson BP. 2006. Reproduction, embryonic development, and maternal transfer of contaminants in an amphibian *Gastrophryne carolinensis*. *Environ Health Persp* 114:661–666.

Jackson BP, Miller W. 1998. Arsenic and selenium speciation in coal fly ash extracts by ion chromatography-inductively coupled plasma mass spectrometry. *J Analyt Atom Spectrom* 13:1107–1112.

Jackson BP, Hopkins WA, Unrine J, Baionno J, Punshon T. 2005. Selenium speciation in amphibian larvae developing in a coal fly ash settling basin. In: Holland JG, Bandura DR, editors, Plasma source mass spectrometry, current trends and future developments. Special Publication No. 301. Cambridge (UK): The Royal Society of Chemistry. p 225–234.

Nagle RD, Rowe CL, Congdon JD. 2001. Accumulation and selective maternal transfer of contaminants in the turtle *Trachemys scripta* associated with coal ash deposition. *Arch Environ Contam Toxicol* 40:531–536.

Punshon T, Jackson BP, Lanzirotti A, Hopkins WA, Bertsch PM, Burger J. 2005. Application of synchrotron x-ray microbeam spectroscopy to the determination of metal distribution and speciation in biological tissues. *Spectros Lett* 38:343–363.

Roe JH, Hopkins WA, Baionno JA, Staub BP, Rowe CL, Jackson BP. 2004. Maternal transfer of selenium in *Alligator mississippiensis* nesting downstream from a coal burning power plant. *Environ Toxicol Chem* 23:1969–1972.

Roe JH, Hopkins WA, Jackson BP. 2005. Species- and stage-specific differences in trace element tissue concentrations in amphibians: implications for the disposal of coal-combustion wastes. *Environ Pollut* 136:353–363.

Roe JH, Hopkins WA, Durant SE, Unrine JM. 2006. Effects of competition and coal combustion wastes on recruitment and life history characteristics of salamanders in temporary wetlands. *Aquat Toxicol* 79:176–184.

Rowe CL, Hopkins WA, Congdon JD. 2002. Ecotoxicological implications of aquatic disposal of coal combustion residues in the United States: a review. *Environ Monit Assess* 80:207–276.

Snodgrass JW, Hopkins WA, Roe JH. 2003. Effects of larval stage, metamorphosis, and metamorphic timing on concentrations of trace elements in bullfrogs (*Rana catesbeiana*). *Environ Toxicol Chem* 22:1597–1604.

Snodgrass JW, Hopkins WA, Broughton J, Gwinn D, Baionno JA, Burger J. 2004. Species-specific responses of developing anurans to coal combustion wastes. *Aquat Toxicol* 66:171–182.

Snodgrass JW, Hopkins WA, Jackson BP, Baionno JA, Broughton J. 2005. Influence of larval period on responses of overwintering green frog (*Rana clamitans*) larvae exposed to contaminated sediments. *Environ Toxicol Chem* 24:1508–1514.

Staub BP, Hopkins WA, Novak J, Congdon JD. 2004. Respiratory and reproductive characteristics of Eastern mosquitofish (*Gambusia holbrooki*) inhabiting a coal ash settling basin. *Arch Environ Contam Toxicol* 46:96–101.

Stepanauskas R, Glenn TC, Jagoe CH, Tuckfield RC, Lindell AH, McArthur JV. 2005. Elevated microbial tolerance to metals and antibiotics in metal-contaminated industrial environments. *Environ Sci Technol* 39:3671–3678.

Unrine JM, Hopkins WA, Romanek CS, Jackson BP. 2007. Bioaccumulation of trace elements in omnivorous amphibian larvae: implications for amphibian health and contaminant transport. *Environ Pollut* 149:182–192.

A5.0 LAKE MACQUARIE, NEW SOUTH WALES, AUSTRALIA

A5.1 INTRODUCTION

Lake Macquarie is a 125-km^2 barrier estuary south of the city of Newcastle on the central coast of New South Wales, Australia. The lake catchment has an area of approximately 622 km^2. It is shallow with an average depth of about 6.7 m. The estuary extends 22 km in a north-south direction with a maximum width of 9 km and is the largest coastal lake in eastern Australia. The lake's narrow entrance results in poor tidal flushing and an intertidal range of approximately 6–15 cm. As a result, most contaminants in the catchment are expected to accumulate in the lake. Lake Macquarie is characteristically marine given the minimal freshwater contribution from the 2 main fluvial inflows.

Lake Macquarie once supported a large commercial and recreational fishery. Approximately 280 species of fish are known to inhabit the waters of Lake Macquarie. Prior to 2006, sea mullet caught commercially had an estimated market value of AU\$ 1.0 million per annum, with a catch in 1991/92 of about 108 tonnes (NSW Fisheries 1995). The lake is now encompassed within a marine park and fishing is limited to recreational licenses.

A5.2 SELENIUM SOURCES

Industrial sources of contamination in the northern section of the lake include the discharges from the lead-zinc smelter that commenced operation in 1897, a steel foundry, collieries (underground coal mines), and sewage-treatment works. The southern section of the lake includes 2 coal-fired power plants at Eraring and Vales Point. For the best part of a century, the ash produced during coal combustion was placed into dams located on the foreshore of the lake. The ash dams and stack emissions from the coal-fired power stations produced large amounts of seleniferous fly ash. The emissions and ash contained Se concentrations of 50–300 mg/kg dw. Peters et al. (1999) showed that Se contamination from fly ash dams has been occurring for a long period, but significant Se contamination of the lake is a phenomenon of the last 30 years. Sediment Se concentrations measured at reference sites at depth (to 30 cm) average 0.3 mg/kg dw, and surficial sediments range from 0.9 to 5.6 mg/kg dw (<100 μm fraction) (Kirby et al. 2001). The highest concentration, 14 mg/kg dw at depth (Batley 1987), was found adjacent to the Vales Point coal-fired power station. Elevated Se concentrations occur at depths well within the reach of common bioturbating organisms.

Lake Macquarie is a marine seagrass–dominated system, which is predominantly lentic due to the poor flushing of the waters with minor currents in the system generated by wind activity. The lake has low turbidity due to the high salinity, low movement, and high iron content in the water. Dissolved Se concentrations (using a 0.45 μm filter) are below detection limits (<0.25 μg/L), and the dominant species is selenite (Jolley unpublished data).

A5.3   FATE AND TRANSFORMATION

Barwick and Maher (2003) provide a detailed food web with associated Se concentrations. Based on these data, the authors conclude that Se biomagnification has occurred and that the elevated Se concentrations in the top carnivores resulted from uptake of Se through benthic food chains. Uptake of Se from sediments by Lake Macquarie benthic species has been confirmed in laboratory studies (Peters et al. 1999).

A5.4   EFFECTS

Selenium concentrations in a number of species exceeded those allowable for human consumption (Barwick and Maher 2003). Effects on the lake's inhabitants have not been confirmed.

A5.5   LESSONS LEARNED

Elevated Se concentrations occur in the food webs of Lake Macquarie. Further research at this site is needed to determine the cumulative effects of Se on higher trophic individuals, populations, and communities. Future studies should consider incorporating aquatic birds into the monitoring regime.

A5.6   REFERENCES

Barwick M, Maher W. 2003. Biotransference and biomagnification of selenium, copper, cadmium, zinc, arsenic and lead in a temperate seagrass ecosystem from Lake Macquarie Estuary, NSW, Australia. *Mar Environ Res* 56:471–502.
Batley GE. 1987. Heavy metal speciation in waters, sediments and biota from Lake Macquarie, New South Wales. *Austr J Mar Freshw Res* 38:591–606.

Kirby J, Maher W, Krikowa F. 2001. Selenium, cadmium, copper, and zinc concentrations in sediments and mullet (*Mugil cephalus*) from the southern basin of Lake Macquarie, NSW, Australia. *Arch Environ Contam Toxicol* 40:246–256.

NSW Fisheries. 1995. A review of the information on the factors affecting the fisheries of Lake Macquarie. Fisheries Research Institute, Australia.

Peters GM, Maher WA, Jolley D, Carroll BI, Gomes VG, Jenkinson AV, McOrist GD. 1999. Selenium contamination, redistribution and remobilisation in sediments of Lake Macquarie, NSW. *Organic Geochem* 30:1287–1300.

A6.0 ELK RIVER VALLEY, SOUTHEAST BRITISH COLUMBIA, CANADA

A6.1 SELENIUM SOURCES

The Elk Valley and surrounding Rocky Mountains are rich in high-grade coal deposits that have been mined since the late 1800s. With the introduction of open-pit mining techniques in the 1950s, the greatly increased disturbance area and subsequent volumes of waste rock led to weathering and release of pyrite-associated Se. The waste rock is transported to extensive dump areas, often valley fills, which leach and drain to the Elk River. There are 5 large mines in the valley with a 2008 production of about 22 million tonnes of coal, and which together contributed in 2006 an estimated load of 6.2 tonnes of Se to the Elk River (Figure A.3).

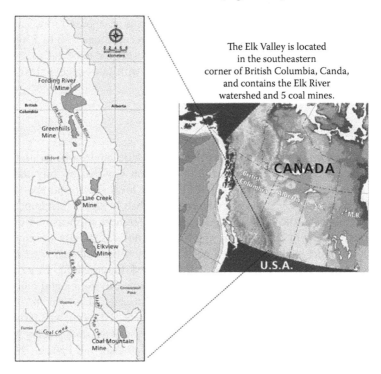

FIGURE A.3 Location of active coal mining areas in the Elk River watershed (from Teck Coal Ltd.).

A6.2 FATE AND TRANSFORMATION

Waste rock leachates drain directly to the Elk River or to the Fording River, which is the major upstream tributary, or may flow into small wetland areas before entering the main river. Selenium is released as selenate (Martin et al. 2008), with effluent concentrations ranging to more than 300 µg/L. Where effluents flow to ponds and marshes, dissolved Se concentrations typically range from 50 to 80 µg/L. Selenium levels in the main Elk River in 2008 ranged from 9.6 µg/L near the mines to 5.8 µg/L at a long-term water quality monitoring site roughly 60 km downstream. Se concentrations in the Elk River have been increasing at about 6% to 8% per year for the past decade (Elk Valley Se Task Force 2008a).

The elevated water-column Se concentrations have translated to elevated concentrations in biota. In 2002, Se concentrations in periphyton in lentic receiving waters averaged about 5 mg/kg dw, and ranged from 3.9 to 12.3 mg/kg dw in rooted macrophytes (Orr et al. 2006). Benthic invertebrates range from 26 to 96 mg/kg dw in lentic habitats and from 2.7 to 9.6 mg/kg dw in lotic habitats. Levels in vertebrates have been measured as well, with concentrations in Columbia spotted frog (*Rana luteiventris*) eggs in Se exposed areas ranging from 10 to 38 mg/kg dw, in eggs of shorebirds (spotted sandpiper, *Actitis macularia*) from 3 to 4 mg/kg dw, and in eggs of marsh birds (red-winged blackbird, *Agelaius phoeniceus*) from 5 to 23 mg/kg dw (Minnow Environmental 2007). Fish tissues (westslope cutthroat trout, *Oncorhyncus clarki lewisi*) are likewise elevated, with muscle Se concentrations in exposed lentic areas up to 76 mg/kg dw and in exposed lotic areas from 4 to 15 mg/kg dw compared to reference area concentrations of 3 to 5 mg/kg dw. Higher levels in lentic areas of the watershed are postulated to be due to production of organo-Se by the detrital microbial community, with subsequent uptake by biota (Orr et al. 2006).

A6.3 EFFECTS

Studies have been conducted in both lentic and lotic environments and species to examine effects of elevated Se and to establish regional toxicity thresholds. Studies of waterbirds (American dipper [*Cinclus mexicanus*] and spotted sandpiper; Harding et al. 2005) showed decreased hatchability in the sandpiper at average egg Se concentrations of 2.2 mg/kg ww. Differences in measured endpoints in American dipper (1.1 mg/kg ww egg Se) were not statistically significant ($p = 0.056$). Amphibian studies showed a positive relationship between egg Se content and development of deformities (Elk Valley Se Task Force 2008). Three attempts (Kennedy et al. 2000; Rudolph et al. 2008; Nautilus Environmental and Interior Reforestation 2009) have been made to establish site-specific Se effects thresholds in cutthroat trout. The first attempt was by Kennedy et al. (2000), which showed no effect even at high Se concentrations (~80 mg/kg dw egg Se). Subsequent work was conducted to address a critique of this study by Hamilton and Palace (2001). The 2 subsequent studies (Rudolph et al. 2008; Nautilus Environmental and Interior Reforestation 2009) showed similar results, with effect thresholds (inferred either as an "effect level" or a calculated EC10) of 19 to 22 mg/kg dw Se. Both of these studies also indicated reproductive failure, through either unviable embryos or unfertilizable eggs, at Se concentrations greater than about 35 mg/kg dw egg Se. Selenium in trout in the most exposed areas of the Elk Valley are at risk of Se toxicity, while levels elsewhere in the valley are currently (2009) lower than the foregoing effect levels.

A6.4 Lessons Learned

The studies conducted in the Elk River Valley highlight the challenge of conducting field-based research on Se-contaminated sites. Apart from the logistical difficulties, there are potential confounding factors related to water chemistry, weather, or hydrology that make it difficult to isolate the effects attributable solely to Se. Selenium accumulation in biota in lentic and lotic environments has been well demonstrated in studies of wetland areas adjacent to the river.

Based on studies conducted to date, current levels of Se in the Elk River Valley are resulting in only localized adverse effects (Elk Valley Se Task Force 2008; Canton et al. 2008).

A6.5 References

Canton SP, Fairbrother A, Lemly AD, Ohlendorf H, McDonald LE, MacDonald DD. 2008. Experts workshop on the evaluation and management of selenium in the Elk Valley, British Columbia, Workshop summary report. http://www.env.gov.bc.ca/eirs/epd

Elk Valley Selenium Task Force. 2008. Selenium status report 2007—Elk River Valley, BC. Sparwood, BC, Canada.

Hamilton SJ, Palace VP. 2001. Assessment of selenium in lotic ecosystems. *Ecotox Env Safet* 50:161–166.

Harding LE, Graham M, Paton D. 2005. Accumulation of selenium and lack of severe effects on productivity of American dippers (*Cinclus mexicanus*) and spotted sandpipers (*Actitis macularia*). *Arch Environ Contam Toxicol* 48:414–423.

Kennedy CJ, McDonald LE, Loveridge R, Strosher MM. 2000. The effect of bioaccumulated selenium on mortalities and deformities in the eggs, larvae, and fry of a wild population of cutthroat trout (*Oncorhynchus clarki lewisi*). *Arch Environ Contam Toxicol* 39: 46–52.

Martin AJ, Wallschläger D, London J, Wiramanaden CIE, Pickering IJ, Belzile N, Chen YW, Simpson S. 2008. The biogeochemical behaviour of selenium in two lentic environments in the Elk River Valley, British Columbia. Technical Paper 8, Proceedings of the 32nd Annual British Columbia Mine Reclamation Symposium, September 15–18, 2008. 2008. Kamloops, BC, Canada.

Minnow Environmental. 2007. Selenium monitoring in the Elk River Watershed 2006. Sparwood (BC, Canada): Elk Valley Selenium Task Force.

Nautilus Environmental and Interior Reforestation. 2009. Evaluation of the effects of selenium on early lifestage development of westslope cutthroat trout from the Elk Valley, BC. Elkford (BC, Canada): Elk Valley Selenium Task Force.

Orr PL, Guiguer KR, Russel K. 2006. Food chain transfer of selenium in lentic and lotic habitats of a western Canadian watershed. *Ecotoxicol Environ Saf* 63:175–188.

Rudolph BL, Andreller I, Kennedy CJ. 2008. Reproductive success, early life stage development, and survival of westslope cutthroat trout (*Oncorhynchus clarki lewisi*) exposed to elevated selenium in an areas of active coal mining. *Environ Sci Technol* 42:3109–3114.

A7.0   AREAS OF THE APPALACHIAN MOUNTAINS AFFECTED BY MOUNTAINTOP MINING AND VALLEY FILLS

A7.1   SELENIUM SOURCES

Large-scale land disturbance is associated with waste rock management at coal mines in the southern and central Appalachian Mountains. Tops of mountain ridges are sheared off, and waste rock is deposited in adjacent valleys (valley fills) and back-stacked to preserve landscape contours. Over 4.8 million hectares, mainly in West Virginia, are scheduled for mining through this technique during the next several years because low-sulfur coal resources are dwindling, forcing mining of near-surface, thin-layered coals of Appalachia. These coal layers are unique in that they are enriched in Se and do not generate as much acid drainage as other coals in the area.

Surface coal mining production (million tonnes) for 1998 was as follows: southern West Virginia, 44.1; eastern Kentucky, 45; Virginia, 7.7; and Tennessee, 1.5. Ninety-five percent of the surface mining in southern West Virginia would be classified as mountaintop mining. Estimated remaining years of surface production in West Virginia is about 50 and in Kentucky is about 100.

A7.2   FATE AND TRANSFORMATION

Most major rivers and tributaries east of the Mississippi River originate in the mountains of the Appalachian regions. Ecoregions in the mountaintop mining area are unique because they combine characteristically northern species with their southern counterparts, and thus boast enormous richness and diversity. Headwater stream populations have the greatest potential for natural selection processes that may result in development of new species/subspecies. The southern Appalachians have one of the richest salamander fauna in the world. Many species of birds, such as the cerulean warbler (*Dendroica cerulea*), Louisiana waterthrush (*Seiurus motacilla*), and Acadia flycatcher (*Empidonax virescens*), depend on large areas of relatively unbroken forest (93% forest cover) and headwater stream habitats. The mountaintop mining area also is unique and important in the evolution and speciation of North American freshwater fishes. Fifty-six species of fish are present in the watersheds, with small headwater streams harboring populations with unique genetic diversity.

A review of available environmental data from Appalachia reveals leaching of Se into streams below valley fills (median, 11.7 µg/L) when compared to streams in non-mined areas (median, 1.5 µg/L) (USEPA 2005). Ponds located at the toes of fills, which provide mitigation habitat for birds and fish, had the highest concentrations (up to 42 µg/L). This mobilization of Se is expected because, as mined materials are exposed, Se is 1) oxidized into mobile selenate; 2) transported regionally within watersheds; 3) biochemically transformed to bioavailable Se; and 4) eventually accumulates in prey, which serve as diet for higher trophic level predators (Presser et al. 2004a,b).

A7.3   EFFECTS

Concentrations of Se in streams exceed Se water-quality criteria for protection of aquatic life (USEPA 2005). Fish tissue Se concentrations exceed toxicity criteria for fish and dietary criteria for protection of wildlife. The Upper Mud River Reservoir in West Virginia, downstream from one of the largest mountaintop removal operations in the United States, recently came under scrutiny because of the identification of deformed juvenile fish (Appalachian Center for the Economy and the Environment et al. 2009). Selenium concentrations in bluegill averaged 28 mg/kg dw (USGS 2008). Streams affected by waste material had lower numbers of total species and benthic species than unmined streams, and mayfly populations have been decimated (Pond et al. 2008).

A7.4 LESSONS LEARNED

The problems in the Appalachians are arising not from coals extraordinarily enriched in Se, but from the mining practices. Fish in reservoirs below valley fills are vulnerable, along with food webs in ponds at toes of valley fills. Conclusions of studies in the phosphate fields of southeast Idaho concerning processes of Se mobilization from waste rock piles are relevant to processes occurring in waste overburden deposited in valley fills (Presser et al. 2004a,b).

A7.5 REFERENCES

Appalachian Center for the Economy and the Environment, Save Our Cumberland Mountains, Ohio Valley Environmental Coalition, Coal River Mountain Watch, Sierra Club. 2009. Toxic selenium: how mountaintop removal coal mining threatens people and streams. Lewisburg (WV, USA) http://www.sierraclub.org/coal/downloads/Seleniumfactsheet.pdf.

Pond GJ, Passmore ME, Borsuk FA, Reynolds I, Rose CL. 2008. Downstream effects of mountaintop coal mining: comparing biological conditions using family- and genus-level macroinvertebrate bioassessment tools. *J N Am Benthol Soc* 27:717–737.

Presser TS, Piper DZ, Bird KJ, Skorupa JP, Hamilton SJ, Detwiler SJ, Huebner MA. 2004. The Phosphoria Formation: a model for forecasting global selenium sources to the environment. In: Hein JR, editor, Life cycle of the Phosphoria Formation: from deposition to the post-mining environment. New York (NY, USA): Elsevier. p 299–319.

Presser TS, Hardy M, Huebner MA, Lamothe PJ. 2004. Selenium loading through the Blackfoot River watershed: linking sources to ecosystems. In: Hein JR, editor, Life cycle of the Phosphoria Formation: from deposition to post-mining environment. New York (NY, USA): Elsevier. p 437–466.

USEPA (U.S. Environmental Protection Agency). 2005. Programmatic environmental impact statement on mountaintop mining and associated valley fills in Appalachia (http://www.epa.gov/region3/mtntop/eis2005.htm).

USGS. 2008. Annual report for West Virginia, Mud River Reservoir data. Charleston (WV, USA): U.S. Geological Survey.

A8.0 KESTERSON RESERVOIR, SAN JOAQUIN VALLEY, CALIFORNIA

A8.1 SELENIUM SOURCE

The arid San Joaquin Valley in the southern Central Valley of California has undergone extensive agricultural development since the 1950s. Regional water supplies were increasingly diverted to agricultural and urban uses and away from natural wetlands within the valley. More than 90% of wetlands in the valley disappeared, severely limiting Pacific Flyway waterfowl use during annual migrations (Ohlendorf et al. 1990).

Beginning in 1949, as part of the Central Valley Project under the direction of the U.S. Bureau of Reclamation, plans were developed to build a large-scale subsurface drainage system to alleviate potential salinization of irrigation croplands. Funding was obtained in the 1970s, and planning and construction began on the San Luis Drain (SLD). The SLD was a canal intended to collect subsurface (tile) irrigation drainwater containing high levels of mineral salts and transport it to a yet-to-be determined discharge point in the San Francisco Bay (California).

Kesterson Reservoir, a series of a dozen shallow interconnected ponds totaling approximately 500 ha, was constructed adjacent to the SLD. As originally envisioned, the reservoir would have controlled flows within the completed drain system. Located at the temporary endpoint of SLD construction, the reservoir became, in effect, a series of terminal flow evaporative ponds. Although using agricultural drainage to supply wetlands in the valley was controversial because of mineral contaminants, the Kesterson National Wildlife Refuge (NWR) was established by cooperative agreement between the U.S. Fish and Wildlife Service and the U.S. Bureau of Reclamation to benefit wildlife populations and related recreational users.

In 1975, funding limitations and environmental concerns stopped construction of the SLD. Kesterson Reservoir was managed as a series of interconnected evaporation ponds and surrounding native grassland (Presser and Ohlendorf 1987; Ohlendorf et al. 1990; Schuler et al. 1990).

A8.2   FATE AND TRANSFORMATION

Elevated concentrations of Se in irrigation drainage entering Kesterson Reservoir were not initially detected because of a flawed sample preparation procedure performed prior to analysis (Presser and Ohlendorf 1987). Without the correct digestion, some forms of Se were lost in the analysis procedure. This delayed the identification of Se as a possible contaminant in drainage and allowed its discharge to continue, loading the ecosystem with an estimated 7900 kg of Se during the period from 1981 to 1985 (Presser and Piper 1998). By 1983, Se concentrations measured in mosquitofish ranged up to 247 mg/kg dw.

Water-column Se concentrations entering Kesterson Reservoir from the SLD in 1983 ranged from 140 to 1400 µg/L, averaging 340 µg/L. These Se concentrations were far in excess of concentrations (<2 µg/L) measured in the nearby Volta Wildlife Management Area, which was not impacted by agricultural drainage (Presser and Barnes 1984; Presser and Ohlendorf 1987). In terms of speciation, 98% of dissolved Se was selenate in the SLD and pond 2, but 20%–30% of dissolved Se was selenite in pond 11, the terminal pond in the evaporation pond series. Sediment Se concentrations in the SLD were exceptionally elevated (up to 210 mg/kg dw). Kesterson Reservoir aquatic organisms consumed by waterfowl, as well as species as diverse as amphibians, reptiles, and mammals utilizing the site, bioaccumulated elevated Se concentrations from incoming drainwater (Ohlendorf et al. 1988). Initial sampling showed Se concentrations moving through the food webs: filamentous algae, rooted plants, and net plankton, 35–85 mg/kg dw; insects and mosquitofish, 22–175 mg/kg dw. Highest mean Se concentrations occurred in midge (Diptera) larvae (139 mg/kg dw), dragonfly (Odonata) nymphs (122 mg/kg dw), damselfly nymphs (175 mg/kg dw), and mosquitofish (170 mg/kg dw) (Ohlendorf et al. 1986). These means were approximately 12 to 130 times those found at Volta Wildlife Management Area (a reference area). Sampling in 1984 showed species mean Se concentrations ranging from 1.5 to 170 mg/kg dw for Kesterson aquatic macrophytes, and from 18.6 to 102 mg/kg dw for aquatic insects (Schuler et al. 1990).

A8.3 Effects

In 1983, a local fish extirpation occurred involving up to 8 warm-water species, with only mosquitofish, a species that is relatively Se tolerant, remaining in the SLD and refuge (Saiki and Lowe 1987; Skorupa 1998). The SLD mosquitofish population, analyzed in 1984–1985, experienced impaired reproductive success, as documented by increased rates of stillborn fry with unabsorbed yolk sacs, and a 70% to 77% survival rate, compared to 97% to 99% survival rate for reference site fish (Saiki and Ogle 1995).

Deformity and death in embryos and hatchlings of aquatic bird populations were widespread; toxicity and immune deficiency contributed to the death of adult aquatic birds (Ohlendorf 1989; Skorupa 1998). Kesterson NWR waterfowl populations exhibited severe reproductive impairment, with over 40% of nests having one or more dead embryos, and nearly 20% having a malformed embryo or chick. Species affected initially included mallard ducks (*Anas platyrhyncos*), American coots (*Fulica americana*), avocets (*Recurvirostra avosetta*), eared grebes (*Podiceps nigricollis*), and black-necked stilts (*Himantopus mexicanus*). Subsequently, malformed young were observed in additional duck species. Developmental abnormalities were of types linked to excessive Se exposure, and included missing or abnormal eyes, wings, legs, feet, and beaks, with frequently more than one abnormality observed per specimen (Ohlendorf 1989; Skorupa 1998). Additionally, malformed major organs such as heart, brains, and liver were documented (Ohlendorf et al. 1986; Ohlendorf et al. 1988; Hoffman et al. 1988). Eared grebe eggs averaged 69.7 mg/kg dw and coot eggs 30.9 mg/kg dw. These 2 species contained the highest frequency of embryotoxicity, with 64% of nest affected. Ducks and stilts had lower frequencies of embryotoxicity (24% of nests affected). Mean egg Se from these species ranged from 6.85 to 28.2 mg/kg dw. By comparison, mean Se concentrations in eggs from the reference Volta Wildlife Management Area were generally less than 2 mg/kg dw, and no abnormal embryos were found.

By 1986, agricultural irrigation inputs were stopped, the SLD closed, and Kesterson NWR was designated as a contaminated site to be remediated. The ponds were filled and capped with clean soil in 1988, substantially limiting the wildlife exposures. The Kesterson experience subsequently led the U.S. Department of Interior to initiate screening at wildlife refuges associated with agricultural irrigation in 13 western U.S. states (Presser et al. 1994; Seiler et al. 2003).

A8.4 LESSONS LEARNED

The Kesterson Reservoir case history greatly expanded our understanding of the geologic origin of Se, the analytical methods necessary for the complete reduction of Se species, and the importance of speciation in determining outcomes in Se-contaminated ecosystems. Specifically, in terms of the San Joaquin Valley, documentation at Kesterson Reservoir led to understanding the ecological risks associated with irrigation practices in arid environments containing seleniferous soils or source rock strata such as marine shales. At Kesterson Reservoir, the following basic steps were identified: 1) Se lost from the water column of the ponds entered the food chain through uptake by biota; 2) biological processes, rather than thermodynamic processes, determined the reduction of Se and its consequent entry into the food web; and 3) the probability of embryo death or deformity was statistically related to Se concentrations in the egg, with both of these factors understood to be influenced by dietary Se levels.

The culmination of Se-associated avian reproductive impairment and abnormalities came to be known as the "Kesterson Syndrome" (Skorupa 1998). The associated phenomenon of biogeochemical translocation of toxic Se levels from source (sedimentary rock formations) via irrigation drainage conveyance to the aquatic ecosystem and ultimately to waterfowl came to be termed the "Kesterson Effect" (Presser 1994). Identification of the components in this biogeochemical pathway led to a reconnaissance effort within the western United States to look for areas where seleniferous agricultural wastewater was being discharged into wetland areas (Presser et al. 1994; Seiler et al. 2003).

Selenium exposures triggering reproductive impairment in waterfowl vary from site to site, which is explained in part by species differences in migration patterns, feeding preferences, and wildlife site fidelity (i.e., the tendency to remain in an area over an extended period or return to an area previously occupied). Site assessments should include relatively sedentary species (e.g., coots and grebes), and sampling should be performed near the end of the nesting season to ensure appropriate consideration of worst-case ecological risk (Ohlendorf et al. 1990; Fairbrother et al. 1999).

A8.5 REFERENCES

Fairbrother A, Brix KV, Toll JE, McKay S, Adams WJ. 1999. Egg selenium concentrations as predictors of avian toxicity. *Human Ecol Risk Assess* 5:1229–1253.

Hoffman DJ, Ohlendorf HM, Aldrich TW. 1988. Selenium teratogenesis in natural populations of aquatic birds in central California. *Arch Environ Contam Toxicol* 17:519–525.

Ohlendorf HM. 1989. Bioaccumulation and effects of selenium in wildlife. In: Jacobs LW, editor, Selenium in agriculture and the environment. Special Publication 23. Madison (WI, USA): American Society of Agronomy and Soil Science Society of America. p 133–177.

Ohlendorf HM, Hoffman DJ, Saiki MK, Aldrich TW. 1986. Embryonic mortality and abnormalities of aquatic birds: apparent impacts of selenium from irrigation drainwater. *Sci Total Environ* 52:49–63.

Ohlendorf HM, Kilness AW, Simmons JL, Stroud RK, Hoffman DJ, Moore JF. 1988. Selenium toxicosis in wild aquatic birds. *J Toxicol Environ Health* 24:67–92.

Ohlendorf HM, Hothem RL, Bunck CM, Marois KC. 1990. Bioaccumulation of selenium in birds at Kesterson Reservoir, California. *Arch Environ Contam Toxicol* 19:495–507.

Presser TS. 1994. "The Kesterson Effect." *Environ Manage* 18:437–454.

Presser TS, Barnes I. 1984. Selenium concentrations in waters tributary to and in the vicinity of the Kesterson National Wildlife Refuge, Fresno and Merced Counties, California. USGS Water-Resources Investigation Report 84-4122. Menlo Park (CA, USA): U.S. Geological Survey.

Presser TS, Ohlendorf HM. 1987. Biogoechemical cycling of selenium in the San Joaquin Valley, California. *Environ Manage* 11:805–821.

Presser TS, Piper DZ. 1998. Mass balance approach to selenium cycling through the San Joaquin Valley: from source to river to bay. In: Frankenberger WT Jr, Engberg RA, editors, Environmental chemistry of selenium. New York (NY, USA): Marcel Dekker. p 153–182.

Presser TS, Sylvester MA, Low WH. 1994. Bioaccumulation of selenium from natural geologic sources in the Western States and its potential consequences. *Environ Manage* 18:423–436.

Saiki MK, Lowe TP. 1987. Selenium in aquatic organisms from subsurface agricultural drainage water, San Joaquin Valley, California. *Arch Environ Contam Toxicol* 16:657–670.

Saiki MK, Ogle, RS. 1995. Evidence of impaired reproduction by western mosquitofish inhabiting seleniferous agricultural drainwater. *Trans Amer Fish Soc* 124:578–587.

Seiler RL, Skorupa JP, Naftz DL, Nolan BT. 2003. Irrigation-induced contamination of water, sediment, and biota in the Western United States—synthesis of data from the National Irrigation Water Quality Program: U.S. Geological Survey Professional Paper 1655. 123 p.

Schuler CA, Anthony RG, Ohlendorf HM. 1990. Selenium in wetlands and waterfowl foods at Kesterson Reservoir, California, 1984. *Arch Environ Contam Toxicol* 19:845–853.

Skorupa JP. 1998. Selenium poisoning of fish and wildlife in nature: lessons from twelve real-world examples. In: Frankenberger WT Jr, Engberg RA, editors, Environmental chemistry of selenium. New York (NY, USA): Marcel Dekker. p 315–354.

A9.0   TERRESTRIAL AND AQUATIC HABITATS, SAN JOAQUIN VALLEY, CALIFORNIA

Kesterson National Wildlife Refuge (NWR) ephemeral pools (1988 to the present) and Reuse Areas (1995 to present)

A9.1   SELENIUM SOURCES

Salt and Se in the soils, subsurface drainage, and groundwater aquifers of the western San Joaquin Valley accumulate as a result of the combination of arid climate, natural and agricultural runoff, and erosion from the marine sedimentary rocks of the California Coast Ranges (Presser and Luoma 2006).

Perhaps the best-known case of Se poisoning in a field environment was found at Kesterson NWR in the San Joaquin Valley of California (Presser and Ohlendorf 1987). There, deformity and death in embryos and hatchlings of aquatic bird populations were widespread, toxicity and immune deficiency contributed to the death of adult aquatic birds, multispecies warm-water fish assemblages disappeared, and a high incidence of still-born fry occurred in tolerant mosquitofish (Ohlendorf 1989; Skorupa 1998).

By the end of 1988, Kesterson Reservoir had been dewatered and the low-lying areas within the evaporation pond system filled with soil to at least 15 cm above the expected average seasonal rise of groundwater. The created terrestrial habitat at Kesterson NWR is used by such bird species as kestrel (*Falco sparverius*), owl (*Tyto alba*), shrike (*Lanius ludovicianus*), stilt (*Himantopus mexicanus*), and killdeer (*Charadrius vociferous*).

Current active management projects in the Valley include 1) conveyance of agricultural drainage to the San Joaquin River (see Grassland Bypass Project Case Study); and 2) pumping of stored drainage to the surface landscape (i.e., collectively called Reuse Areas). Most options present opportunities for wildlife use and human exposure. Over 42 species of birds have been found to use drainwater pilot Reuse Areas.

A9.2 FATE AND TRANSFORMATION

Kesterson NWR is now a mosaic of primarily terrestrial habitats that are less contaminated than the aquatic habitat they replaced (Ohlendorf and Santolo 1994). However, ephemeral pooling creates aquatic habitat in wet years. Agricultural drainage Reuse Areas exist on a pilot-scale level in the valley but are an important component of recently proposed management plans (Presser and Schwarzbach 2008). Planned Reuse Area expansion in the valley ranges from 3000 to 7700 ha. This would represent up to a 10-fold scale-up from the existing pilot project of 730 ha. The Reuse Areas integrate Se-contaminated drainage with terrestrial habitats by exposing the landscape to an agricultural waste stream. In addition to concern about day-to-day management of such large-scale operations, there is also the potential for adverse effects related to Se-contamination with ponding and creation of transitory aquatic habitats 1) during rainfall events and 2) when management actions call for maintaining flow.

Because these habitats are transitory in nature, data sets characterizing the ecosystem are not complete. Monitoring of bird eggs does take place, and data for water-column Se concentrations are collected from time to time, driven mainly by rainfall events. In 2000, ephemeral pools at Kesterson NWR showed water-column Se concentrations from 15 to 247 µg/L (USBR 2000). Available food items appearing in these temporary pools and their Se concentrations included the following: aufwuchs (biofilm), 11 to 190 mg/kg dw; *Daphnia* (*Cladocera*), 35 to 77 mg/kg dw; corixids (*Heteroptera*), 8 to 21 mg/kg dw; beetles (*Coleoptera*) 14 to 149 mg/kg dw; and backswimmers (*Notonectidae*), 2.3 to 24 mg/kg dw.

The Red Rock Ranch integrated drainage management pilot project, a Reuse Area, includes traditional cropland, tree plantations, halophyte cropland, and a solar evaporator basin (Skorupa 1998). In 1996, a sample of the water impounded in the solar evaporator contained more than 11,000 µg/L Se. Water entering the halophyte plot during the spring of 1996 averaged approximately 1,600 µg/L. In 2004, the Panoche Drainage District included an accidentally flooded pasture within the Reuse Area (Presser and Schwarzbach 2008). Here, concentrated agricultural drainage (60 to 200 µg/L Se) was being managed to reduce the amount of Se discharged to the San Joaquin River.

A9.3 EFFECTS

Aquatic dietary items for birds at Kesterson NWR exceeded 6 mg/kg dw, a level considered a high risk for reproductive impairment (USDOI 1998). Risk assessments took into account such factors as 1) birds being able to feed on a mixture of aquatic invertebrates, terrestrial invertebrates, or terrestrial plants; 2) the transitory nature of ponding; and 3) the proportion of aquatic habitat to terrestrial habitat available for bird use. Bird egg Se concentrations ranged from 1.6 to 7.4 mg/kg dw for kestrel, swallow (*Hirundinidae*), starling (*Sturnidae*), killdeer, king bird (*Tyrannidae*), and shrike (*Laniidae*) (USBR 2000). No evidence of toxicological impact on these species of birds was observed at Kesterson Reservoir during monitoring from 1988 to 2000.

In 1996, ponding of drainage water inflow at the solar evaporator and adjacent halo-phyte plot at the Red Rock Ranch pilot project was great enough to attract breeding shorebirds (Skorupa 1998). During the short period of ephemeral ponding and spring nesting, a viable food web was formed, birds nested and fed on contaminated dietary items, and eggs developed. Nonviability of eggs was 67% and the level of teratogenic-ity (56.7%) surpassed the level found at Kesterson Reservoir during the 1980s.

In 2004, in the Panoche Drainage District pilot project, the average Se concen-tration in avocet and stilt eggs was 58 mg/kg dw. A reduction of hatchability and deformities of bird embryos were expected to occur at these concentrations (USDOI 1998; Skorupa 1998). From a compilation of data from 2003 through 2006 from the Panoche Drainage District Reuse Area, Se concentrations in bird eggs were consis-tently above concentrations associated with Se toxicity to embryos during those 4 years (Presser and Schwarzbach 2008). Selenium concentrations in avocets and stilts in 2006 exceeded 90 mg/kg dw, higher than during the flooding event of 2003.

A9.4 LESSONS LEARNED

Ephemeral pools and aquatic habitats form efficiently in agricultural areas affected by a high water table. Data taken during accidental flooding events illustrate the fact that even small ponds are inviting to aquatic birds in the San Joaquin Valley given the limited habitat opportunities. To put the magnitude of contamination in context, Table A.1 compares the mean bird egg Se concentrations for the Panoche Reuse

TABLE A.1
Comparison of the Mean Egg Se Concentrations (mg/kg dw bird egg) at the Panoche Reuse Area to Those at Kesterson Reservoir (1983–1985)

Panoche *Reuse Area*			
Species	2003	2004	2005	2006
Killdeer	12.5	13.1	15.9	22.8
Avocets & Stilts	39.0	15.3	35.3	23.0
Blackbirds	5.9	6.0	N/A	8.8

Kesterson Reservoir			
Species	1983	1984	1985
Killdeer	N/A	33.1	46.4
Avocets & Stilts	16.1	20.9	34.6
Blackbirds	N/A	6.0	N/A

Area, 2003–2006, to the mean bird egg Se concentrations for Kesterson Reservoir, 1983–1985 (Ohlendorf et al. 1986; Presser and Schwarzbach 2008).

Note that, except for killdeer, the results for the Panoche Reuse Area are effectively no different from the results for Kesterson Reservoir (see Kesterson Reservoir, California, case study). The timing of appearance of food webs in the ponds and immediate nesting also illustrate the need for vigilance on the part of management for this type of habitat. Selenium concentrations in bird eggs from the majority of reference sites sampled were also highly elevated, suggesting a landscape effect larger than the Reuse Area as management and storage of concentrated drainwater takes place over several years.

A9.5   References

Ohlendorf HM. 1989. Bioaccumulation and effects of selenium in wildlife. In: Jacobs LW, editor, Selenium in agriculture and the environment. Special Publication 23. Madison (WI, USA): American Society of Agronomy and Soil Science Society of America. p 133–177.

Ohlendorf HM, Santolo GM. 1994. Kesterson Reservoir—Past, present, and future: an ecological risk assessment. In: Frankenberger WT Jr, Engberg RA, editors, Environmental chemistry of selenium. New York (NY, USA): Marcel Dekker, p 69–117.

Ohlendorf HM, Hoffman HJ, Saiki MK, Aldrich TW. 1986. Embryonic mortality and abnormalities of aquatic birds: apparent impacts of selenium from irrigation drainwater. *Sci Total Environ* 52:49–63.

Presser TS, Ohlendorf HM. 1987. Biogeochemical cycling of selenium in the San Joaquin Valley, California. *Environ Manage* 11:805–821.

Presser TS, Luoma SN. 2006. Forecasting selenium discharges to the San Francisco Bay-Delta Estuary: ecological effects of a proposed San Luis drain extension. Professional Paper 1646. Menlo Park (CA, USA): U.S. Geological Survey. Available at http://pubs.usgs.gov/pp/p1646.

Presser TS, Schwarzbach SE. 2008. Technical analysis of in-valley drainage management strategies for the Western San Joaquin Valley, California: Open-File Report 2008–1210. Menlo Park (CA, USA): U.S. Geological Survey.

Skorupa JP. 1998. Selenium poisoning of fish and wildlife in nature: lessons from twelve real-world examples. In: Frankenberger WT Jr, Engberg RA, editors, Environmental chemistry of selenium. New York (NY, USA): Marcel Dekker. p 315–354.

[USBR] U.S. Bureau of Reclamation. 2000. Ecological risk assessment for Kesterson Reservoir. Report prepared by CH2MHill and Lawrence Berkeley Laboratory, San Francisco, CA, USA.

[USDOI] U.S. Department of the Interior. 1998. Guidelines for interpretation of the biological effects of selected constituents in biota, water, and sediment. Irrigation Water Quality Program Report 3. Denver (CO, USA): National Bureau of Reclamation, Fish and Wildlife Service, Geological Survey, Bureau of Indian Affairs.

A10.0   GRASSLAND BYPASS PROJECT (SAN JOAQUIN VALLEY), CALIFORNIA

A10.1   Selenium sources

Sources of Se in the San Joaquin Valley, California, are ancient marine sediments of the Coast Ranges to the west of the valley (Presser et al. 1994; Schwartz et al. 2003). Selenium in the alluvium is mobilized via the naturally eroding lithology during rain events in the wet season (October–May). Selenium moves with the flood water through the few streams, such as Panoche Creek, of the arid foothills to the hydrologically altered agricultural fields, and into ditches, canals, and sloughs, finally draining into the San Joaquin River.

The 38,800-ha Grassland Drainage Area is to the west of the Kesterson National Wildlife Refuge (see Kesterson Reservoir case study, Section A8) and to the north of the agricultural area that disposed of drainage into the San Luis Drain and consequently Kesterson Reservoir. Irrigated agriculture in the Grassland Drainage Area requires the use of subsurface drains to move water away from water-logged crop root zones.

Prior to 1995, drainage from the agricultural fields was sent into the Grassland Ecological Area (GEA), a wetland of worldwide importance, part of the San Luis National Wildlife Refuge complex in the vicinity of the San Joaquin River. The wastewater provided a source of water for the wetlands, saving irrigation water for other uses. The wastewater was alternated with fresh water flowing through the wetland channels (e.g., Mud and Salt Sloughs). Selenium bioaccumulation and trophic transfer were documented in organisms inhabiting the GEA (USBR et al. 2004-2005).

In 1995, the discharge was consolidated into a 45-km segment of the original San Luis Drain (renamed the Grassland Bypass) in order to reduce contaminated water supplies, but discharge to the San Joaquin River via Mud Slough remains (Figure A.4) (CSWQCB 2001). Drainage is greatest throughout the long growing season (January through October).

A10.2   FATE AND TRANSFORMATION

The San Joaquin Valley is an important region for migratory birds, although most of the wetland areas in the valley have been drained and converted to agricultural and urban uses. Duck (hunting) clubs and wildlife refuges provide habitat for birds on the Pacific Flyway. As the Grassland Bypass Project is near to the Kesterson National Wildlife Refuge, it is understood that species affected include aquatic and aquatic-dependent species.

Selenium in this system is predominantly selenate (Presser and Luoma 2006). As a flowing water system, the San Joaquin River may be less sensitive to Se effects (especially if selenate dominates inputs) than adjacent riparian wetlands, where residence times and biogeochemical transformations of selenate are more likely. Wherever there is ponded water, there is the possibility of the accumulation of inorganic Se in a more biologically available form.

A10.3   EFFECTS

Since the Grassland Bypass Project began in 1995, physical parameters (pH, conductivity, temperature, and flow) and Se have been monitored in the San Luis Drain, Mud and Salt Slough, the channels leading into the GEA, and the San Joaquin River (Figure A.4) (USBR et al., 1998 and ongoing). Selenium is also measured in sediment accumulating in the drain and in organisms (invertebrates, fish, bird eggs, and vascular plants). Toxicity tests (short-term chronic testing of 4 or 7 days using survival,

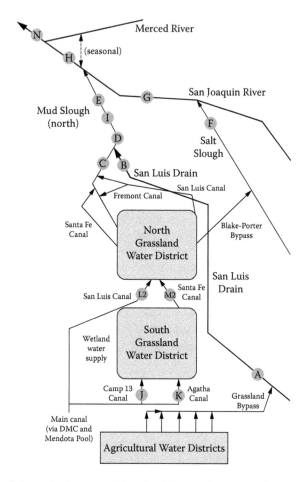

FIGURE A.4 Schematic diagram of Grassland Bypass Project moving agricultural drain-water to the San Joaquin River, CA (Adapted from USBR et al. 1998 and ongoing). Letters indicate sampling locations throughout the project area.

growth, or reproduction endpoints) are conducted with water from all sampling sites using an alga, an invertebrate, and a fish. Selenium concentrations in water in the inlet channels to the GEA and in water and organisms in Salt Slough have decreased to a "moderate hazard to wildlife" Se level (Lemly 1993), but water and organisms in Mud Slough and downstream to the San Joaquin have risen to above a "hazard to wildlife" Se level (USBR et al. 1998 and ongoing, 2004–2005).

The San Joaquin River is now being restored, but the degraded conditions as a result of agricultural drainage discharges to the river over time have led to a predominance of invasive fish species over native fish species.

A10.4   LESSONS LEARNED

Once Se-laden inputs were removed from the GEA, the associated wetlands began to recover. Elevated Se concentrations occur in the surrounding area, however, including Mud Slough. By the time the Grassland Bypass Project ends, considerable data will be available regarding the rates of recovery of this wetland-and-slough system.

The Grassland Bypass Project pioneered a new regulatory structure for agricultural drainage in the United States. The Se load discharged through the Grassland Bypass Project is regulated both by a contract with the Bureau of Reclamation and by a Waste Discharge Requirement from the California Central Valley Regional Water Quality Control Board (CSWQCB 1998a). This is the first case in the United States where quantitative load limits have been imposed on contaminants contained in agricultural drainage (a "nonpoint source discharge"). As a result of this project and its associated regulatory structure, Se loads from the Grassland Bypass Project have decreased from approximately 3636 kg/yr to 1818 kg/yr. Future load restrictions are designed to meet the water-quality criteria for the protection of aquatic life for the San Joaquin River (5 µg/L Se) and for the GEA (2 µg/L; CSWQCB 1998b, 2001; USEPA 1992).

A10.5   References

[CSWQCB] California State Water Quality Control Board. 1998a. Waste discharge requirements for San Luis and Delta-Mendota Authority and US Department of Interior Bureau of Reclamation Grassland Bypass Channel Project. Order No. 98-171. Sacramento (CA, USA): Central Valley Regional Water Quality Control Board.

[CSWQCB] California State Water Quality Control Board. 1998b. The water quality control plan (Basin Plan) for the Sacramento River Basin and San Joaquin River Basin. 4th edition. Sacramento (CA, USA): Central Valley Regional Water Quality Control Board.

[CSWQCB] California State Water Quality Control Board. 2001. Waste discharge requirements for San Luis and Delta-Mendota Authority and US Department of Interior Bureau of Reclamation Grassland Bypass Channel Project. Order No. 5-01-234. Sacramento (CA, USA): Central Valley Regional Water Quality Control Board.

Lemly AD. 1993. Guidelines for evaluating selenium data from aquatic monitoring and assessment studies *Environ Monitor Assess* 28:83–100.

Presser TS, Luoma SN. 2006. Forecasting selenium discharges to the San Francisco Bay-Delta Estuary: ecological effects of a proposed San Luis drain extension. Professional Paper 1646. Menlo Park (CA, USA): US Geological Survey. Available at http://pubs.usgs.gov/pp/p1646.

Presser TS, Sylvester MA, Low WH. 1994. Bioaccumulation of selenium from natural geologic sources in the Western States and its potential consequences. *Environ Manage* 18:423–436.

Schwartz H, Sample J, Weberling KD, Minisini D, Moore JC. 2003. An ancient linked fluid migration system: cold-seep deposits and sandstone intrusions in the Panoche Hills, California, USA. *Geo-Mar Lett* 23:340–350.

[USBR] United States Bureau of Reclamation et al. 1998 and ongoing. Grassland Bypass Project monthly, quarterly, and annual reports. Mid-Pacific Region, Sacramento, CA, USA. http://www.sfei.org/grassland/reports/gbppdfs.htm.

[USBR] United States Bureau of Reclamation et al. 2004-2005. Grassland Bypass Project annual report for 2004-2005. Mid-Pacific Region, Sacramento, CA, USA. http://www.sfei.org/grassland/reports/gbppdfs.htm.

[USEPA] United States Environmental Protection Agency. 1992. Rulemaking, water quality standards, establishment of numeric criteria for priority toxic pollutants, States' compliance: Final Rule, 57 FR 60848 (December 22, 1992).

A11.0   SAN FRANCISCO BAY-DELTA ESTUARY, CALIFORNIA

A11.1   SELENIUM SOURCES

Major sources of Se in the San Francisco Bay-Delta Estuary (Bay-Delta) (Figure A.5) are 1) discharges of irrigation drainage conveyed from agricultural lands of the western San Joaquin Valley into the San Joaquin River; 2) effluents from North Bay refineries, which refine crude oil from the western San Joaquin Valley, along with crude oil from other sources; and 3) Sacramento River inflows that dominate the freshwater contribution (high water volume) to the estuary (Presser and Luoma 2006).

The volume of water flowing into the Bay-Delta is determined by climate and water management. On a broad scale, the Bay-Delta watershed is characterized by a distinct seasonal cycle of high inflows from the Sacramento and San Joaquin rivers in January through June (high flow season), followed by lower inflows through the last

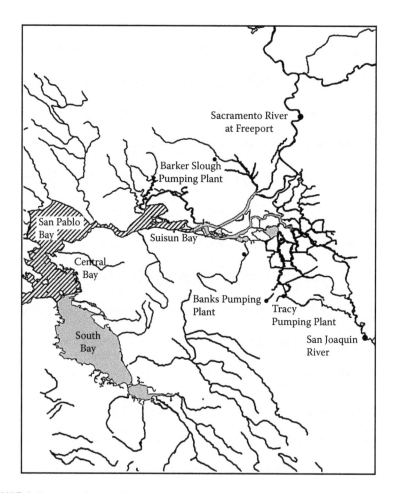

FIGURE A.5   Map of North San Francisco Bay. (From Tetra Tech 2008.)

6 months of the calendar year (low flow season). Riverine influences also depend on water year type (wet year or dry year).

A11.2   FATE AND TRANSFORMATION

The San Joaquin-Sacramento River system and the internal refinery inputs move into an estuarine environment. In the Bay-Delta in the 1980s, speciation differed among source waters (Cutter and San Diego-McGlone 1990). The Sacramento River inflow was 30% to 70% selenate, depending on season; organo-Se was the other main component. The San Joaquin River inflow was 70% selenate and 22% organo-Se. Refinery wastewaters averaged 62% selenite. In the late 1980s, during low flow in the North Bay, as much as 50% of the Se was selenite, reflecting the predominance of refinery inputs. In the late 1990s, studies showed less selenite in the North Bay after refinery inputs were reduced because of regulatory limits on both the Se concentration and load and the installation of additional treatment technology. However, organo-Se increased in abundance, reflecting recycling of the metalloid in the estuary. Selenite plus organo-Se constituted 60% of the mass of Se (Cutter and Cutter 2004).

A11.3   EFFECTS

Selenium contamination documented from 1982 to the mid-1990s was sufficient to threaten reproduction in key species within the Bay-Delta estuary ecosystems (Presser and Luoma 2006). The aquatic or aquatic dependent wildlife species in San Francisco Bay with the greatest potential to have been affected included white sturgeon (*Acipenser transmontanus*), starry flounder (*Platichthys stellatus*), Dungeness crab (*Cancer magister*), diving ducks (scoters [*Melanitta* spp.], and scaups [*Aythya* spp.]).

Human health advisories were posted based on Se concentrations in tissues of diving ducks. From 1989 to 1990 in the North Bay, average Se concentrations in surf scoter liver tissue and sturgeon flesh exceeded reproductive toxicity guidelines (Heinz 1996; Lemly 2002; Presser and Luoma 2006). Analyses from 1982 through 1996 showed that the animals with the highest Se tissue concentrations from the North Bay ingested bivalves (*Corbicula fluminea* prior to 1986 and *Potamocorbula amurensis* in subsequent samplings) as a major component of their diet. Selenium concentrations in the predominant bivalve in the Bay-Delta were higher in the mid-1990s than in 1977 through 1990 (Linville et al. 2002; Presser and Luoma 2006). Selenium concentrations in *P. amurensis* reached 20 mg/kg dw in the North Bay in October 1996, exceeding by 6-fold the USDOI (1998) dietary guideline (>3 mg/kg dw) of concern.

A11.4   LESSONS LEARNED

Prior to refinery cleanup, Se concentrations in the Bay-Delta were well below the most stringent recommended water quality criterion (1 μg/L). Enhanced biogeochemical transformation to bioavailable particulate Se and efficient bioaccumulation by bivalves characterize the Bay-Delta ecosystem (Luoma et al. 1992; Stewart et al. 2004; Presser and Luoma 2006). The results of this research are being incorporated into the regulatory limits for Se discharges to the estuary (http://www.waterboards .ca.gov/sanfranciscobay/water_issues). A first-of-its-kind aquatic-dependent wildlife criterion for Se also is under development for this region as a result of the risk Se poses to estuarine systems (Beckon and Maurer 2008).

A11.5 REFERENCES

Beckon WN, Maurer TC. 2008. Species at risk from selenium exposure in the San Francisco estuary. Final Report to the U.S. Environmental Protection Agency. Inter-Agency Agreement No. DW14922048-01-0. Sacramento (CA, USA): US Fish and Wildlife Service.

Cutter GA, San Diego-McGlone MLC. 1990. Temporal variability of selenium in fluxes in San Francisco Bay. *Sci Total Environ* 97/98:235–250.

Cutter GA, Cutter LS. 2004. Selenium biogeochemistry in the San Francisco Bay estuary: changes in water column behavior: *Estuar Coast Shelf Sci* 61:463–476.

Heinz GH. 1996. Selenium in birds. In: Beyer WN, Heinz GH, Redmon-Norwood AW, editors, Environmental contaminants in wildlife: interpreting tissue concentrations. New York (NY, USA): Lewis. p 447–458.

Lemly AD. 2002. Selenium assessment in aquatic ecosystems: a guide for hazard evaluation and water quality Criteria. New York (NY, USA): Springer-Verlag.

Linville RG, Luoma SN, Cutter LS, Cutter GA. 2002. Increased selenium threat as a result of invasion of the exotic bivalve *Potamocorbula amurensis* into the San Francisco Bay-Delta. *Aquat Toxicol* 57:51–64.

Luoma SN, Johns C, Fisher NS, Steinberg NA, Oremland RS, Reinfelder J. 1992. Determination of selenium bioavailability to a benthic bivalve from particulate and solute pathways. *Environ Sci Technol* 26:485–491.

Presser TS, Luoma SN. 2006. Forecasting selenium discharges to the San Francisco Bay-Delta Estuary: ecological effects of a proposed San Luis drain extension. Professional Paper 1646. Menlo Park (CA, USA): U.S. Geological Survey. Available at http://pubs.usgs .gov/pp/p1646.

Stewart RS, Luoma SN, Schlekat CE, Doblin MA, Hieb KA. 2004. Food web pathway determines how selenium affects ecosystems: a San Francisco Bay case study. *Environ Sci Technol* 38:4519–4526.

Tetra Tech. 2008. Technical memorandum #4: Conceptual model of Se in North San Francisco Bay. Prepared for the California Regional Water Quality Control Board. Alameda, CA, USA.

[USDOI] U.S. Department of the Interior. 1998. Guidelines for interpretation of the biological effects of selected constituents in biota, water, and sediment. Irrigation Water Quality Program Report 3. Denver (CO, USA): National Bureau of Reclamation, Fish and Wildlife Service, Geological Survey, Bureau of Indian Affairs.

A12.0 PHOSPHATE MINING IN THE UPPER BLACKFOOT RIVER WATERSHED, IDAHO

A12.1 SELENIUM SOURCES

The Meade Peak Member of the Phosphoria Formation extends throughout southeast Idaho, and adjacent areas of Wyoming, Montana, and Utah (McKelvey et al. 1959, 1986; Presser et al. 2004a). Outcrops occur over a vast part of that area as a result of folding, faulting, uplift, and subsequent erosion of younger deposits. The area supports phosphate mining, livestock grazing, fishing, and hunting. Over the last half of the 20th century, mining in Idaho provided approximately 4.5% of world demand for phosphate, used mainly in fertilizer (USDOI and USGS 2000). About 49% of the total production has occurred since 1985. This tonnage represents approximately 15% of the estimated 1 billion tonnes accessible to surface mining within the Phosphoria Formation (USDOI and USDA 1977). Out of 19 mining sites in Idaho, 4 are presently active (Dry Valley Mine, Enoch Valley Mine, Rasmussen Ridge Mine, and Smoky Canyon Mine) and 2 are categorized as existing mine operations (Maybe Canyon Mine and Lanes Creek Mine) (Presser et al. 2004a). The Phosphoria Formation also is estimated to have generated about 3×10^{10} tonnes of oil.

The Meade Peak Member contains Se concentrations of up to 1200 mg/kg dw. Average Se concentrations range from 48 to 560 mg/kg dw in westernmost Wyoming and in southeast Idaho (McKelvey et al. 1986; Piper et al. 2000). Selenium is dispersed throughout the deposit, but achieves its highest concentration in a waste-shale zone between 2 major phosphate-ore zones of the Meade Peak Member. The waste-shale beds are phosphate lean but enriched in organic carbon compared to the ore zones (Herring and Grauch 2004). The lower-ore zone is about 12 m thick, the waste zone 27 m thick, and the upper-ore zone 5 m thick, each of which approximately maintains its thickness over 21,500 km^2.

The upper Blackfoot River watershed in southeast Idaho receives drainage from 11 of 16 phosphate mines that have extracted ore from the Phosphoria Formation (Presser et al. 2004a,b). Ten of the active and legacy mines are now Superfund sites (http://www.greateryellowstone.org/media/pdf/phosphate-superfund-map.pdf).

A12.2   FATE AND TRANSFORMATION

Mining removes phosphate-rich beds and exposes organic carbon-rich waste rock to subaerial weathering (Presser et al. 2004a). Waste rock is generated at a rate of 2.5 to 5 times that of mined ore (USDOI and USDA 1977). Individual dumps contain 5-65 million tonnes of waste rock that is either contoured into hills, used as cross-valley fill, or used as back-fill in mine pits. In terms of Se chemistry, when Se hosted by organic matter in source rocks is exposed to the atmosphere and surface and groundwater, Se is oxidized from relatively insoluble selenide and elemental Se to soluble oxyanions of selenite and selenate (Piper et al. 2000).

The cross-valley fills at the Smoky Canyon mine (45 million tonnes) and South Maybe Canyon mine (27 million tonnes) are stabilized with under-drains (Presser et al. 2004a). Discharges from these drains are source waters for Pole Creek and Maybe Creek. Concentrations of Se in these 2 drains, as well as in a dump seep at the inactive Conda Mine, were equal to or exceeded 1000 µg/L.

In the upper Blackfoot River watershed, where the majority of mines are located, Se concentrations in streams draining both active and inactive mines contained up to 400 µg/L. Waste-rock dump seeps and surface streams exhibit annual cycles in Se concentration that peak during the spring period of maximum flow (Montgomery Watson 1998, 1999, 2000, 2001a,b; Presser et al. 2004a,b). In 1998, the stream Se concentration maximum for all samples collected in May was 260 µg/L, whereas it was 32 µg/L in September. In May 2000, the Se concentration in East Mill Creek was 400 µg/L and was 19 µg/L in September 1999.

Samples collected in 2001 at 2 mining areas (Mackowiak et al. 2004) showed mean Se concentrations in forage plants (legume and grass) grown on waste-rock dumps exceeded thresholds of dietary toxicity for horses (5–40 mg/kg dw) and sheep (5–25 mg/kg dw). Location within a dump site, as well as the species of plant, were factors in determining Se concentrations in vegetation. For example, in 2001, alfalfa bioaccumulated Se to a greater extent (mean 150 mg/kg dw, maximum 952 mg/kg dw) than grasses (mean 27 mg/kg dw, maximum 160 mg/kg dw). Maximum Se concentrations in grass and mean and maximum Se concentrations in legume would qualify the plant material itself, regardless of dietary considerations, as hazardous based on the criterion for a hazardous Se solid waste (100 mg/kg ww, or 143 mg/kg dw at 30% moisture) (USDHHS 1996).

A12.3 EFFECTS

Livestock deaths from Se toxicosis have been documented, including deaths of horses and sheep (Presser et al. 2004a). Permits for grazing have been suspended for some mine-disturbed areas (Idaho Department of Environmental Quality 2002). Embryo deformities in birds have occurred, most notably in the American coot and the Canada goose; a Se concentration of 80 mg/kg dw is the highest reported Se concentration among coot eggs (Presser et al. 2004a; Skorpa et al. 2002). Aquatic invertebrate samples in the area also contained record elevated amounts of Se (up to 788 mg/kg dw) (Skorupa et al. 2002).

Deer, elk, and moose populations also may be at risk based on studies of Se concentrations in livers donated by hunters in the areas near the phosphate waste dumps (there is a direct correlation between elevated Se concentrations in the liver versus distance of the harvested elk from the nearest mine site) (Idaho Department of Environmental Quality 2002).

Selenium concentrations in submerged macrophytes reached a maximum of 56 mg/kg dw in spring and 44 mg/kg dw in fall (Montgomery Watson 1998, 1999, 2000, 2001a,b). Benthic macroinvertebrates showed a maximum Se concentration of 150 mg/kg dw in spring and 63 mg/kg dw in fall. Based on tissue, whole-body Se concentrations in forage fish (suckers, sculpins, minnows, and salmonids <15 cm) exceeded the USDOI (1998) risk threshold of 6 mg/kg dw for growth and survival during both seasons (maxima: 35 mg/kg dw in spring and 11 mg/kg dw in fall). These values also exceeded the USDOI (1998) risk threshold for diet, if these fish were eaten by larger fish. Concentrations of Se in gamefish (>15 cm) showed a maximum skin-on fillet concentration of 33 mg/kg dw in spring and 17 mg/kg dw in fall. Depending on the conversion factor used (Piper et al. 2000), whole-body Se concentrations in game-fish would reach 55–77 mg/kg dw in spring. These included Yellowstone cutthroat trout (*Oncorhynchus clarki bouvieri*), brook trout (*Salvelinus fontinalis*), and rainbow trout (*Oncorhynchus mykiss*). Trout have disappeared from the most Se-contaminated streams (e.g., East Mill Creek, dissolved Se > 20 ug/L; whole-body cutthroat trout 52 mg Se/kg dw) (Montgomery Watson 2001a,b; Hamilton and Buhl 2004). The proportion of trout above the 2 mg/kg ww Se guideline for human consumption of fish flesh (USDHHS 1996; USDOI 1998) increased from <1% in fall to 30% in spring.

In 1999, a significant larval salamander die-off (at least 250 individuals) at a pit lake at the Gay Mine was confirmed as Se poisoning (Green 2001). A Se concentration of 126 mg/kg dw measured in salamander tail tissue is a record for that type of tissue. Samples of dead larval salamanders from the Smoky Canyon Mine were collected in 2000. Two important diseases were diagnosed: chronic selenosis and iridovirus infection (Green and Miller 2001). It was concluded that Se contributed directly to their deaths by poisoning or indirectly by causing damage to the immune system.

Two human health advisories for Se have been issued for activities in the phosphate mining areas (Idaho Department of Environmental Quality 2003). The first was a hunter's advisory that recommended limited consumption of elk liver by area hunters (Idaho Department of Fish and Game 2000; Idaho Bureau of Environmental Health and Safety 2001). The second advisory recommended limited consumption of fish from East Mill Creek by children based on elevated Se concentrations observed in edible fish tissue from this stream (ATSDR 2002).

A12.4   LESSONS LEARNED

Large-scale land disturbance in southeast Idaho leaves behind waste rock that provides many routes of Se exposure for fish, birds, and livestock. The average Se concentration of the Meade Peak Member of the Phosphoria Formation is an order of magnitude higher than those of other exploited marine shales that have been linked to incidences of Se toxicosis via oil refining and irrigation in the western United States. The addition of phosphorites as a category of Se-containing rocks to that of other carbon-rich source rocks enabled a forecast of global Se sources (Presser et al. 2004b).

A12.5   REFERENCES

[ATSDR] Agency for Toxic Substances and Disease Registry. 2002. Health consultation: Selenium in fish streams of the upper Blackfoot River watershed, southeast Idaho selenium project, Soda Springs, Caribou County, Idaho. Atlanta, GA, USA.

Green DE. 2001. Final diagnostic report for Case #16322, 001-022: Gay Mine tiger salamander larvae. Madison (WI, USA): National Wildlife Health Center, Biological Resources Division, U.S. Geological Survey.

Green DE, Miller KJ. 2001. Final diagnostic report for Case #16947, 001-003: Smoky Canyon Mine tiger salamander larvae and garter snake. Madison (WI, USA): National Wildlife Health Center, Biological Resources Division, U.S. Geological Survey.

Hamilton SJ, Buhl KJ. 2004. Selenium in water, sediment, plants, invertebrates and fish in the Blackfoot River drainage. *Water Air Soil Pollut* 159:3–34.

Herring JR, Grauch RI. 2004. Lithogeochemistry of the Meade Peak Phosphatic Shale Member of the Phosphoria Formation, southeast Idaho. In: Hein JR, editor, Life cycle of the Phosphoria Formation: from deposition to the post-mining environment. New York (NY, USA): Elsevier p 321–366.

Idaho Bureau of Environmental Health and Safety. 2001. Health consultation: evaluation of selenium in beef, elk, sheep and fish in the Southeast Idaho Phosphate Resource Area. Pocatello (ID, USA): Idaho Department of Health and Welfare.

Idaho Department of Environmental Quality. 2002. Area wide human health and ecological risk assessment and related memorandum. Report prepared by R Clegg. Boise (ID, USA): Tetra Tech EM Inc.

Idaho Department of Environmental Quality. 2003. Area wide risk management plan, area wide investigation: Southeast Idaho Phosphate Mining Resource Area. Soda Springs, ID, USA.

Idaho Department of Fish and Game. 2000. Elk liver consumption advisory, Unit 76. Pocatello, ID, USA.

Mackowiak CL, Amacher MC, Hall JO, Herring JR. 2004. Uptake of selenium and other elements into plants and implications for grazing animals in Southeast Idaho. In: Hein JR, editor, Life cycle of the Phosphoria Formation: from deposition to the post-mining environment. New York (NY, USA): Elsevier. p 527–555.

McKelvey VE, Williams JS, Sheldon RP, Cressman ER, Cheney TM, Swanson RW. 1959. The Phosphoria, Park City, and Shedhorn Formations in the western phosphate field. Professional Paper 313-A. Denver (CO, USA): U.S. Geological Survey.

McKelvey VE, Strobell JD, Slaughter AL. 1986. The Vanadiferous Zone of the Phosphoria Formation in western Wyoming and southeastern Idaho. Professional Paper 1465. Denver (CO, USA): U.S. Geological Survey.

Montgomery Watson. 1998. Fall 1997 interim surface water survey report, Southeast Idaho phosphate resource area, Selenium Project. Steamboat Springs, CO, USA.

Montgomery Watson. 1999. Final 1998 Regional investigation report, Southeast Idaho phosphate resource area, Selenium Project. Steamboat Springs, CO, USA.

Montgomery Watson. 2000. 1999 Interim investigation data report, Southeast Idaho phosphate resource area, Selenium Project. Steamboat Springs, CO, USA.

Montgomery Watson. 2001a. Draft 1999–2000 regional investigation data report for surface water, sediment and aquatic biota sampling activities, May–June 2000, Southeast Idaho phosphate resource area, Selenium Project. Steamboat Springs, CO, USA.

Montgomery Watson. 2001b. Draft 1999–2000 regional investigation data report for surface water, sediment and aquatic biota sampling activities, September 1999, Southeast Idaho Phosphate resource area, Selenium Project. Steamboat Springs, CO, USA.

Piper DZ, Skorupa JP, Presser TS, Hardy MA, Hamilton SJ, Huebner M, Gulbrandsen RA. 2000. The Phosphoria Formation at the Hot Springs Mine in southeast Idaho: a source of trace elements to ground water, surface water, and biota. Open-File Report 00-050. Menlo Park (CA, USA): U.S. Geological Survey.

Presser TS, Piper DZ, Bird KJ, Skorupa JP, Hamilton SJ, Detwiler SJ, Huebner MA. 2004a. The Phosphoria Formation: a model for forecasting global selenium sources to the environment. In: Hein JR, editor, Life cycle of the Phosphoria Formation: from deposition to the post-mining environment. New York (NY, USA): Elsevier, p 299–319.

Presser TS, Hardy MA, Huebner MA, Lamothe P. 2004b. Selenium loading through the Blackfoot River watershed: linking sources to ecosystems. In: Hein JR, editor, Life cycle of the Phosphoria Formation: from deposition to the post-mining environment. New York (NY, USA): Elsevier. p 437–466.

Skorupa JP, Detwiler SJ, Brassfield R. 2002. Reconaissance survey of selenium in water and avian eggs at selected sites within the Phosphate Mining Region near Soda Springs Idaho: May–June 1999. Sacramento (CA, USA): U.S. Fish and Wildlife Service. Available at http://wwwrcamnl.wr.usgs.gov/Selenium/library.htm.

[USDHHS] United States Department of Health and Human Services. 1996. Toxicological profile for selenium. Altanta (GA, USA): Agency for Toxic Substances and Disease Registry.

[USDOI] United States Department of the Interior. 1998. Guidelines for interpretation of the biological effects of selected constituents in biota, water, and sediment. Irrigation Water Quality Program Report 3. Denver (CO, USA): National Bureau of Reclamation, Fish and Wildlife Service, Geological Survey, Bureau of Indian Affairs.

[USDOI and USDA] United States Department of the Interior and US Department of Agriculture. 1977. Final environmental impact statement: development of phosphate resources in southeastern Idaho. Vol. I. Washington (DC, USA): US Government Printing Office.

[USDOI and USGS] United States Department of the Interior and US Geological Survey. 2000. Minerals yearbook, metals and minerals. Vol. I. Washington (DC, USA): US Government Printing Office.

Appendix B: Commentary: Persistence of Some Fish Populations in High-Se Environments*

Steven Canton

CONTENTS

B1.0 INTRODUCTION

Selenium (Se) is an essential micronutrient required by most aquatic and terrestrial species in order to maintain metabolic functions (USEPA 2004). Selenium occurs at trace concentrations in virtually all environmental media, including rocks, soils, water, and living organisms. Anthropogenic activities, such as irrigation of

---

* This appendix comprises material presented by the author to the workshop attendees that is referenced in Chapter 6 but that does not itself form part of the consensus (the main text of this book) arising from this workshop.

seleniferous soils, creation of waste-rock drainage at coal and phosphorus mines, fly-ash pond discharges from coal-fired power plants, and discharges from oil refineries, can result in elevated Se concentrations in aquatic ecosystems (Lemly 1997a). However, "elevated" Se concentrations also occur naturally in areas of marine shale deposits, especially where those shales occur near shallow groundwater and surface water flows (ARCADIS 2006).

Given that Se is an essential micronutrient, aquatic organisms readily assimilate organic forms of Se (e.g., selenomethionine), yet frequently are not able to excrete Se at the same rate of consumption at elevated concentrations. As noted in deBruyn et al. (2008), direct toxic effects have been measured in adult organisms via decreased survival or growth and in young by decreased survival (reproductive success), growth, or increased occurrences of larval deformities. Additionally, the margin between required concentrations and those that may become toxic is narrow, perhaps as low as 1 order of magnitude for some vertebrate species, and highly variable within and between species, making toxic thresholds difficult to define (DeForest 2008).

While a tissue-based standard is likely more relevant than a water-column-based value (deBruyn et al. 2008), this approach raises a potential issue: Are chronic Se threshold tissue concentrations predictive of impacts in the field? "Ground-truthing" of Se-effects research is not a common analysis and has rarely been examined with stream fish communities (Canton and Baker 2008).

Relevant questions when evaluating how laboratory-based thresholds might predict impacts to natural fish populations include:

- What are the effects of prior exposure on fish population responses?
- Compensatory response involves functional adaptations in the organisms under toxicological stress (Hayes 2008). Thus, is Se-related mortality compensatory; i.e., is Se-related mortality "masked" by natural population-level mortality rates among the various life stages?
- Are Se-related impacts buffered or mitigated by migration (immigration and emigration) from unexposed populations?
- Can increased Se sensitivity of juveniles versus adults be found in natural populations through examination of life history tables or age class evaluations?
- Can Se impacts be separated from potential effects of other contaminants occurring in concert with Se?
- Can population-related impacts from Se be masked by non-Se factors, such as density-related effects? Changes in inter-specific interactions? Seasonal effects? Habitat quality? Hydrologic flow regime?

Obviously, no single study can address all of these issues. The majority of field-based research to date has documented severe population-level Se impacts (elimination of fish populations) in lakes and reservoirs receiving coal fly-ash discharges (Bryson et al. 1984; Garrett and Inman 1984; Lemly 1985; Chapter 3; Appendix A).

There are few population-level evaluations of fish from lakes or streams with naturally elevated Se inputs. Population-level effects have been more clearly documented in lentic than in lotic habitats (Orr et al. 2006). This may be a result of more organic-rich sediments in lentic systems, thereby promoting conversion to organic Se compounds (Van Derveer and Canton 1997). Such organoselenium compounds rapidly accumulate up the food chain (Brix et al. 2005). In highly oxic states common to flowing waters, selenate is less likely to be reduced to forms that partition to sediments, thus reducing the likelihood of organoselenium compound formation (Brix et al. 2005).

One other issue with field evaluation of tissue Se thresholds is related to conclusions that egg tissue is the best predictive tissue for Se toxicity (deBruyn et al. 2008; DeForest 2008). Unfortunately, few data exist on egg or ovary concentrations for multiple fish species from field studies. As such, the analyses provided in this commentary necessarily rely on comparisons of fish population metrics to whole-body Se concentrations—which is consistent with USEPA (2004).

In summary, this commentary focuses on Se-related trends observed in a portion of the Arkansas River in south-central Colorado that maintains a warm-water stream fish community in an area with natural geologic Se sources and elevated Se concentrations (GEI 2007). Although this study does not provide definitive "answers" as to whether tissue thresholds can predict field effects, it does provide useful insights as to how warm-water stream fish populations behave in areas with naturally elevated Se concentrations—and how this response is potentially tempered by habitat, flow, interaction between species, and possibly other factors not measured or considered.

B2.0 ARKANSAS RIVER—STUDY BACKGROUND

The objectives of this study were to collect aquatic biological data, determine background Se concentrations, and define physical habitat characteristics of a portion of the Arkansas River and nearby tributaries in the vicinity of the city of Pueblo, Colorado (Figure B.1), to evaluate relationships between Se concentrations and fish populations (GEI 2007). Data were collected from fall 2004 to fall 2006 from the Arkansas River, Fountain Creek, Wildhorse Creek, and the St. Charles River (Figure B.1). We combined these data to evaluate potential population-level effects due to elevated Se.

Data collected to accomplish the study goals included:

1) Seasonal sampling over a period of 2 to 3 years of fish and macroinvertebrate populations for species composition and relative abundance of aquatic organisms at 10 sites showing a wide range of Se concentrations;
2) Collection of whole-body fish tissue, composite macroinvertebrate tissue, and sediment samples for the analysis of Se bioaccumulation pathways with whole-body fish tissue Se concentrations converted to dry weight using sample-specific percent solids data;
3) Evaluation of fish collected at all sites for potential Se-related deformities, including abnormal spots, fin erosion, tail erosion, spinal deformities, lesions, and red fins.
4) Collection of monthly water column samples by City of Pueblo personnel as part of the larger Se load allocation study; all samples were analyzed for total Se and dissolved sulfate.

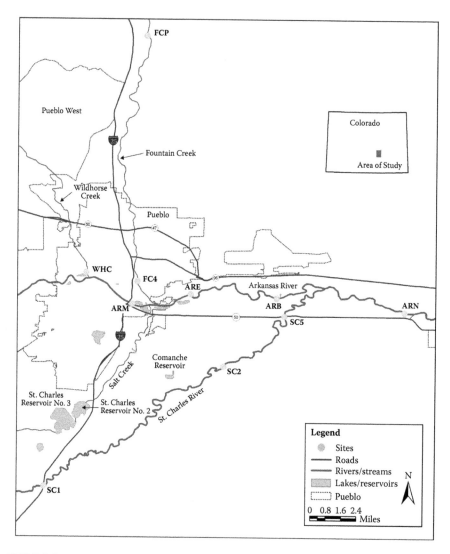

FIGURE B.1 Location of study sites on the Arkansas River, Wildhorse Creek, Fountain Creek, and the St. Charles River, 2004–2006.

5) Physical habitat measurements, including sediment particle size, to determine relationships between the biota and their environment. In addition, habitat types, number of habitat units, total length, average wetted area width, average depth, total length of eroded bank, average percentage of permanent vegetation, and flow conditions were also measured (GEI 2007).

Not all sites could be sampled for all data during all sampling episodes due to changes in flow, access, site location, scope, or changes resulting from review of past data. Those sites or dates where data are not available are noted.

An accompanying analysis had also been conducted to evaluate the sources of Se within the reach of the Arkansas River from upstream of the city of Pueblo to the confluence with the Huerfano River (ARCADIS 2006). The purpose of that study was to determine whether irrigation of the Se-rich lands along the Arkansas River was contributing significantly to the elevated Se levels noted within this reach. Analysis of surface and groundwater sources of Se clearly indicated that the amount of Se discharged into the Arkansas River from nonpoint sources, including groundwater, as a result of irrigation return flows was small (<10%) relative to other natural sources (ARCADIS 2006). Wildhorse Creek and Fountain Creek were the largest contributors of Se to this portion of the Arkansas River. The Pueblo Water Reclamation Facility, a permitted source, was the seventh largest contributor of Se. However, Se from this source enters the facility through the infiltration of groundwater into the sanitary sewer system and is, therefore, also considered a natural source of Se (ARCADIS 2006).

In addition, groundwater Se concentrations in shale-influenced zones were estimated as being 100 times higher than concentrations in the alluvial zones, regardless of irrigation practices in those areas (ARCADIS 2006). The results of this study indicated that Se in this reach of the Arkansas River and tributaries occurs from natural sources, with insignificant human-induced additions to total loading. This is an important consideration in the evaluation of effects of elevated Se to fish populations. Since elevated concentrations of Se are naturally occurring at the study sites, this study provided a unique evaluation of Se impacts to fish populations that have been exposed naturally to elevated Se for potentially thousands of years.

B2.1   SUMMARY OF FISH POPULATIONS

B2.1.1   Arkansas River Mainstem Sites

Twenty-one species of fish and 1 hybrid fish were collected in the Arkansas River during the 2005 and 2006 sampling efforts (Table B.1). The number of fish species collected at each site ranged from 9 to 14. Only white suckers were collected from every site during both years. Density estimates at Arkansas River sites ranged from 238 fish/acre to 1,300 fish/acre (Figure B.2). No single fish species dominated in terms of density across all the Arkansas River sites, but central stonerollers, flathead chubs, red shiners, sand shiners, and smallmouth bass were numerically dominant at one or more sites and sampling times and were generally common in relatively large numbers throughout this reach of the Arkansas River (GEI 2007).

Length-frequency data indicated multiple age groups were present for most of the species found at the sites. "Condition" parameters (Fulton condition factor K and relative weight Wr), also showed that the fish collected were generally healthy in regards to being an appropriate weight for their length (GEI 2007).

The few abnormalities noted for fish from the Arkansas River sites (caudal lesions and black spotted fins) were probably due to parasitic infection, not Se exposure (Lagler 1956; Herman 1990). A single central stoneroller collected from Site ARM in 2006 had a deformed tail and a single fathead minnow collected from Site ARE in 2005 was missing one eye. This low percentage of deformities (< 6%) is likely related to factors other than Se (Lemly 1997b).

TABLE B.1

Fish Density Estimates for Sites on the Arkansas River Mainstem, Fall 2005 and 2006. Site Locations Per Figure B.1

| Fish Species | | | Density (Number/Acre) | | | | | | | |
| | | | 2005 Sampling | | | | 2006 Sampling | | | |
Family	Common Name	Scientific Name	ARM	ARE	ARB	ARN	ARM	ARE	ARB	ARN
Catostomidae	Longnose sucker	*Catostomus catostomus*	110	—	12	7	—	—	—	5
	White sucker	*Catostomus commersoni*[a]	186	208	186	64	30	50	49	30
Centrarchidae	Bluegill	*Lepomis macrochirus*	135	—	—	—	50	—	20	—
	Green sunfish	*Lepomis cyanellus*[a]	219	6	—	6	30	—	10	—
	Green sunfish × bluegill hybrid	Not applicable	—	—	—	—	30	—	—	—
	Largemouth bass	*Micropterus salmoides*	203	19	—	39	—	50	59	43
	Orangespotted sunfish	*Lepomis humilus*[a]	203	—	—	—	10	—	—	—
	Smallmouth bass	*Micropterus dolomieu*	228	47	35	32	110	—	—	—
	White crappie	*Pomoxis annularis*	8	—	—	13	—	—	—	—
Cyprinidae	Central stoneroller	*Campostoma anomalum*[a]	—	13	244	6	380	—	118	—
	Common carp	*Cyprinus carpio*	—	—	47	13	—	—	—	15
	Fathead minnow	*Pimephales promelas*[a]	—	32	221	26	20	60	—	—
	Flathead chub	*Platygobio gracilis*[a]	—	231	23	26		460	196	5
	Longnose dace	*Rhinichthys cataractae*[a]	8	—	35	—	60	10	20	—
	Red shiner	*Cyprinella lutrensis*[a]	—	33	81	286	—	420	118	101
	Sand shiner	*Notropis stramineus*[a]	—	334	93	102	—	—	69	10
Fundulidae	Plains killifish	*Fundulus zebrinus*[a]	—	—	35	—	—	10	—	—

TABLE B.1 (CONTINUED)

Fish Density Estimates for Sites on the Arkansas River Mainstem, Fall 2005 and 2006. Site Locations Per Figure B.1

| | Fish Species | | Density (Number/Acre) | | | | | | | |
| | | | 2005 Sampling | | | | 2006 Sampling | | | |
Family	Common Name	Scientific Name	ARM	ARE	ARB	ARN	ARM	ARE	ARB	ARN
Ictaluridae	Black bullhead	*Ameiurus melas*	—	—	—	—	30	10	—	—
	Channel catfish	*Ictalurus punctatus*[a]	—	—	12	—	—	10	—	—
	Yellow bullhead	*Ameiurus natalis*[a]	—	—	—	6	—	—	—	—
Percidae	Saugeye	*Stizostedion canadense*	—	—	—	—	—	10	—	10
Poeciliidae	Mosquitofish	*Gambusia affinis*		—	—	13	—	—	10	19
Total density			1,300	923	1,024	638	750	1,090	669	238
# of native species			4	7	9	8	6	7	7	4
Total # of species			9	9	12	14	10	10	10	9

[a] Native to the Arkansas River basin.

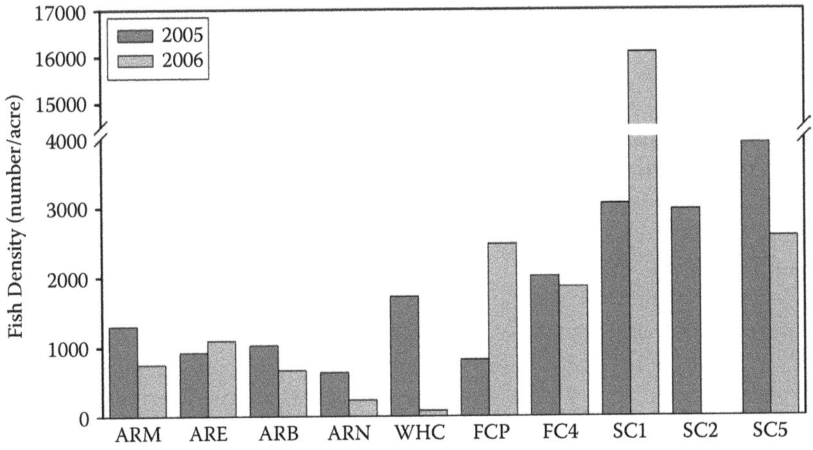

FIGURE B.2 Fish density estimates (number/acre) at sites on the Arkansas River, Wildhorse Creek, Fountain Creek, and St. Charles River in fall 2005 and 2006 (* = no data collected). Site locations per Figure B.1.

Several fish species collected in this reach of the Arkansas River appear to have established viable, reproducing populations, with juveniles and adults present at one or more of the Arkansas River sites during the study (GEI 2007). However, others were much more variable over the 2 years of sampling, with some species present in high numbers in 1 year and either absent or in low numbers in the other year (Table B.2). These data suggest that some species possibly moved out of these sites or were temporarily eliminated from them over the 2-year span of the study, while other species became established. The hydrologic differences between years may have made the habitat more homogenous and less hospitable to certain fish species or displaced organisms.

B2.1.2 Tributary Sites

Wildhorse Creek: Two fish species, the central stoneroller and white sucker, were collected in 2005 and 2006 from the site on Wildhorse Creek (WHC). Both of these species were common in the Arkansas River sites. Central stonerollers were numerically dominant during both years of sampling (Table B.2). Density estimates at Site WHC were 20 times higher in 2005 than they were in 2006 (Figure B.2). Both species contributed to this decrease, with central stonerollers decreasing by 91% and white suckers by 99% between 2005 and 2006. Despite the large differences between years, the age class distribution of central stonerollers was similar between years, indicating a reproducing population that included both juvenile and adult fish in both years (GEI 2007).

The 2 fish species at Site WHC were in fair condition based on their length-weight ratios (GEI 2007) and within the ranges observed at the Arkansas River sites. No abnormalities were observed for fish collected in 2005, but an eroded caudal fin and spotted fins were observed on 1 central stoneroller each in 2006. As with the Arkansas River sites, the differences between the 2 years of study in Wildhorse Creek are likely correlated to the increased water levels and accompanying habitat changes that occurred at this site, which appear to have had a more pronounced effect at Site WHC, possibly due to the smaller stream size (GEI 2007).

Fountain Creek: Nine species were collected from the 2 Fountain Creek sites in 2005 and 2006, all of which were species that had also been collected from the Arkansas River sites (Table B.2). Densities at these 2 sites were generally higher than densities at the Arkansas River sites (Figure B.2). Flathead chubs dominated the fish populations numerically at both sites, composing 66 to 85% of the total density. Length-frequency analysis of the flathead chubs indicated that the populations were reproducing, with juvenile and older adult fish present in relatively high numbers at both sites and years (GEI 2007). Fish condition index values indicated that most species of fish present in Fountain Creek were in only fair condition (GEI 2007), possibly indicating that food resources were limited for some species. The only abnormality or injury observed at these sites was a tail lesion noted on a single flathead chub, potentially indicating a parasitic infection.

St. Charles River: Eleven species of fish were collected from the St. Charles River sites between 2005 and 2006 (Table B.2). All of the species present in the St. Charles River sites were also present in one or more of the Arkansas River sites. The number of species collected at each site ranged from 5 to 10. No trend toward higher or lower

TABLE B.2
Fish Density Estimates for Sites on Wildhorse Creek, Fountain Creek, and the St. Charles River, Fall 2005 and 2006. Same Format as Table B.1

| | | | Density (Number/Acre) | | | | | | | | | | |
| | Fish Species | | 2005 Sampling | | | | | | 2006 Sampling | | | | |
Family	Common Name	Scientific Name	WHC	FCP	FC4	SCI	SC2	SC5	WHC	FCP	FC4	SCI	SC5
Catostomidae	White sucker	*Catostomus commersoni*[a]	794	74	15	51	—	780	5	124	11	833	—
Centrarchidae	Bluegill	*Lepomis macrochirus*	—	—	—	—	—	—	—	—	—	83	
	Green sunfish	*Lepomis cyanellus*[a]	—	—	—	—	—	8	—	—	—	83	—
	Largemouth bass	*Micropterus salmoides*	—	—	—	—	—	16	—	—	—	—	—
Cyprinidae	Central stoneroller	*Campostoma anomalum*[a]	926	107	15	1,061	20	764	81	94	198	7,250	95
	Fathead minnow	*Pimephales promelas*[a]	—	8	—	667	2,495	228	—	—	55	1,333	27
	Flathead chub	*Platygobio gracilis*[a]	—	598	1,330	61	202	1,433	—	2,093	1,385	—	14
	Longnose dace	*Rhinichthys cataractae*[a]	—	8	—	—	—	220	—	51	11	167	—
	Red shiner	*Cyprinella lutrensis*[a]	—	25	15	242	—	323	—	—	99	3,500	770
	Sand shiner	*Notropis stramineus*[a]	—		618	889	40	134	—	114	77	2,333	1,662
Fundulidae	Plains killifish	*Fundulus zebrinus*[a]	—	—	—	81	212	8	—	—	22	500	14
Ictaluridae	Black bullhead	*Ameiurus melas*[a]	—	—	15	—	—	—	—	—	—	—	—
Total density			1,720	819	2,010	3,052	2,969	3,914	86	2,476	1,858	16,082	2,582
# of native species			2	6	6	7	5	9	2	5	8	8	6
Total # of species			2	6	6	7	5	10	2	5	8	9	6

[a] Native to the Arkansas River basin.

species richness was observed between years. Densities were higher than those seen at the Arkansas River sites and other tributary sites (Figure B.2). No single species numerically dominated all of the St. Charles River sites, but central stonerollers, fathead minnows, flathead chubs, and sand shiners were dominant at sites for one or both years (Figure B.2). Length-frequency analysis of the fish populations at the St. Charles River sites indicated that sites had reproducing populations of central stonerollers, fathead minnows, and sand shiners, with juvenile and adult fish collected during both years (GEI 2007). Compared to sites on the other streams, more abnormalities and injuries were observed in the St. Charles River. Black-spotted fins were observed on several of the minnow species, likely the result of parasitic infections rather than Se effects (Lagler 1956; Herman 1990). Additionally, a shortened opercule, a deformed jaw, and a growth on the underside of the mouth (potential Se-related deformities) were noted in 1 flathead chub each in 2005 (GEI 2007). These 3 fish composed less than 1% of the fish collected at this site.

B2.1.3   Fish Population Summary

Composition of fish populations in the Arkansas River, Wildhorse Creek, Fountain Creek, and the St. Charles River within the study area were similar, with the same species present in the tributary sites as were found in the Arkansas River mainstem sites. Population parameters such as density and biomass varied substantially between sites and years. Such changes are likely linked to the higher stream flows present in 2006 and significant habitat changes due to beaver activity at some sites. Variable population parameters are not uncommon in plains streams with highly variable flow regimes and habitat conditions (Schlosser 1987). When observed, the few abnormalities noted at the sites were more consistent with signs of parasitic infections than increased Se levels (Lagler 1956; Herman 1990). There is also the potential for "survivor bias" with regard to deformities that is inherent in analyses of field populations—i.e., Se-related deformity effects could be occurring, but masked by mortality (Lemly 2002a).

B2.2   Habitat Evaluation Summary

Habitat at the sites on the Arkansas River and Fountain Creek were more similar to each other than the sites on Wildhorse Creek and the St. Charles River (GEI 2007). The segments of the Arkansas River and Fountain Creek within the study area were wider and had higher velocities than sites on the other 2 streams. They also had higher amounts of bank cover and some sites had backwater areas, which provide increased habitat diversity for fish (GEI 2007). The sites on Wildhorse Creek and the St. Charles River had more limited cover and less habitat complexity.

Substrate composition has many significant impacts on the aquatic biota in streams. Heterogeneous substrate provides a variety of microhabitats for organisms to colonize, resulting in a diverse macroinvertebrate community. Fish feeding, spawning, and habitat requirements are likewise affected by the types of substrate available in streams. Substrate in the Arkansas River, Wildhorse Creek, Fountain Creek, and St. Charles River was relatively homogenous and tended to be dominated by sand at most sites (Figure B.3). Streambeds composed largely of fine substrates,

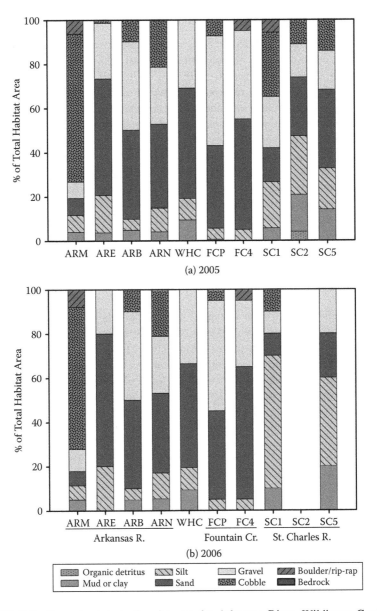

FIGURE B.3 Percent substrate for sites on the Arkansas River, Wildhorse Creek, St. Fountain Creek, and the St. Charles River in fall of a) 2005 and b) 2006.

such as sand and silt, are typical for plains streams and are generally less stable (Cross and Moss 1987). The 2 sites on the St. Charles River were the only sites to show substantial changes in substrate composition over the 2 study years. As previously discussed, beaver activity altered the habitat noticeably at both sites between 2005 and 2006. As a result, the percentage of substrate composed of silt was over twice as high at those 2 sites in 2006 in comparison to 2005.

B2.3 SELENIUM CONCENTRATION ANALYSES

Sediment Se: Fine sediment mean Se concentration was generally up to 2 times greater than coarse sediment Se concentration at all sites (GEI 2007), with values in fine sediments reaching over 20 mg/kg dw. When all sites were included in a regression analysis, a positive relationship between total mean sediment Se concentration (mean includes coarse and fine sediment concentrations), mean dissolved Se water column concentration, and mean sediment total organic carbon (TOC) (mean includes coarse and fine sediment percentages) was observed (Figure B.4), similar to that reported by Van Derveer and Canton (1997).

Water column Se: Total Se water column concentrations were generally elevated throughout the study area, with only the upper reaches on the St. Charles River and Fountain Creek having mean Se concentrations below the current USEPA chronic standard of 5 µg/L Se (Table B.3). Se concentrations measured during the biological study at the Wildhorse Creek site were more than 20 times greater than all of the other biological sampling locations, with a mean concentration of 418 ± 115 µg/L. Even the minimum concentration measured at Site WHC (315 µg/L) was 7 times greater than the maximum Se concentration measured at other study sites (44 µg/L at SC5).

The mean Se concentration increased in the Arkansas River downstream of the confluence with Fountain Creek. Mean Se concentrations were generally greater in the St. Charles River than the Arkansas River (Table B.3). The farthest downstream site (SC5) had the second highest mean Se concentration of 20 ± 13 µg/L. Although mean Se concentrations were high, variability in concentrations was also high (Table B.3), which suggests high Se exposure to aquatic life via the water column may not be constant.

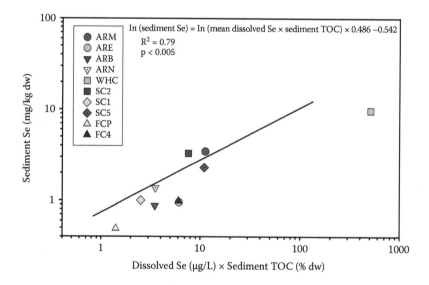

FIGURE B.4 Relationship between mean total sediment Se concentration and the mean dissolved Se-sediment total organic carbon (TOC) interaction.

TABLE B.3
Total Se Concentration Summary for Paired Surface Water and Biological (GEI 2007; Figure B.1) Sampling Locations

Water Quality Sampling Location	GEI Sampling Location	Start	End	n	Total Se (µg/L)		
					Min	Max	Mean ± SD
Arkansas River							
AR-17 Arkansas River at Moffat	ARM	1/4/05	8/24/06	15	2.33	14.1	7.05 ± 3.69
AR-19 Arkansas River at CS-10	ARE	1/6/05	2/25/06	9	4.48	15.5	10.6 ± 4.06
AR-18 Offtake at Excelsior Ditch	ARB	1/13/05	10/15/05	7	3.02	13.4	8.72 ± 4.00
AR-27 Arkansas River at Nyberg	ARN	1/6/05	10/15/05	8	4.64	11.7	8.81 ± 2.85
Tributaries							
AR-09 WP-118 Wildhorse near mouth	WHC	1/4/05	6/8/06	17	315	715	418 ± 115
FC-05 Fountain Creek above Pinon	FCP	1/5/05	2/25/06	9	1.72	4.93	3.43 ± 1.05
FC-09 Fountain Creek above confluence	FC4	1/5/05	6/8/06	12	7.49	21.1	12.1 ± 4.34
AR-36 St. Charles River at I-25	SCI	1/22/05	6/8/06	6	1.43	4.75	3.09 ± 1.37
AR-22 St. Charles River below Bessemer	SC2	1/14/05	6/8/06	11	2.64	19.8	11.7 ± 6.22
AR-23 St. Charles River at Hwy. 50	SC5	1/6/05	6/8/06	13	3.74	43.6	20.3 ± 13.0

Invertebrate Tissue Se: Mean invertebrate Se tissue concentrations varied by season, year, and site and ranged from 6.0 to 45.5 mg/kg dw (GEI 2007). Individual tissue composite sample concentrations ranged from less than the minimum detection limit to 90.4 mg/kg dw (Figure B.5). Although mean macroinvertebrate Se concentrations varied by site and watershed, only the mean macroinvertebrate Se concentration for Site WHC was significantly different (higher) than other sites ($p = 0.001$; Figure B.5).

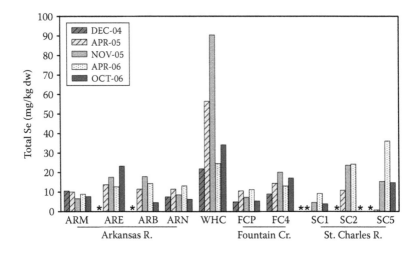

FIGURE B.5   Total Se concentration of composite benthic macroinvertebrate tissues ($n = 1$) for each sampling date and location (* = no data collected).

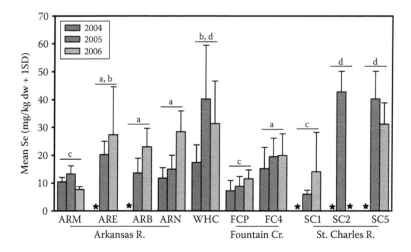

FIGURE B.6 Mean whole-body fish tissue Se concentration by site and year. Means equal the arithmetic mean of all species and replicates sampled that year. Different letters indicate significant differences between site means (1-way ANOVA on site means followed by Bonferroni, $p < 0.005$) (* = no data collected).

Fish tissue Se: Mean Se concentrations in whole-body fish tissue generally increased from the most upstream sampling location to the most downstream location within each watershed. Statistical differences between sites and years were observed (1-way ANOVA, $p < 0.001$), with lower fish Se concentrations observed during the 2004 sampling than both 2005 and 2006 (GEI 2007). No statistical differences were observed between the farthest upstream sites in the Arkansas River, Fountain Creek, and St. Charles River (Figure B.6). Furthermore, the Arkansas River sites downstream of the confluence with Fountain Creek were not significantly different from the Fountain Creek site located immediately upstream from the confluence (FC4).

The greatest whole-body fish tissue concentrations were collected from the 2 downstream St. Charles River sites. Equally high Se concentrations were measured in fish collected from Wildhorse Creek (Figure B.7). Fish tissue Se concentrations at the farthest upstream sites on the Arkansas River, Fountain Creek, and the St. Charles River were all statistically similar ($p > 0.05$; Figure B.7) and were generally among the lowest fish tissue concentrations observed.

Se tissue concentrations varied noticeably by fish family. Se tissue concentrations were measured for 3 cyprinids (central stoneroller, sand shiner, and red shiner), 1 catostomid (white sucker), and 3 centrarchids (green sunfish, smallmouth bass, and largemouth bass) (Figures B.7 and B.8). Mean concentrations in all cyprinids were greater (21.1 mg/kg dw; SE = 1.4) than either centrarchids (19.7 mg/kg dw; SE = 1.3) or catostomids (17.5 mg/kg dw; SE = 1.5). However, there was no statistical difference between the overall mean for the 3 families ($p = 0.19$). Most mean

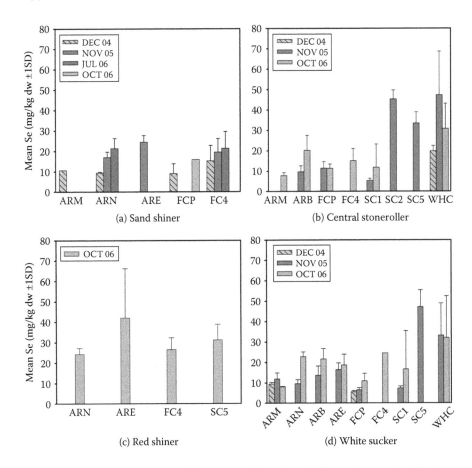

FIGURE B.7 Mean whole-body Se concentration of cyprinid species sampled for tissue analyses, including a) sand shiner, b) central stoneroller, and c) red shiner and 1 catostomid represented by the d) white sucker. Absence of a bar indicates samples were not collected during that sampling event.

whole-body Se concentrations were well above the USEPA (2004) chronic tissue criterion of 7.9 mg/kg dw.

B.2.3.1   Cross-Media Se Relationships

Media with measured Se concentrations and sediment TOC collected simultaneously with fall fish population sampling were analyzed in an all possible regression analysis (Hintze 2000) to determine the parameters potentially affecting bioaccumulation pathways for Se throughout the study area. For the fall data, site-mean fish tissue Se concentrations (all species combined) were most influenced by macroinvertebrate Se tissue concentrations, followed by coarse sediment Se

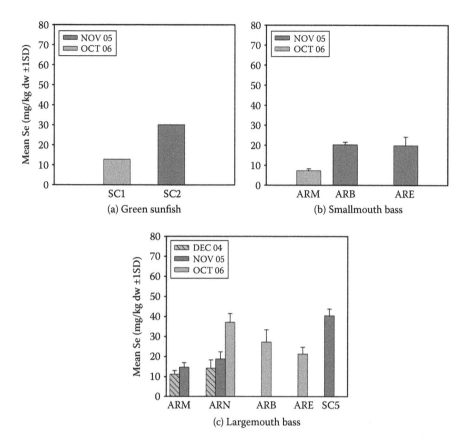

FIGURE B.8 Mean whole-body Se concentration of centrarchid species sampled for tissue analyses, including a) green sunfish, b) smallmouth bass, and c) largemouth bass. Absence of a bar indicates samples were not collected during that sampling event.

concentration and sediment TOC. Of these 3 variables, only macroinvertebrate Se had a significant positive relationship with whole-body fish tissue concentration ($R^2 = 0.38$; $p = 0.001$) for the fall data (Figure B.9). Although these 3 variables were most influential, together they explained only 45% of the variability in fish tissue Se concentrations.

Fish population metrics were also evaluated relative to fish tissue Se concentration. Fish taxa richness appeared to have a negative relationship with increasing mean tissue Se concentration (Figure B.10, Part A). Likewise, fish density also appeared to have a marginally significant ($p = 0.10$; Figure B.10, Part C) negative trend with increases in fish tissue Se concentration, although when fish density was weighted by habitat availability, this relationship was not even marginally significant ($p = 0.69$; Figure B.10, Part B).

Rather than strong relationships with tissue Se concentrations, substrate conditions, represented by percent silt and boulder/rip rap, explained most of the variability in total fish density weighted by habitat ($R^2 = 0.44$, 2-parameter model).

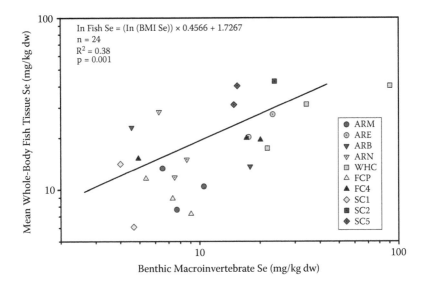

FIGURE B.9   Relationship between benthic macroinvertebrate (BMI) Se tissue concentration and mean whole-body fish tissue Se concentration.

Of the two significant parameters, percent silt had the most influence on total density and exhibited a slightly significant positive relationship (R^2 = 0.37; Figure B.11). Although silt by itself likely does not directly contribute to greater fish densities, high silt may be an indicator of higher primary production that could influence food availability or may be related to increasingly diverse habitat that contains slow backwater or eddy refugia.

This trend was also evaluated for individual fish families. There was a slight negative relationship between densities of cyprinid species and their whole-body Se concentrations (Figure B.12, Part A), although this relationship was not statistically significant (R^2 = 0.04; p = 0.40). However, when habitat parameters, substrate characteristics, and Se data were included in an all possible regressions analysis, the strongest 2-parameter model again suggests percent silt and whole-body Se together have the greatest influence on cyprinid densities (Figure B.12). A 3-parameter model of percent silt, whole-body Se, and percent sand explains 57% of the variability in the densities of cyprinid species, indicating a potential habitat-related confounding factor in the evaluation of potential Se effects.

Grouping species-specific data by the family Catostomidae results in trends that are similar to trends observed with cyprinids (Figure B.13). Catostomid densities (equivalent to white sucker density) were not significantly correlated with whole-body Se concentrations (R^2 = 0.02; p = 0.65). Rather, a positive relationship was once again observed with percent silt (R^2 = 0.25; p = 0.05).

A significant negative relationship was observed between whole-body Se and species-specific densities for the family Centrarchidae (R^2 = 0.53; p = 0.02; Figure B.14, Part A). However, densities of centrarchid species were also significantly correlated with primary habitat average depth (R^2 = 0.74; p < 0.001;

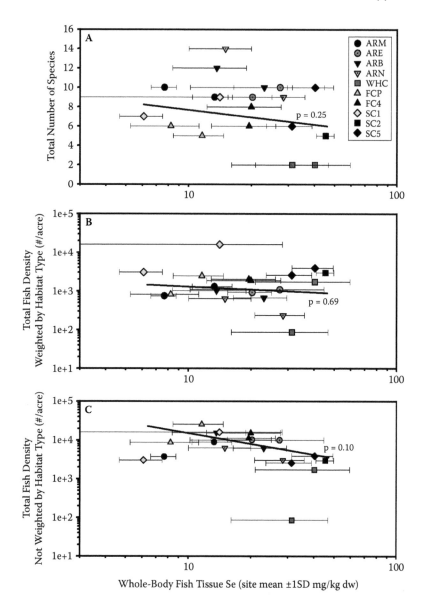

FIGURE B.10 Relationships between mean whole-body fish tissue Se concentration (mean includes all species) and a) total number of species identified at the site, b) total fish density weighted by habitat, and c) total fish density not weighted by habitat.

Figure B.14, Part B); greater population densities were associated with greater habitat depths. A 3-parameter model identified primary habitat depth, percent sand, and percent cobble as having the greatest influence on centrarchid density. Together, these 3 parameters explain 90% of the variability in centrarchid density. These results and the significant negative relationship between mean primary

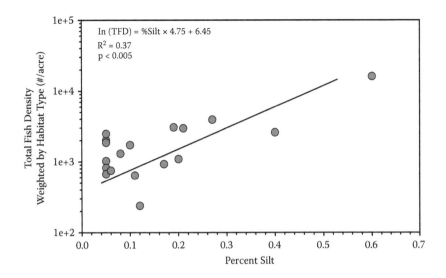

FIGURE B.11 The relationship between total fish density (TFD) weighted by habitat type and site percent silt.

habitat depth and centrarchid Se ($R^2 = 0.43$; $p = 0.04$; Figure B.14, Part C) suggest population differences between sites are more related to habitat parameters than whole-body Se concentrations.

Despite a wide range of Se concentrations in the water, sediments, invertebrates, and fish, no significant relationships were observed between total fish densities or fish species richness and site mean fish tissue Se concentrations. When individual fish families are considered, it does appear there is a significant relationship between species-specific whole-body Se concentrations and densities within the family Centrarchidae (bass and sunfish). No significant relationships were observed in the other 2 dominant families in the study sites, Cyprinidae or Catostomidae.

However, it is important to note that attempts to determine Se effects were highly confounded by significant correlations with habitat. In fact, habitat parameters, in general, consistently explained more of the variability in fish parameters than whole-body fish tissue Se, suggesting that tissue concentrations were not driving populations despite the wide range of tissue levels.

B3.0   POTENTIAL MECHANISMS ACCOUNTING FOR PATTERNS OBSERVED

The previous discussion describes a warm-water fish community existing in an environment naturally elevated in Se. There are some potential relationships between fish tissue levels and fish community health that would appear to support the concept of field validation of tissue threshold effects. However, the tissue thresholds that might be found vary significantly between fish families—and may actually be more closely tied to habitat availability, stream size, and/or other mechanisms, such as interaction with other water quality parameters or population/community ecology factors.

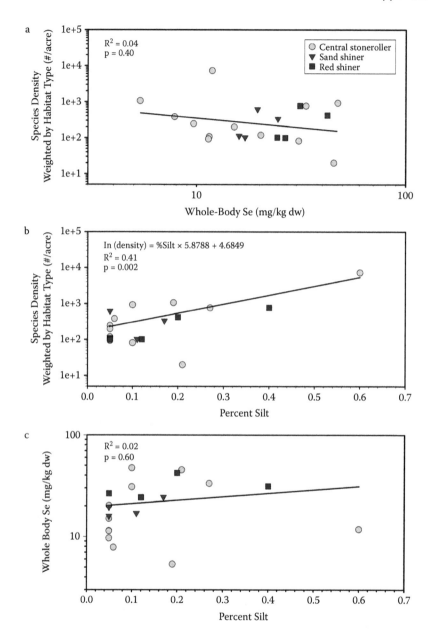

FIGURE B.12 Relationships between cyprinid species-specific a) densities and whole-body Se, b) densities and percent silt, and c) whole-body Se and percent silt. Only data with paired Se tissue and population data were included in analyses.

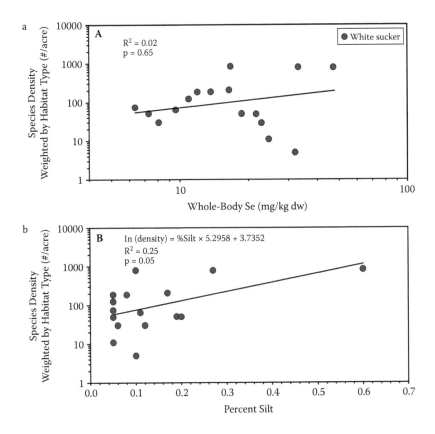

FIGURE B.13 Relationships between catostomid (represented by white suckers) species-specific a) densities and whole-body Se and b) densities and percent silt. Only data with paired Se tissue and population data were included in analyses.

B3.1 SULFATE

Persistence of fish populations in this naturally high Se environment may be related to other site-specific conditions that affect Se toxicity. For example, many researchers have noted the interaction and ameliorating effect of sulfur on Se. Specifically, sulfur has been shown to reduce Se bioaccumulation and uptake rates for a number of aquatic organisms (Bailey et al. 1995; Hansen et al. 1993; Ogle and Knight 1996; Riedel and Sanders 1996), as well as reduced acute toxicity from the selenate form of Se (Brix et al. 2001).

The concurrent Se loading studies show that in this portion of the Arkansas River basin, Se concentrations appear to be autocorrelated with sulfate concentrations (ARCADIS 2006). Donnelly and Gates (2005) also found positive relationships between Se water concentrations with sulfate and nitrate concentrations ($R^2 = 0.55$) in the surface and groundwater of the Arkansas River Valley system. Sulfate was also elevated in sediments (ARCADIS 2006); if sediments are a potential primary

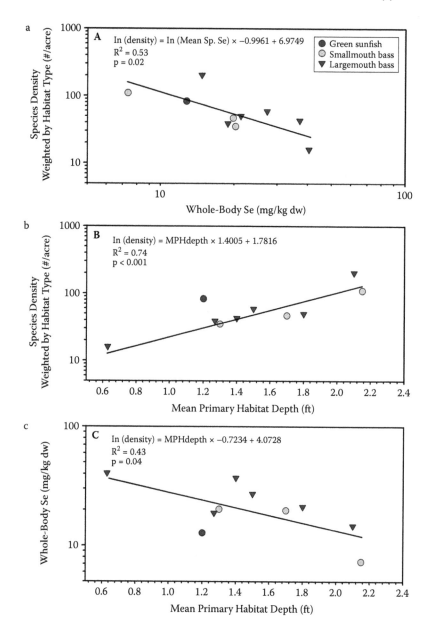

FIGURE B.14 Relationships between centrarchid species-specific a) densities and whole-body Se, b) densities and mean primary habitat (MPH) depth, and c) whole-body Se and MPH depth. Only data with paired Se tissue and population data were included in analyses.

pathway to the food web/dietary uptake (as noted above), then sulfate may play a role in the persistence of these fish.

B3.2   CENTRAL STONEROLLER SE TOLERANCE

Central stonerollers were abundant at a number of sites in the Arkansas River basin. These fish persisted and thrived, even in Wildhorse Creek (which appeared Se stressed, given the limited fish species richness) where Se concentrations in surface water, sediment, and invertebrate tissues were extremely high, and central stoneroller tissue concentrations reached over 40 mg/kg dw (Figure B.15). Interestingly, tissue Se concentrations in central stonerollers from sites on the lower St. Charles River are similar to those on Wildhorse Creek, despite significantly the lower concentrations of Se in other media from the lower St. Charles River (Figure B.15).

So, what makes central stonerollers special, with relation to Se? As noted earlier, fish tissue Se was strongly correlated with diet, in the form of invertebrate tissue Se concentrations (Figure B.9). However, the diet of the central stoneroller is primarily comprising plants, detritus, and algae/diatoms, with only occasional ingestion of small insects (Becker 1983). Such a diet could potentially reduce exposure to Se from an invertebrate diet, although central stoneroller tissue Se concentrations were generally elevated at all sites. This may explain the general similarity in central stoneroller tissue concentrations despite substantially different invertebrate Se between Wildhorse Creek and the St. Charles River. That difference could possibly also explain the lack of other fish species in Wildhorse Creek whose primary diet does consist of invertebrates (such as other minnow species). Still, many of those species are found in the St. Charles River, despite elevated Se in invertebrate tissues.

Could central stonerollers persist as a result of their reproductive strategies? Spawning of central stonerollers in Colorado begins in May (Woodling 1985), with reports in Wyoming of April to June (Baxter and Stone 1995). Cross and Moss (1987) report that central stonerollers spawn in early spring before evapotranspiration

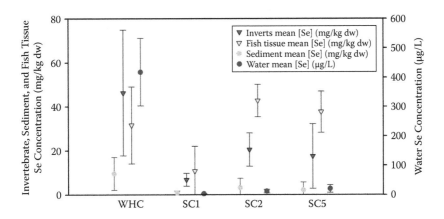

FIGURE B.15   Concentration of Se in various media (only central stoneroller data included for "fish") at sites on Wildhorse Creek and the St. Charles River.

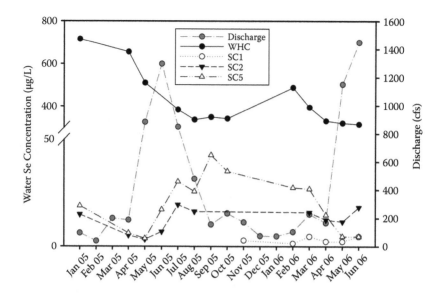

FIGURE B.16   Seasonal patterns in water column Se for sites on the St. Charles River and Wildhorse Creek, with comparisons to flow data from the gaging station at Moffat Street on the mainstem Arkansas River.

reduces groundwater influx. It is unclear whether they spawn repeatedly during late spring and summer, like other minnow species.

Would spawning in late spring–early summer explain their pattern of differential Se tissue levels? Potential Se exposure through seasonal changes in water column Se and flow were compared between the 2 streams (Figure B.16). Again, St. Charles River Se water concentrations are much lower than in Wildhorse Creek, and they almost appear to have opposite seasonal trends. Wildhorse Creek Se is high in winter and early spring and lower in summer/fall, whereas St. Charles River sites are high in summer and fall, with declines to lower values in spring.

Wildhorse Creek Se is high when flows are low in spring, whereas St. Charles River sites are high on the descending limb of the hydrograph in summer and in fall. Thus, it is possible that central stoneroller spawning may be coinciding with the times of year of relatively lower Se doses for Wildhorse Creek and higher doses for St. Charles River.

Seasonal patterns suggest that perhaps, in some respects, naturally derived Se could be behaving more like a "pulse exposure" versus "constant exposure." It has been reported that Se concentrations in fish can have a relatively short half-life (Hamilton 2004), which suggests pulse exposure may not be as much of a concern. However, Hamilton also cites many cases where, after Se inputs have stopped, Se concentrations remained high. Still, Bennett et al. (1986) reported the biological half-life of food-derived Se in the larvae of fathead minnows was 28 days. Hoang and Klaine (2008) found that, when Se exposure was pulsed instead of constant in an experiment with *Daphnia magna*, individuals were able to recover after the removal of pulses. Not only that, but the reproduction of surviving *D. magna* was not affected by pulsed Se exposure, since they exhibited recovery after removal of the pulses (Hoang and Klaine 2008). Nonetheless, even if we consider the Se exposures to be pulse-related in Wildhorse Creek and the St. Charles River, tissue Se levels are still elevated and in the case of Wildhorse Creek, fish species are still being precluded from the stream.

Could tolerance of the central stoneroller be related to producing large numbers of eggs, compared to larger species such as centrarchids? Lam and Wang (2006) evaluated transgenerational retention and maternal transfer of Se in *D. magna*. Approximately 19%–24% of dietary Se accumulated in daphnids was transferred maternally to the next generation, but maternally derived Se was transferred less efficiently to the next generation—i.e., successive generations were less affected than the first generation. In addition, the authors showed that food availability affected Se maternal transfer through reproduction, where lots of food resulted in lots of offspring and subsequently a lower dose of Se per young (low food → few offspring → high dose of Se per young). If this pattern holds for central stonerollers in these streams, then perhaps high availability of food resources affects dosage level to their young, such that lots of food leads to lots of offspring having a low Se dose and, thus, no population effect. Of course, this does not explain the lack of other minnow species in Wildhorse Creek, even though they would have a similar reproductive strategy.

Persistence of central stonerollers (and other minnow species) in these higher Se systems may also be reflecting "species selectivity," postulated by Hamilton (2004) as a "shift from large, long-lived native species to small, short-lived non-native species, except for tolerant species like carp and black bullhead." This lends credence to central stonerollers persisting in a high Se environment.

Could persistence of central stonerollers be a result of acclimation/adaptation/genetic mutation/etc.? Kennedy et al. (2000) speculated that the lack of toxic response for trout in their study could be due to an "evolved tolerance" to high Se concentrations for a fish population living in a seleniferous environment. De Rosemond et al. (2005) also speculated about evolved tolerance to Se in their study on white sucker larval development. Hamilton (2004) notes that others have speculated an evolved tolerance, but states no one has given any evidence—which is still true.

Although Se thresholds were exceeded at many sites, widespread reproductive impairment for minnows was not observed. The calculated chronic level for whole-body Se (40 mg/kg dw) to protect fathead minnows (GEI 2008) was not exceeded at any of the sites (based on mean whole-body concentrations), perhaps explaining the lack of observable population effects—i.e., tissue levels are below effects levels for Se on minnows, assuming fathead minnows are an appropriate surrogate for other members of the cyprinid family. One aspect of interest of that study was the substantial variability observed in the Se concentrations of eggs from females of roughly similar whole-body Se concentrations. Similar variability has been reported by Breitburg et al. (2001), who postulated that the high variability in Se concentrations among clutches and females may actually contribute to the ability of some fishes to maintain viable populations in waters with elevated Se. When combined with high reproductive output and lower Se accumulation in tissues than that found for some highly susceptible species, this high variability in Se concentrations among clutches and females may contribute to the ability of some fishes to maintain viable populations in waters contaminated with Se. Could this be the "evolutionary" or "acclimation" mechanism that would allow fish to maintain viable populations in waters with moderately high Se concentrations?

Lastly, could this phenomenon simply result from compensatory mortality? For example, perhaps population-level effects in situations like the Arkansas River basin (except for extreme cases of high Se concentrations) are not occurring because low survival rates at hatch and development of larvae due to Se exposure are compensated by higher survival rates among those unaffected. Van Kirk and Hill (2007) used population modeling to evaluate the effects of toxicity-related juvenile mortality on population size. While toxicity-related juvenile mortality was the primary cause of population declines, at whole-body Se concentrations below 7 to 10 mg/kg dw, toxicity-induced pre-winter mortality was compensated for by decreased density-dependent winter mortality. Until individual-level toxicity-induced mortality exceeded 80%, greater response was observed at the individual level than the population level. Compensatory mortality has also been demonstrated in some sport fish populations, where mortality due to angling in crappie (*Pomoxis* spp.) populations was compensated by a lower natural mortality rate in the survivors (Allen et al. 1998). Of course, this is a slightly different mechanism, because angling mortality would preferentially be on adults (thereby reducing predation), rather than on embryos or larvae. Still, low survival rates of larvae may allow for increased survival of other young, which would have less competition for more resources. Could this ultimately result in selectivity for more resistant fish?

B3.3   Where Are the Deformities?

As noted in this study, robust fish populations with extremely few deformities were observed despite elevated Se concentrations. Hamilton (2004) postulates that fish larvae with deformities would have a better chance of surviving with no predators present—specifically larger predatory fish that have already succumbed to Se. This concept was also postulated by Lemly (2002b); a lack of predation in Belews Lake may have allowed many deformed nonpiscivorous individuals to survive to juvenile and early adult life stages.

If this is a valid hypothesis, then why are there not deformed fish in the Arkansas River tributaries? As a result of small stream size and shallow depths, there naturally would not be significant predaceous fish populations in these systems. Even the bass and larger sunfish found in the mainstem Arkansas River are a result of escapees from upstream reservoirs. So, if Se were having an effect, there should be a higher incidence of deformities in tributaries where Se concentrations are high and there are no predators. Yet, deformities were not observed in these streams—although it is possible all deformed fish are dying by means other than predation.

Out of the almost 3,000 fish observed during the study, only 13 fish showed deformities. The most common deformity was "black spots," which is not considered Se related. Regarding the observed lack of deformities, de Rosemond et al. (2005) also found lower levels of deformities than expected in white sucker, based on the egg

Se concentrations and a lack of correlation with embryo Se concentration. Kennedy et al. (2000) also observed this result, with high Se in cutthroat trout eggs and a lack of developmental effects (deformities). Hamilton et al. (2002) also reported very high Se concentrations in razorback sucker eggs and much lower levels of deformities than expected. The presence of deformities may not be a meaningful endpoint for evaluation of Se effects in natural fish population studies.

B3.4 Could the Form of Se Explain Persistence of Fish in High Se Streams?

In waters receiving Se from natural discharges (i.e., drainage and runoff from marine shale deposits), selenate is the predominant form of Se present (GEI 2007). In contrast, selenite was the primary form of Se found in sites with measurable Se impacts, such as waters receiving coal fly-ash discharges (Lemly 2002b). The form of Se may be one other factor helping to explain persistence of fish populations in elevated Se environments, along with the physical habitat conditions of those systems. From an acute toxicity perspective, it is apparent that selenite is more toxic than selenate, especially when considering the sulfate effect (USEPA 2004). In addition, reducing environments that promote the formation of organic forms of Se may also affect persistence of fish, since selenomethionone has been shown to bioaccumulate preferentially compared to selenite or selenate (Besser et al. 1989). Thus, organoselenium compounds may contribute disproportionately to Se bioaccumulation and toxicity (e.g., Besser et al. 1993).

The preponderance of selenate forms, combined with the stream environments (with primarily low organic content sand substrates), may produce an environment that influences Se toxicity to these fish communities. Of course, while these factors do exist in the Arkansas River basin, there were still considerably elevated fish tissue Se concentrations for many species of fish at many of the sites—again, without substantial evidence of Se impairment. So, the form of Se present at the site also appears to be of limited use in explaining the observed patterns.

B4.0 CONCLUDING REMARKS

Can Se impacts actually be measured in the field? The short answer is yes, as found in the classic Se incidents in Belews Lake and Hyco Reservoir (Lemly 1993; Garrett and Inman 1984; Chapter 3; Appendix A). In addition, the example discussed herein indicates effects in the form of reduced fish species richness in Wildhorse Creek. However, these all seem to be the extremes. In the case of Belews and Hyco, there was a sudden input of high Se concentrations that appears to have immediately entered the food web, again at very high concentrations, and quickly impacted the fisheries. In Wildhorse Creek, reduced fish species were found in a stream with mean water column Se concentrations in excess of 400 µg/L.

So, without extreme exposure, can Se impacts actually be measured in the field? The long answer is "maybe … maybe not." I am unaware of any other stream sites where measureable impacts from Se have been documented unconditionally (i.e., without a possible caveat of confounding influences). In this case study, it appears on the surface that increased Se concentrations in water and tissues are related to at least a reduced number of centrarchids in these systems. The trend seems real—until two issues are considered: 1) the trend in centrarchid abundance was more strongly influenced by habitat availability than Se levels, and 2) the centrarchids showing the greatest reductions are not native to the system (e.g., largemouth bass) and their populations are maintained by escapees from upstream reservoirs, not spawning, which limits the potential to measure impacts through "reproductive success."

Perhaps there are simple reasons noticeable impacts are not found. As postulated in the literature and evaluated above, it could be that some varying combination of all the factors above work together to provide Se protection. And, although this evaluation provides detailed data from at least 2 years of sampling efforts, more sensitive tools may be necessary to detect Se impacts in the field. Or perhaps we cannot distinguish Se impacts in the field due to ameliorating effects of co-occurring sulfate, or compensatory mechanisms, or differential predator–prey interactions, or advantageous reproductive strategies—which means determining risk from Se exposure in the field is perhaps far from a simple matter.

B5.0 ACKNOWLEDGMENTS

I thank Norka Paden and Stephanie Baker, along with Jason Mullen and Susan Meyer, for their tremendous help in pulling this document together. Their assistance and insight improved the manuscript. It was also helped by key reviews from David DeForest and Peter Chapman, and other members of the steering committee. The data summarized in this commentary were from a comprehensive study of Se in the middle Arkansas River, funded by the City of Pueblo; thanks go to Gene Michael and Nancy Keller for their seemingly endless support!

B6.0 REFERENCES

Allen MS, Miranda, LE, Brock RE. 1998. Implications of compensatory and additive mortality to the management of selected sportfish populations. *Lakes Reserv: Res Manage* 3:67–79.

[ARCADIS] ARCADIS G&M, Inc. 2006. Sources and occurrence of selenium in the Arkansas River and Fountain Creek near Pueblo, Colorado. Prepared for the City of Pueblo (CO, USA).

Bailey FC, Knight AW, Ogle RS, Klaine SJ. 1995. Effect of sulfate level on selenium uptake by *Ruppia maritima*. *Chemosphere* 30:579–591.

Baxter TG, Stone MD. 1995. Fishes of Wyoming. Cheyenne (WY, USA): Wyoming Game and Fish Department.

Becker GC. 1983. Fishes of Wisconsin. Madison (WI, USA): University of Wisconsin Press.

Bennett WN, Brooks AS, Brooks ME. 1986. Selenium uptake and transfer in an aquatic food chain and its effects on fathead minnow larvae. *Arch Environ Contam Toxicol* 15:513–517.

Besser JM, Canfield TJ, La Point TW. 1993. Bioaccumulation of organic and inorganic selenium in a laboratory food chain. *Environ Toxicol Chem* 12:57–72.

Besser JM, Huckins JN, Little EE, La Point TW. 1989. Distribution and bioaccumulation of selenium in aquatic microcosms. *Environ Pollut* 62:1–12.

Breitburg DL, Riedel GF, Pacey CA. 2001. Variation in selenium concentration in eggs produced by fathead minnows experimentally exposed to food containing elevated selenium concentrations. Selenium cycling and impact in aquatic ecosystems: defining trophic transfer and water-borne exposure pathways. Palo Alto (CA, USA): EPRI. pp 6-1 to 6-12.

Brix KV, Volosin JS, Adams WJ, Reash RJ, Carlton RC, McIntyre DO. 2001. Effects of sulfate on the acute toxicity of selenate to freshwater organisms. *Environ Toxicol Chem* 5: 1037–1045.

Brix KV, Toll JE, Tear LM, DeForest DK, Adams WJ. 2005. Setting site-specific water-quality standards by using tissue residue thresholds and bioaccumulation data. Part 2. Calculating site specific selenium water-quality standards for protecting fish and birds. *Environ Toxicol Chem* 24:231–237.

Bryson WT, Garret WR, Mallin MA, Macpherson KA, Partin WE, Woock SE. 1984. Roxboro Steam Electric Plant 1982 Environmental Monitoring Studies, Volume II, Hyco Reservoir Bioassay Studies. Environmental Technology Section. Carolina Power and Light Company. Raleigh (NC, USA).

Canton SP, Baker S. 2008. Field application of tissue thresholds: potential to predict fish population or community effects in the field. In: GEI Consultants, Inc., Golder Associates, Parametrix, and University of Saskatchewan Toxicology Centre. Selenium tissue thresholds: tissue selection criteria, threshold development endpoints, and field application of tissue thresholds. Washington (DC, USA): North American Metals Council-Selenium Working Group; www.namc.org/selenium

Cross FB, Moss RE. 1987. Historic changes in fish communities and aquatic habitats in plains streams of Kansas. In: Matthews WJ, Heins DC, editors, Community and evolutionary ecology of North American stream fishes. Norman and London: University of Oklahoma Press. p 155–165.

deBruyn A, Hodaly A, Chapman P. 2008. Tissue selection criteria: selection of tissue types for development of meaningful selenium tissue thresholds in fish. In: GEI Consultants, Inc., Golder Associates, Parametrix, and University of Saskatchewan Toxicology Centre. Selenium tissue thresholds: tissue selection criteria, threshold development endpoints, and field application of tissue thresholds. Washington (DC, USA): North American Metals Council-Selenium Working Group; www.namc.org/selenium

DeForest DK. 2008. Threshold development endpoints: review of selenium tissue thresholds for fish—evaluation of the appropriate endpoint, life stage, and effect level, and recommendation for a tissue-based criterion. In: GEI Consultants, Inc., Golder Associates, Parametrix, and University of Saskatchewan Toxicology Centre. Selenium tissue thresholds: tissue selection criteria, threshold development endpoints, and field application of tissue thresholds. Washington (DC, USA): North American Metals Council-Selenium Working Group; www.namc.org/selenium

de Rosemond SC, Liber K, Rosaasen A. 2005. Relationship between embryo selenium concentration and early life stage development in white sucker (*Catostomus commersoni*) from a northern Canadian lake. *Bull Environ Contam Toxicol* 74:1134–1142.

Donnelly JP, Gates TK. 2005. Assessing irrigation-induced selenium and iron in the lower Arkansas River Valley in Colorado. Proceedings of the 2005 World Water and Environmental Resources Congress. Anchorage (AK, USA).

Garrett GP, Inman CR. 1984. Selenium-induced changes in fish populations of a heated reservoir. *Proc Annu Conf SE Assoc Fish Wildl Agencies* 38:291–301.

[GEI] GEI Consultants, Inc. 2007. Aquatic biological monitoring and selenium investigations of the Arkansas River, Fountain Creek, Wildhorse Creek, and the St. Charles River. Technical Report prepared on behalf of the City of Pueblo WRP. Littleton (CO, USA).

[GEI] GEI Consultants, Inc. 2008. Maternal transfer of selenium in fathead minnows, with modeling of ovary tissue-to-whole body concentrations. Technical Report prepared on behalf of Conoco-Phillips. Littleton (CO, USA).

Hamilton SJ. 2004. Review of selenium toxicity in the aquatic food chain. *Sci Total Environ* 326:1–31.

Hamilton SJ, Holley KM, Buhl KJ, Bullard FA, Weston LK, McDonald SF. 2002. Impact of selenium and other trace elements on the endangered adult razorback sucker. *Environ Toxicol* 17:297–323.

Hansen LD, Maier KJ, Knight AW. 1993. The effect of sulfate on the bioconcentration of selenate by *Chironomus decorus* and *Daphnia magna*. *Arch Environ Contam Toxicol* 25:72–78.

Hayes AW. 2008. Principles and methods of toxicology. 5th edition (NY, USA): Taylor and Francis Group.

Herman RL. 1990. The role of infectious agents in fish kills. In: Meyer FP, LA Barclay, editors, Field manual for the investigation of fish kills. Resource Publication 177. Washington (DC, USA): United States Department of the Interior, Fish and Wildlife Service.

Hintze JL. 2000. NCSS 2001 statistical software system for Windows. Kaysville (UT, USA): Number Cruncher Statistical Systems.

Hoang TC, Klaine SJ. 2008. Characterizing the toxicity of pulsed selenium exposure to *Daphnia magna*. *Chemosphere* 71:429–438.

Kennedy CJ, McDonald LE, Loveridge R, Strosher MM. 2000. The effect of bioaccumulated selenium on mortalities and deformities in the eggs, larvae, and fry of a wild population of cutthroat trout (*Oncorhynchus clarki lewisi*). *Arch Environ Contam Toxicol* 39:46–52.

Lagler KF. 1956. Freshwater fishery biology. 2nd edition. Dubuque (IA, USA): William C Brown Company.

Lam IK, Wang WX. 2006. Transgenerational retention and maternal transfer of selenium in *Daphnia magna*. *Environ Toxicol Chem* 25:2519–2525.

Lemly AD. 1985. Toxicology of selenium in a freshwater reservoir: implications for environmental hazard evaluation and safety. *Ecotoxicol Environ Safety* 10:314–338.

Lemly AD. 1993. Teratogenic effects of selenium in natural populations of freshwater fish. *Ecotoxicol Environ Safety* 26:181–204.

Lemly AD. 1997a. Environmental implications of excessive selenium: a review. *Biomed Environ Sci* 10:415–435.

Lemly AD. 1997b. A teratogenic deformity index for evaluating impact of selenium on fish populations. *Ecotoxicol Environ Safety* 37:259–266.

Lemly AD. 2002a. Selenium assessment in aquatic ecosystems: a guide for hazard evaluation and water quality criteria. New York (NY, USA): Springer-Verlag.

Lemly AD. 2002b. Symptoms and implications of selenium toxicity in fish: the Belews Lake case example. *Aquat Toxicol* 57:39–49.

Ogle RS, Knight AW. 1996. Selenium bioaccumulation in aquatic ecosystems: 1. Effects of sulfate on the uptake and toxicity of selenate in *Daphnia magna*. *Arch Environ Contam Toxicol* 30:274–279.

Orr PL, Guiger KR, Russel CK. 2006. Food chain transfer of selenium in lentic and lotic habitats of a western Canadian watershed. *Ecotoxicol Environ Safety* 63:175–188.

Riedel GF, Sanders JG. 1996. The influence of pH and media composition on the uptake of inorganic selenium by *Chlamydomonas reinhardtii*. *Environ Toxicol Chem* 15:1577–1583.

Schlosser IJ. 1987. A conceptual framework for fish communities in small warmwater streams. In: Matthews WJ, DC Heins, editors, Community and evolutionary ecology of North American stream fishes. Norman (OK, USA): University of Oklahoma Pr. p 17–24.

[USEPA] U.S. Environmental Protection Agency. 2004. Draft aquatic life water quality criteria for selenium—2004. EPA-822-D-04-001. Washington (DC, USA): Office of Water.

Van Derveer WD, Canton SP. 1997. Selenium sediment toxicity thresholds and derivation of water quality criteria for freshwater biota of western streams. *Environ Toxicol Chem* 16:1260–1268.

Van Kirk RW, Hill SL. 2007. Demographic model predicts trout population response to selenium based on individual-level toxicity. *Ecol Model* 206:407–420.

Woodling J. 1985. Colorado's little fish: a guide to the minnows and other lesser known fishes in the state of Colorado. Denver (CO, USA): Colorado Division of Wildlife.

Index

Page references in *italics* refer to figures.
Page references in **bold** refer to tables.

Other Titles from the Society of Environmental Toxicology and Chemistry (SETAC)

Genomics in Regulatory Ecotoxicology: Applications and Challenges
Ankley, Miracle, Perkins, Daston, editors
2007

Population-Level Ecological Risk Assessment
Barnthouse, Munns, Sorensen, editors
2007

*Effects of Water Chemistry on Bioavailability and Toxicity of Waterborne Cadmium,
Copper, Nickel, Lead, and Zinc on Freshwater Organisms*
Meyer, Clearwater, Doser, Rogaczewski, Hansen
2007

Ecosystem Responses to Mercury Contamination: Indicators of Change
Harris, Krabbenhoft, Mason, Murray, Reash, Saltman, editors
2007

Freshwater Bivalve Ecotoxicology
Farris, Van Hassel, editors
2006

*Estrogens and Xenoestrogens in the Aquatic Environment:
An Integrated Approach for Field Monitoring and Effect Assessment*
Vethaak, Schrap, de Voogt, editors
2006

*Assessing the Hazard of Metals and Inorganic Metal Substances
in Aquatic and Terrestrial Systems*
Adams, Chapman, editors
2006

Perchlorate Ecotoxicology
Kendall, Smith, editors
2006

Natural Attenuation of Trace Element Availability in Soils
Hamon, McLaughlin, Stevens, editors
2006

*Mercury Cycling in a Wetland-Dominated Ecosystem:
A Multidisciplinary Study*
O'Driscoll, Rencz, Lean
2005

SETAC

A Professional Society for Environmental Scientists and Engineers and Related Disciplines Concerned with Environmental Quality

The Society of Environmental Toxicology and Chemistry (SETAC), with offices currently in North America and Europe, is a nonprofit, professional society established to provide a forum for individuals and institutions engaged in the study of environmental problems, management and regulation of natural resources, education, research and development, and manufacturing and distribution.

Specific goals of the society are

- Promote research, education, and training in the environmental sciences.
- Promote the systematic application of all relevant scientific disciplines to the evaluation of chemical hazards.
- Participate in the scientific interpretation of issues concerned with hazard assessment and risk analysis.
- Support the development of ecologically acceptable practices and principles.
- Provide a forum (meetings and publications) for communication among professionals in government, business, academia, and other segments of society involved in the use, protection, and management of our environment.

These goals are pursued through the conduct of numerous activities, which include:

- Hold annual meetings with study and workshop sessions, platform and poster papers, and achievement and merit awards.
- Sponsor a monthly scientific journal, a newsletter, and special technical publications.
- Provide funds for education and training through the SETAC Scholarship/Fellowship Program.
- Organize and sponsor chapters to provide a forum for the presentation of scientific data and for the interchange and study of information about local concerns.
- Provide advice and counsel to technical and nontechnical persons through a number of standing and ad hoc committees.

SETAC membership currently is composed of more than 5000 individuals from government, academia, business, and public-interest groups with technical backgrounds in chemistry, toxicology, biology, ecology, atmospheric sciences, health sciences, earth sciences, and engineering.

If you have training in these or related disciplines and are engaged in the study, use, or management of environmental resources, SETAC can fulfill your professional affiliation needs.

All members receive a newsletter highlighting environmental topics and SETAC activities and reduced fees for the Annual Meeting and SETAC special publications.

All members except Students and Senior Active Members receive monthly issues of Environmental Toxicology and Chemistry (ET&C) and Integrated Environmental Assessment and Management (IEAM), peer-reviewed journals of the Society. Student and Senior Active Members may subscribe to the journal. Members may hold office and, with the Emeritus Members, constitute the voting membership.

If you desire further information, contact the appropriate SETAC Office.

1010 North 12th Avenue	Avenue de la Toison d'Or 67
Pensacola, Florida 32501-3367 USA	B-1060 Brussels, Belgium
T 850 469 1500 F 850 469 9778	T 32 2 772 72 81 F 32 2 770 53 86
E setac@setac.org	E setac@setaceu.org

www.setac.org
Environmental Quality Through Science®